TECHNOLOGY
IN AMERICA
A BRIEF HISTORY

ALAN I MARCUS
Iowa State University

HOWARD P. SEGAL
University of Maine

Harcourt Brace Jovanovich College Publishers
Fort Worth   Philadelphia   San Diego
New York   Orlando   Austin   San Antonio
Toronto   Montreal   London   Sydney   Tokyo

For Jean and for Deborah

ISBN: 0-15-589762-4
Library of Congress Catalog Card Number: 88-80640
Printed in the United States of America

5 6 7 8 9 016 9 8 7 6 5 4

Introduction

Since the mid-1950s scholars have demonstrated an increased awareness of the importance of technology in the American past. Members of Science, Technology, and Society (STS) and Science, Technology, and Human Values programs have written monographs and essays detailing technology's role in fostering social and environmental change. Others in more traditional history of technology and engineering programs have investigated technological development and chronicled changes in material production processes. These efforts have been supplemented by numerous studies by American social, economic, business, and political historians. This book is testament to the many contributions of these diverse groups.

The maturity of this aforementioned scholarship has given us the confidence to attempt a brief, single-volume synthesis of the history of technology in America. We are indebted to both our predecessors and our contemporaries for showing us that technology is a multifaceted phenomenon, and that its history involves both material productions—techniques, structures, and apparatuses—and the organization of human activities. The history of technology is both the hardware and that hardware's relationships to American society and culture. This book acknowledges those precepts, but also seeks to break new ground by concentrating its examination on the impact of American society and culture on technology, rather than vice versa.

Such an attempt must begin with an understanding of American social and cultural ideas. We have something of a precedent to follow in that regard. Historians have long searched for the "essence" of the American experience and endeavored to delineate America's national character. Scholars have found America to be a nation of immigrants, or a land of abundance, and have suggested that Americans have been marked by the presence of a frontier, a frontier mentality, or a Puritan ancestry. These and other assessments have been offered to account for what historians have alternately, and with frequent contradiction, characterized as the American people's pervasive individualism, manifest pluralism, or quest for material gain. While each of these all-embracing designations contains something to commend itself (and admitting that America has been blessed with an abundance of material and human resources, and that its citizens have been infatuated with material gain), we have taken, we think, a more fruitful approach. Rather than ascribe to a solitary cause or characterization the panorama of events that constitutes American history, we have identified specific, dominant cultural notions and social themes for different eras in the American past. In this regard, we have drawn our inspiration from more contemporary scholars who have demonstrated that culture is not simply a phenomenom of people and place. Cultures rarely emerge from the settlement of geographic areas and persist inviolate; instead, a culture has distinctive epochs, each characterized by a particular set of assumptions locked in time.

It is these cultural notions and social themes that make a series of years a distinct era. The identification of cultural notions and social themes during the periods in which they predominate strikes us as a particularly useful and important way to understand America's technological past. After all, an era's social and cultural desires, opinions, ideas, themes, and notions help explain public acceptance or rejection of specific technologies. Those same factors also dictate how technologies are conceived and ultimately employed, as well as circumscribe such relative concepts as efficiency and profitability. In that sense, what makes and has made technology in America distinct from technology someplace else is and was the cultural epochs through which America is passing and has passed.

These latter considerations are reflected in the book's organization. The trendier designations of preindustrial, industrializing, industrial, and postindustrial America prove too limiting. Products of the impact-of-technology-on-society question, they lack the flexibility required for a systematic discussion of the influence on technology of a particular period's prevailing cultural notions and social themes. A relatively conventional American history periodization instead provides the necessary latitude. It facilitates, for example, the examination of colonial technology with reference to mercantile theory and community practice (especially the closed commercial corporation), and the investigation of late–nineteenth- and early–twentieth-century American technology in the context of that period's obsession with system, efficiency, standardization, centralization, nationalism, and the like.

The richness of America's technological past, in conjunction with our complex agenda for this book, have forced us to make some hard choices. The book covers some technologies in depth over time—manufacturing, agriculture, and transportation, for example—highlights aspects of others that were central during limited periods, and occasionally investigates less-familiar technologies in depth to demonstrate their representativeness or importance to a specific period. A few get only a passing nod. Technologies associated with the petroleum industry are perhaps the most neglected. Also, we have varied the complexity with which we discuss specific technologies, trusting that the occasional more technical explanation is written in such a manner as to serve as a model for and a guide to complementary textbook sections, and thus will satisfy those readers with some technical training without hampering those lacking a similar background.

We have decided to emphasize the years after the 1830s. That decision stemmed from the fact that the century after the 1830s marked America's technological heyday; with that era's extraordinary technological developments came profound social and cultural changes. Few pre-1830s technologies still seem vital today. That cannot be said of technologies fostered after those years: Americans in the 1980s continue to live with and employ variants of technologies first developed more than a hundred years ago.

ACKNOWLEDGMENTS

Many people deserve acknowledgment for their contributions to this book. Graduate students Matthew McCright, Paul Melde, Gregory Sanford, Maureen Ogle, Mark Finlay, Kent Mutchler, and Judith Lopez, all in Iowa State University's Ph.D. program in the history of technology and science; and undergraduates Charlotte Chiu and Gregory Welch of Harvard University, and Andrew Ward and Robert Bradford of the University of Maine all provided invaluable research assistance. Robert Bradford also prepared the index. The Vigilantes, an informal group of Iowa State University's history faculty, provided helpful comments; and fellow historians of technology W. David Lewis, Bruce Sinclair, Carroll Pursell, and Melvin Kranzberg offered encouragement and timely suggestions.

History Department secretaries Kathie Moring and Carol Rickards of the University of Maine, and Audrey Burton and Carole Kennedy of Iowa State University, expertly and cheerfully prepared several drafts of the text. In addition, a Faculty Summer Research Grant from the University of Maine expedited completion of the final draft.

Equally important, the editorial and production staff at Harcourt Brace Jovanovich—Drake Bush, acquisitions editor; Chuck Allen, free-lance editor; Kay Kaylor, production editor; Kim Turner, production manager; Merilyn Britt, designer; and Vicki Kelly, art editor—were all consummate professionals from the book's earliest stages until its last ones.

Finally, our families—Jean, Jocelyn, and Gregory Marcus, and Marjorie Marcus Schulz, Deborah Rogers, and Robert Segal—gave the emotional support necessary to see us through this enterprise.

Contents

Introduction   iii

Part One
FROM THE OLD WORLD TO THE NEW: 1607 TO THE 1870s

1   Manufacturing America: 1607 to 1800   3

The Old World in the New   3
   Survival Through Utilization of Resources   4
   Mercantilism and Colonial Governance   5
The Search for Community Self-sufficiency: The Seventeenth
   Century   9
   The Ubiquitous Mill   9
   The Colonial Master–Apprentice System   12
   Providing Necessities   13
   Larger-Scale Enterprise   21
Beyond Subsistence: The Eighteenth Century   28
   Signs of New Wealth   29
Technology and the Revolution, 1763–1783   31
   Printing as a Catalyst for the Revolution   31
   Manufacture of Arms   34
Thirteen Commercial Nations, 1776–1789   37
   Interstate Commerce   37
Framing and Implementing a Commercial Pact, 1787–1800   41
   The Constitution: A Commercial Document   42
   Hamilton's Call for Federal Assistance   43
   Private Development of Manufacturing   46

Notes   49
For Further Reading   50

**2   Young America and Individual Opportunity:
1800 to the 1830s**   52

Beyond Mercantilism   52
  Extending Europe   53
  The Ascendancy of the Individual   54
Building the Nation's Infrastructure   55
  Building Turnpikes   57
  Building Bridges   59
  Improved River Travel—Building Steamboats   59
  Building Canals   62
  Building Railways   66
Changes in Manufacturing and Labor   69
  Machine Shops as Adjuncts to Mills   70
  Use of Machinery in Textiles   72
  Reducing Labor Costs   77
Preparing America's Yeomanry   79
  Mechanics' Institutes   81
  Agricultural Societies   82
  Technical Education for the Yeomanry   82
Machinery, Production, and Democracy   84

Notes   85
For Further Reading   86

3   America as a Social Unit: the 1830s to the 1870s   88

The Search for a Uniquely American Character   88
  The Break from European Society   89
  Using Technology to Find America—Photography   91
Providing a Common Material Experience   93
  Setting the Stage   93
  Railroads Ascendant   96
  Furnishing Uniform Goods   106
Uniform Agriculture   115
  Equipment for a Uniform Agriculture   117
The Creation of International Markets   124
  Mid-Century Shipping   124
  The International Industrial Exhibition   125
  Transoceanic Telegraphy   127
The Civil War: A Mass Experience   129

Notes   129
For Further Reading   130

Part Two
SYSTEMATIZING AMERICA:
THE 1870s TO THE 1920s

4   Systematizing Power, Communications, and the Power to Communicate   135

The Great Centennial Exhibition: Late–Nineteenth-Century American Notions in Miniature   135
The Idea of System: The Pervasive Late–Nineteenth-Century American Notion   139
The Electrification of America   141
   The Development of Generators   142
   The Advent of Electric Lighting   143
   The Growth of Electric Power   150
The Development of Telephones   158
   Telephone and Radio   163
Systematizing the Technologists   165
   Engineering Organizations   165
   Systematizing Technical Education   169
   Systematizing Technical Research   173

Notes   177
For Further Reading   178

5   Systematizing the Fabric of American Life: the 1870s to the 1920s   180

Systematizing the Physical Environment   181
   The New City   181
   Political Control of the Cities   185
   The New Agriculture   187
   Systematizing Nature   194
Systematizing Leisure   202
   Sports and the Growth of the Middle Class   202
   Bicycles   207
   Automobiles   208
   Motion Pictures   212

Notes   216
For Further Reading   217

6   Systematizing Workers and the Workplace   220

The System in the Home   220
  New Household Machinery   221
Systematizing Factory Work   225
  Building Systematic Factories   226
  Systematizing Factory Administration   226
  Systemization of Large-Scale Production   230
Systematizing the Human Element   234
  Scientific Management   234
  Industrial Social Welfare and Social Science   236
  New Relations between Labor and Owners   239
World War I: A System of Thought's Last Battle   243
  Marshaling Resources for War   244
  New War Technology on Land   245
  New War Technology at Sea   246
  New War Technology in the Air   247
  Production of Munitions During the War   249
  The End of a System of Thought   250

Notes   252
For Further Reading   253

Part Three

FROM INDUSTRIAL AMERICA TO POSTINDUSTRIAL AMERICA:
THE 1920S TO THE PRESENT

7   Technology as a Social Solution: the 1920s to the 1950s   257

Emergence of a New Notion of Systems   257
  The Hawthorne Experiments   257
  The New Notion of Systems   260
The Government and Social Engineering   261
  Hoover and National Planning   261
  Roosevelt and the New Deal   262
  The Administration of New Deal Technologies   264
Revitalizing Rural America   270
  Balancing Agriculture and Industry   270
Changes in Production Techniques   275
  Changes in Agriculture after 1920   276
  Changes in Industrial Manufacturing after 1920   283
New Marketing and Delivery Technologies   289
  Using Radio to Reach a National Market   289
  The Railroads after 1920   291

New Delivery Technologies—Trucks and Buses   291
The Growth of Commercial Aviation   293
Mid-Century High Tech   295
  The Development of Computers   295
  The Development of Transistors   300
Technology in World War II: Solving the Problem of
  Totalitarianism   301
  Creation of the Atomic Bomb   302
  The Development of Radar   305
  The Development of Sonar   306
  The Development of Proximity Fuses   307
  The Role of Aviation in World War II   308
American Use of Technology in the Postwar World: Solving the
  Problem of Communism   309

Notes   311
For Further Reading   312

8   **Technology as a Social Question: the 1950s to the Present**   315

Assaulting the Establishment and Established Notions   315
  Technology as the Driving Force Behind Modern Life   316
The Nuclear Age   317
  National Defense and Nuclear Debate   318
  The Rise and Fall of Nuclear Power   319
Spaceflight   323
  The Origins of Spaceflight   324
  The Space Race   324
  Decline of Support for the Space Program   326
Television and the Fabric of America   328
  Criticism of Television Programming   328
The Advent of High-tech Electronics   330
  Development of the Microchip   332
  Development of Computers—The Information Age   333
The New American Manufacturing   334
  The Growth of High-tech Industry   336
  An Example of High-tech Industry—The Silicon Valley   337
  Robotics and Labor Relations   339
  Transportation for the New Manufacturing   340
High-tech Agriculture   341
  *Some Results of Pesticides, Mechanization,
    and Hybrid Crops*   342
The Growth of Biotechnology   344
The Environment and Engineers   346

The Response of Engineers to Criticism   347
The Technology of Policy   350
  Systems Analysis   351
  Technology Assessment   352
  Futurecasting   353
  Growth of the Ideology of Appropriate Technology   356
Americans and Their Technologies   359

Notes   361
For Further Reading   362

Illustration Credits   365

Index   367

P_{art}
O_{ne}

From the Old World to the New

1607 TO THE 1870s

Technologies were instrumental in sustaining the first permanent English settlements in North America. Established as mercantilistic ventures, the colonies mined and modified the New World's abundant resources, first to maintain themselves, and eventually for export. The success with which the colonists plied their skills, many brought with them from the Old World, seemed by the 1760s to place them in direct competition with England. Although the American Revolution broke formal connections with England, it neither lessened the New World's reliance on mercantilism nor separated America from Europe.

The United States Constitution was a product of mercantilistic notions, but Americans soon recognized that their country was a land of seemingly limitless possibilities. They endeavored to fulfill the promise of a democratic republic by rejecting aristocracies of wealth and heredity and by expanding the right to vote. By the first decade of the nineteenth century, Americans had abandoned mercantilism, concentrating instead on exploring the New World's possibilities and providing the opportunity for citizens to tap them.

In this milieu, technologies became equated with the provision of individual opportunity. Governments sponsored or built turnpikes, canals, bridges, and railroads to facilitate commerce, and created an unprecedented number of corporations to take advantage of these internal improvements. A shortage of skilled and inexpensive labor plagued many of these enterprises. Entrepreneurs responded by encouraging emigration from Europe of individuals possessing special skills, by copying European production machines, and by replacing workers with machinery. They also established institutions to provide learning opportunities and to join capital with new inventions. That manufactories quickly dotted the nation's landscape was an indication that opportunity had been provided.

These technological initiatives were understood in a different way after the mid-1830s. As Americans defined themselves as a people exhibiting a distinctive and unique character and further distanced their country from Europe, the technologies of the previous decades were reconceptualized as providing a means of unifying and homogenizing the nation. They were bolstered by new technologies as the technology of individual opportunity gave way to the technology of uniformity and coherence. Mid-nineteenth-century Americans adopted those technologies that measured the distinctly American character, furnished the nation's citizens with a common material experience, or enhanced the country's standing within the world of nations.

1

Manufacturing America: 1607 to 1800

THE OLD WORLD IN THE NEW

Modern vessels brought the first European settlers to the New World. Brightly colored and round-sterned, these carvel-built ships generally had several masts, a combination of square and lateen sails, and were steered by adjusting the sails or by turning a rudder-attached tiller. They were among the most sophisticated products of European technology.

Though the early colonists traveled to North America in relative comfort and ease on truly seaworthy ships, they faced very different circumstances once they landed. They had left all amenities behind. Inclement weather, death, and disease were now their constant companions. Unlike the Spanish conquistadors, these transplanted Europeans discovered neither gold nor silver, nor did they encounter an indigenous urban work force, like the Incas or Aztecs, to do their bidding. Nor could these men (and, later, women) depend on the prolonged support of the Old World or on friendly Indians. Relations with Indians often were poor, while European shipping was frequently disrupted. Shipwrecks, pirates, and hostile natives made transport of commodities between continents unreliable, and communications difficult. The first British colonists had to rely on their technical skills just to stay alive.

Survival Through Utilization of Resources

Survival of the settlements demanded development of New World resources, and colonists quickly brought to bear their Old World technologies. Greeted by an abundance of fertile land, much of it forest, they cleared it and placed it under cultivation. An impediment to settlement while standing, when felled, trees served as the primary building material and fuel, and as a source of naval stores. The continent's plentiful rivers and streams provided water sufficient for drinking—and waste disposal—and facilitated transportation, enabling pole boats, barges, and ships to travel inland. These waterways also furnished an important source of motive power; waterwheels came to rival windmills, animals, and men as prime movers. Through the judicious use of gears, cams, and triphammers, waterwheels powered bellows, and saw, grist, paper, grinding, rolling, slitting, crushing, and fulling mills. In the colonists' able hands, sand became glass and ores metals. Animal hides were converted into leather for shoes and apparel. Animal hair was felted or woven. Processed skins trimmed hats. Potash was leached from wood ashes, mixed with animal fat, and transformed into soap. Evaporation of ocean water yielded salt. Flax and hemp became linen cloth and rope. Tallow and whale oil served to make candles.

But perhaps the most significant resource was the colonists themselves. Every boatload of settlers crossing the Atlantic brought knowledge of European technological processes and skills. While English people and practices were most heavily represented, colonists also emigrated from France, Spain, the Netherlands, Poland, Sweden, the German principalities, Scotland, Wales, and Northern Ireland. They all carried with them expertise obtained in their native lands. Not all of their technical knowledge and skill was suitable to the new setting; though abundant in many raw materials, the North American environment was different, thus limiting the possibilities of the first settlers. Colonists needed to reject some European-acquired information and adapt other aspects to their new situation. This flexibility, coupled with a familiarity with European techniques, proficiency in putting them to use, and hard work, enabled these men and women to fashion an existence in a harsh, sometimes antagonistic environment.

Mercantilism and Colonial Governance

Europeans carried more to the New World than practical knowledge, adaptability, and know-how. They also brought a common understanding of the nature of colonization. They generally recognized that settling North America was a corporate enterprise. Despite the physical burdens they bore, participants were unified in the endeavor by prospects of improving their situation. These expectations provided a compelling reason to accept less-than-idyllic terms and conditions. Colonization was an investment in the near future; as a consequence, men and women endured hardships, economic and otherwise, and often put themselves in subservient positions, as temporary expedients to a grand prize. Each European state risked funds or granted individuals and groups exclusive privilege to settle or trade in an area in hope of either securing a percentage of the profits or collecting taxes. Individual entrepreneurs or those acting collectively as joint-stock companies also were immediately involved. For a proportion of the settlements' anticipated profits, they covered colonists' transportation costs, supported them in the New World for specified periods, and sometimes gave them land. Usually in return for passage to North America, beneficence of their sponsors, and promise of future rewards, colonists entered into agreements—sometimes tacit but always in some ways restrictive—that mandated their responsibilities and obligations to their benefactors.

Mercantilism and the Colonial Social Structure   Colonization, then, entailed a network of well-defined relationships and duties. Seventeenth- and eighteenth-century men and women accentuated the entire web, not any single aspect. They supposed that all parties—nation states, entrepreneurs, and colonists—needed to benefit in order for any one party to derive benefit. Though they recognized that there would be inequities, they felt that it was not necessary for each party or for each member of each party to benefit equally. Emphasis on the whole process over, and almost to the exclusion of, the individuals that comprised it provided an attractive perspective during a period marked by an awareness that the total wealth of the world—both old and new—was limited. Pooling funds, property, and interests in return for, and according to, an established set of commitments, including guarantees of exclusivity implied in grants of monopoly, spread the risk and the burden of colonization. It allowed European states and their subjects to attempt collectively what none could effectively do separately. Success was determined predominantly by the venture's intrinsic viability, and participants rested secure in the knowledge that a successful venture meant that they would all reap benefits.

It bears repeating that an essential feature of this social formulation was that the world's wealth was finite. The idea of wealth in the seventeenth and eighteenth centuries, moreover, was not akin to more modern notions of capital (a nineteenth-century concept, capital seemed to have the power to reproduce itself). Wealth was not created per se, but rather acquired, gathered, captured, or corralled. It was the sole purpose of seventeenth- and eighteenth-century governments in both the Old and New World to accrue for their citizens as much wealth as possible, and to guard wealth previously amassed. Government by its very definition was to restrain competition for wealth within its jurisdiction. That sort of competition seemed wasteful, potentially a destroyer of wealth. Rather than permit each citizen to pursue his own destiny, which would presumably result in a "condition of war of everyone against everyone," governments were created and endowed with great and arbitrary powers, including the power to grant monopolies, to prevent such an occurrence. Indeed, life without government was thought to be hopeless: "solitary, poor, nasty, brutish, and short." Only through the formation of "commonwealths," as seventeenth- and eighteenth-century governments were sometimes known, could the common wealth—the wealth that citizens had corralled through collective activity—be protected most efficiently.

Given the government aim to acquire additional wealth, a technology–government nexus became the standard manner of capturing more of the world's limited wealth. Conversion of natural resources into products was the first step, but new wealth could not be brought into a commonwealth merely through production. Only through export and sale of products to a different nation-state could a commonwealth gain wealth (the British colonies were part of the English commonwealth, not a separate entity). Conversely, imports reduced wealth by removing it beyond a commonwealth's jurisdiction. As a consequence, governments rigorously encouraged exportation and just as strenuously discouraged importation of any kind.

Political economists call these beliefs *mercantilism* or *mercantile theory*. The notions upon which mercantilism rested—especially the idea that the whole, not the individual parts, stood as the key determinant—were woven tightly within the fabric of colonial thought. Indeed, colonists had transplanted European assumptions about the relationship among things and the nature of things to the North American soil. These suppositions were fortified by the arrival of subsequent settlers and reinforced by the colonial connection to Europe. They found expression in many areas. The nature of local government—plantations, towns, and cities—was among the clearest expressions of mercantile thought.

The Colonial Style of Local Government Colonial cities, towns, and plantations, including the patroon manors of New Netherlands, drew upon traditional European governmental practice, and received charters from the crown, proprietors, or colonial legislatures that granted the right of local government. These charters were an extraordinary privilege in the seventeenth and eighteenth centuries. Since they controlled the sites around which the large-scale production and distribution of products revolved, local governments regulated virtually all aspects of commercial and manufacturing life. Almost everyone engaged in those activities had some government-related task. For example, a partial list of governmental posts in Boston around 1700 included

> constable, town clerk or recorder, treasurer, assessors, collectors, surveyors of highways, clerks of the market, fence viewers, hog reeves, pound keepers . . . , common drivers . . . , water bailiffs, cow keepers, town drummers and teachers of town drummers, tithingmen, perambulators, ringers and yokers of swine, sealers of weights and measures, keepers of ordinaries, town bellmen, cullers of staves, measurers of corn and of boards, corders of wood and overseers of wood corders, overseers of chimneys and chimney-sweepers, overseers of almshouses, gaugers, viewers, surveyors of casks of tar, fire wards, town criers, informers of offenders against the license laws, scalers of leather, licensors and inspectors of brick makers, cullers of fish, inspectors of hides for transportation, measurers of salt, packers of flesh and fish, inspectors of the killing of deer, deer reeves, school wardens, school teachers, truckmasters, brewers, rebukers of boys, sizers of meadows, warners of town meetings, scavengers, viewers of lands, lot layers, judges

of delinquents at town meetings, judges of boundary disputes, branders of cattle, pinders, jurymen, town cannoneers, commissioners for equalization of assessment, town fishers, town deputies, town doctors, town grubbers, and persons to keep dogs out of church.[1]

And Boston's population at the time stood at only 6,700.

Most of these positions required only a small outlay of time. They merely suggest the scope of local government's remarkable involvement and authority in commerce and manufacturing. The local government owned the docks, markets, counting houses, warehouses, and often the vessels. It required idle persons to work. It regulated who might enter the locality, how long they might remain, and what occupations or trades they could pursue. It certified when apprentices could establish their own shops. It set prices for goods and services, fixed wages for workers, and mandated the number of persons allowed to practice each trade.

Local government was essentially a closed commercial corporation (as was any governmental unit), for its powers eliminated economic competition within its jurisdiction. Regulation of wages, prices, and number of workers placed the welfare of the plantation, city, or town above the desires and ambitions of individuals residing and working there. At the same time, these restrictions guaranteed that all citizens would share—not necessarily equally—in whatever gain resulted.

A local government's influence typically extended far beyond its legal boundaries to the production in surrounding areas. Authority to determine who could enter the locality and participate in commercial endeavors enabled local government to control access to both the largest local market—the city, town, or plantation—and the European trade. Most rural artisans or agriculturists lacked the means to ship their products to Europe and therefore had to seek assistance of those closed commercial corporations. And that meant agreeing to abide by that local government's stipulations.

Channeling Colonial Economies   Rather than merely regulate trade and production, however, colonial governments aimed to direct or channel them. Each colonial government strove to encourage its citizens to produce goods and services essential to the colony's survival, and to increase its wealth. The latter ambition produced attempts to decrease importation of certain commodities from other colonies and European states, and to expand exports to those entities. The desire to approach self-sufficiency, so as not to squander wealth through importation, and to tap export markets to corral additional wealth—two keystones of mercantilism—led colonial governments to favor some industries, places, and people over others. The colony as a whole, not just some of its individuals, was to benefit.

These manipulations of economies took several forms. For example, the Virginia Assembly in the 1660s presented premiums for the manufacture of linen thread, prizes for the best woolen cloth, and rewards based on tonnage for ship building. As a method of stimulating local manufacture, it also prohibited importation of salt. Some colonies fostered production by levying duties on imported articles. Pennsylvania in 1704 laid an import tax on hops, and Rhode Island in 1731 placed a similar duty on beer and ale. Other colonies granted land. Connecticut offered land in return for establishment of a sawmill, Rhode Island for a grist mill, and Maryland for a flour mill. Monopolies were common practice. Massachusetts gave a twenty-one-year monopoly to the proprietor of the Boston dry dock and Rhode Island provided a fourteen-year guarantee of exclusiveness to the owner of the Providence sawmill. Outright grants of cash were sometimes employed. Massachusetts used a bounty to encourage investors to develop a papermill. Other enticements included exemptions from taxes for specified periods.

THE SEARCH FOR COMMUNITY SELF-SUFFICIENCY: THE SEVENTEENTH CENTURY

The Ubiquitous Mill

That several colonial governments encouraged mill establishment suggests the mill's New World significance. Mills were crucial for large-scale production, but on a smaller scale they also were among the first institutions erected in pioneering settlements and served as foci of rural communities. Unlike the Old World, which faced scarce material resources and labor abundance, the New World possessed an abundance of land and minerals but a shortage of labor. New wealth could be acquired by tapping natural resources and cultivating lands; governments and joint-stock companies profited by enticing men and women to use their technical skills to develop new areas and to produce additional products. Success in attracting colonists to unsettled regions improved markedly when mills were established to ease work burdens.

Mills provided colonists with power generated by continuous motion. Animals or people turned the simplest mills by walking in circles or on treadmills. Windmills were especially common in the early 1600s. Neither type of mill proved ideal, however; the former demanded investment in labor or in beasts of burden, while the wind's capriciousness limited the latter. Watermills were more reliable and powerful, and settlers capitalized on the New World's plentiful streams and rivers.

Colonists used three basic, ancient watermill forms that relied on differently positioned waterwheels. In the horizontal or Norse mill, the wheel lay flat, immersed in the water. Horizontal mills were the least powerful; their chief virtues were cheapness of construction and resulting plane of rotation. No gearing was necessary to accomplish tasks on a plane parallel to the river. Wheels of the other two watermills—undershot and overshot waterwheels—stood perpendicular to the river and delivered motion on that plane. By means of wooden or metal gearing, that motion could be directed where it was needed, even translated to a different plane. Current turned undershot waterwheels by running beneath them; only their lower parts lay in water. Overshot wheels not only decreased the friction generated by having a portion of the wheel constantly immersed, but they also used gravity to help provide power. These wheels were placed below water courses; thus water fell only upon the wheel's top flutes. Such an arrangement usually required situating the wheel in a trough and digging a raceway to channel the river there. Capital investment was greatest for overshot watermills, but they provided the greatest power and were therefore employed in many large-scale enterprises.

An extraordinary variety of mills dotted the colonial landscape. Gristmills appeared on plantations as early as 1617, in Jamestown in 1621, in Massachusetts in 1628, and in virtually every community, including small rural communities, soon thereafter. The earliest colonists and those lacking access to gristmills used querns * to grind grain into meal, a laborious, thankless task taking at least two hours to meet each family's daily needs. Quern-ground meal yielded a coarse, unpalatable bread. Gristmills were so highly prized for the time and labor saved, as well as for the quality of bread produced, that remote settlers willingly traveled several miles to the nearest gristmill to have their corn, rye, or wheat ground. They gave millers a portion of the milled product as payment.

These rural gristmills were generally small and usually powered by horizontal waterwheels. A structure above the wheel housed

---

* *querns:* primitive handmills used to grind grain.

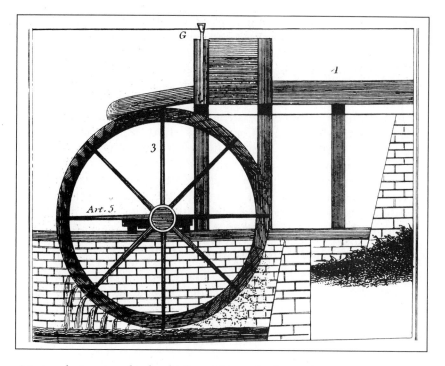

An overshot waterwheel, which combines the force of gravity with water current to increase power

two grinding stones. The top stone was attached to the wheel by a rod in its center. The rod passed unobstructed through a hole in the middle of the lower stone, which was fixed. Movement of the top stone against the stationary lower stone milled the grain.

Rural sawmills proved nearly as desirable as gristmills. Land-clearing provided colonists a ready supply of logs for fuel and shelter, and sawmills speeded and eased the cutting of logs for firewood and timbers. Most colonial dwellings and other structures were framed with oak timbers, not formed of logs chinked with wood chips. Sawmills (typically horizontal watermills) fashioned rough framing timbers, which were sometimes finished by broadax and adz before the frames were assembled on the ground and mortised a side at a time. Corner braces and vertical studs were added to provide the frames stability when raised. Thick clapboards were then nailed to the studs as siding to protect inhabitants from the often harsh New World climate. Roof shingles generally were not sawmill cut, but split by froes * and tapered by drawknives. †

Colonial mills numbered in the thousands. Their importance in the colonies' life and economy marked those who built and repaired them as the New World's most treasured artisans. These millwrights combined knowledge of the various wood-building crafts with smithing and surveying, a rigorous combination of skills that further contributed to their extremely short supply. Millwrights immigrating to the New World filled only a small portion of colonial need and carried with them the latest European practice; colonial governments repeatedly attempted to entice other European millwrights to follow. But immigration and inducements failed to remedy the New-World millwright shortage.

The Colonial Master–Apprentice System

Like most other colonial artisans, many New World millwrights learned their special skills and gained knowledge of their trade through the master–apprentice system. This system was of European origin; colonists adapted it to fit a world characterized by abundant resources and critical artisan and labor shortages. They blurred the lines among apprentices, journeymen, and masters, and generally accelerated and informalized the apprenticeship process. In its most pristine and time-honored European form, the system created bonds between masters—those known for proficiency in a craft or with a process—and apprentices. Apprentices were generally boys in their pre- or early teens who were commended to a master's service by their parents for up to seven years. While in service, apprentices assisted masters. In addition to instructing their charges in their art, masters often were obligated to clothe, shoe, house, and feed apprentices, to guard their morals, and to teach them to read and write. When the apprentice period ended, masters provided the boys with letters certifying that they had served admirably and testifying to their adeptness in the rudiments of the art.

Rarely did these graduates move directly into the ranks of masters. They lacked money to purchase tools and to rent or buy a facility in which to work and live. To secure necessary funding and

---

* *froes:* ax-like cleaving implements for splitting logs.

† *drawknife:* two-handled bladed tool for shaving surfaces by pulling the device toward the operator.

hone their new skills, these young men took positions as journeymen in a master's shop. Journeymen earned wages for their labors; after a period of years, when the opportunity presented itself, they could claim for themselves the mantle of master.

In Europe, guilds or local governments set limits on the number of journeymen and apprentices a master could accept at one time. These restrictions did not apply, however, to the training of kin. Sons and sons-in-law, brothers and brothers-in-law, nephews and grandchildren generally were welcomed into a relative's home to learn an art. Their relationship to the master eased their burdens in a number of ways. They incurred a less rigid set of obligations than did formal apprentices. They also often found themselves able to decrease the duration of, or even to skip, the journeyman stage. Masters sometimes lent relatives money to start their own operations.

The informality of the master–apprentice relationship in the colonies transcended even that accorded kin in Europe. Demand for artisans, especially in rural and pioneering settlements, so far outstripped supply that individuals possessing only rudimentary training and skills sometimes were looked on by desperate settlers as valuable commodities. Lack of consistent, formal stewardship produced great variation in practice among a trade's colonial artisans, which occasionally fostered beneficial experimentation and modification but more often resulted in inferior techniques.

Providing Necessities

Colonial Agriculture   For much of the colonial period, joint-stock companies and proprietors armed with royal charters unsuccessfully attempted to transplant forms of the European manorial system to the New World. Colonial America's abundant, available land undermined these repeated efforts: with alternatives manifestly present, few colonists chose to live as peasants. Unlike most European rural peoples, the overwhelming majority of colonists lived on their own or rented land, grew their own food, and made their own cloth. Moreover, the virgin New-World soil was forgiving and seemed to require neither regular fertilization nor careful rotation of crops.

But it would be a mistake to conclude either that each colonial farmstead was an island unto itself or that the structure of agriculture and the tenor of agricultural life was similar in the North and South. Plantations dominated the southern agricultural scene by 1700 (though never were they the most numerous form—small farms far outnumbered them), while small farms, usually from 10–200 acres, marked the Middle Atlantic and New England colonies. In the North, settlements and towns formed the basis of community; plantations served a similar function in the South. Towns and settlements provided northern agriculturists access to markets, goods, amenities, and services, while each plantation supplied its residents, as well as small farms nearby, the same advantages as towns. A full complement of artisans, many of them black slaves, lived on plantations. Most plantations were located on waterways, and owners purchased or built ships to transport goods to other colonies and Europe. Dozens of people lived on each plantation, but only a handful—usually a family and perhaps a few hired hands, indentured servants, or slaves—resided on a typical northern farm. Although each plantation generally grew agricultural staples sufficient to feed its workforce, the plantation economy relied on slave labor or indentured servants and on volume production, exporting tobacco, rice, and indigo. Northern farms primarily produced grains, beef, pork, flax, and poultry; they were much less dependent on the export trade.

Northern and southern agriculture differed from each other and from England, but agricultural practices and implements of all three areas were fundamentally similar. Blacksmiths, the most common colonial metalworkers, used charcoal-fired hearths, bellows, tongs, anvils, and hammers to fashion a wide assortment of agricultural tools. Earliest settlers did not have plows to break the soil and instead used mattocks,* hoes, and spades. Plows appeared regularly from the 1640s on, but many northern farmers could not afford to own one. Their lands often were tilled by other farmers who had purchased a plow made in the colonies—some farmers even worked as blacksmiths—or from England. Towns sometimes subsidized plowing by paying bounties to plow owners for preparation of local fields. Plows manufactured in Europe were well-crafted articles, but domestic plows were rather crude affairs. Shares were almost always iron, and moldboards usually wood. To slow moldboard deterioration, farmers frequently had blacksmiths cover the wood with thin iron strips. These awkward implements required the force of several yoked oxen to pull them. Only in the mid-eighteenth century did colonists adopt a plow that cut turf more facilely. The shovel plow, its share shaped like a shovel with the convex side

---

mattocks: similar to pickaxes.

turned outward, cut a shallow furrow and could be pulled by a horse or mule. Southern plantations used it extensively.

Plowing was not colonial agriculture's only labor-intensive facet. Animals pulled brush harrows—small tree limbs sometimes weighed down by chains—to break cloddy soil and smooth fields. Farmers broadcast grains and covered newly sown seed with another pass of the brush harrow. Peas and corn were planted by dibble stick and hoe: the farmer poked a hole in the soil with the stick, dropped in several seeds, and hoed dirt over them. Corn and pea seedlings were later thinned by hand, and hoes were used for weeding. Settlers harvested crops with iron-bladed and wooden-handled sickles. Stalks were raked together with hand rakes, bound into bundles on the ground, and carried to barns for threshing.

Threshing was the only agricultural activity that did not depend on the weather. The simplest threshing was carried out indoors with a flail, two wooden sticks joined together by a piece of leather. Farmers placed stalks on floors and beat them with flails to free grain from heads. They then collected grain and chaff and set them aside for winnowing. This threshing method worked well on small farms but proved much too slow for large grain-producing farms such as those of the mid-Atlantic colonies. There colonists used horses to separate the grain by spreading stalks in a circle and repeatedly walking their animals on top of them. Straw was then brushed away, leaving only chaff and grain, which was winnowed either by sieve or wind. For wind-winnowing, farmers placed chaff and grain in a basket and tossed them in the air outdoors or in well-ventilated barns. The heavier grain would fall straight down, while the lighter chaff would blow away.

Threshing yielded seeds for next year's crops and grain for bread. It also provided the raw material for beer, which reigned as the universal beverage for most of the seventeenth century. Colonists had numerous recipes for the brew; flavoring it with spruce, ginger, or molasses was most popular. Beer was regularly consumed in the home. Even the Puritan fathers approved of the drink. In fact, they converted the Great House, the official residence of Governor John Winthrop, into a tavern in the 1640s. Increase Mather best summed up the Puritan view. He argued that "drink was in itself a good creature of God, and to be received with thankfulness," but that "the abuse of drink" came "from Satan";[2] alcohol was acceptable, but drunkenness was not.

Colonial beer-making was based on traditional European practice and required no special skill or apparatus. Home brewers needed only three barrels and a pot, and large-scale production was only slightly more complicated. Commercial manufacturers, who in the seventeenth century served taverns and settlements, took barley, soaked it in water until it sprouted—germination converted some of the starch into sugar—and dried it in the sun or a kiln to produce malt. The malt was then ground or crushed in a mash tub, and heated or boiled in a copper vessel called a cooker. The brewers next passed the mash through a strainer and transferred the liquid—called wort—first to a cooler and finally to a fermenting tub. Yeast was added, and the batch was left to sit. After a period of weeks or even months, the yeast was strained off and the beer was transferred to wooden kegs. The yeast from an especially flavorful batch of beer was saved and used again.

Colonial Textiles   Though beer helped colonists cope with the New World's sometimes harsh climate, clothing offered a more sustained warmth. Toward this end, colonists transformed plant and animal products into cloth and leather. Until the late eighteenth century, North Americans and rural Europeans employed virtually identical processes. Both men and women participated in the time-consuming clothing-making processes; men commonly sheared sheep, skinned animals, and gathered fibrous material from flax, while women characteristically cleaned wool and flax fibers, and spun yarn and thread. Both sexes operated the family-owned hand loom, but in the colonies itinerant weavers were occasionally employed. Fulling was the only cloth-making operation ever done outside the home; woolen fabric was sometimes sent to a fulling mill for finishing. Unlike cloth, however, leather goods were made entirely by artisans.

New Englanders favored woolen articles to protect them from the elements, and Southerners opted for cool linens (cotton did not become important until the late eighteenth century), but neither employed either fabric exclusively. Use of both materials required colonials to master two quite different processes. Woolen yarn-making began with washing wool in tubs with soap to free dirt and grease, drying it in the shade, and beating it to rid fibers of vestiges of foreign material. Colonial women then carded the wool by hand, using leather paddles studded with fine wire. After disentangling, sorting, and straightening the wool, they spun fibers into yarn either on a distaff and spindle or on a spinning wheel. Colonial wheels were usually hand powered, and consisted of a cord that passed around the spindle and over the large wooden wheel. They enabled spinners to revolve their spindles more evenly and quickly, which improved yarn quality and decreased spinning time. The spinner's left hand undertook the more difficult tasks, freeing the right for turning and regulating wheel speed. Carded wool was held in the left hand and attached to the spindle. It was then drafted and elongated. The wheel was put in motion while the yarn was twisted to bind the fibers, and wound on the spindle.

Preparation of linen thread or yarn from flax took longer and demanded strength. A woody stalk, which required removal, surrounded the fibrous material of flax that would become linen. Men retted flax, that is, removed the fiber from the stalk, by either soaking it in water for two weeks or laying it on the ground for as much as a month to putrify. The softened flax stalk was then dried and forcibly cracked in a wooden beam press called a flaxbrake. A wooden knife was then used to separate much of the stalk from the flax fibers. Then a pair of thick-wired wooden paddles were used to remove the rest of the woody material. After washing and bleaching, the fibers were ready to be spun.

Spinning linen bore little resemblance to spinning wool. To make linen thread or yarn, women wrapped the raw material on a distaff and separated its fiber filaments by hand. They then fed the fibers to a foot-operated flax wheel.

Each colonial family spun its own thread and yarn, and then wove it into cloth on wooden hand looms. Like spinning wheels, these looms remained similar in design to those built centuries earlier in Europe. Weaving involved crossing and interlacing two distinct series of threads at right angles to one another. Warp threads were fixed on the loom and provided the frame for the cloth piece to be produced. The weft was wound in single threads on small spools set one after another in a movable shuttle, which was perpendicular to warp threads. The shuttle was passed alternately over and under warp threads from one side of the piece to the other, then returned in the same fashion, but on reverse sides of warp threads. This time-consuming process continued until the entire length of warp threads had become a woven piece of cloth.

Local carpenters built looms and colonists often situated these large devices in barns or sheds rather than houses. Loom frames consisted of four upright posts joined to form a rectangle by cross beams at the top, middle, and bottom. Looms also had cross pieces at the top to suspend heddles, movable parallel cords equipped with eyes and used to separate and guide warp threads. Pulleys and cords connected heddles to pedals called treadles. Warp threads were attached to the back center beam—the warp beam—and passed through the appropriate heddle eyes and connected to the front center

Most colonial cloth was made in the home. An itinerant weaver works the loom as a spinning wheel rests in the foreground.

beam—the cloth beam. Depressing a treadle moved the corresponding heddles and opened space in the warp threads to pass the shuttle. The loom's final major piece, the batten or beater, pressed weft threads into place after the shuttle's passage. A movable frame attached at the loom's top just in front of the heddles, the hand-operated batten was pulled forward sharply after each pass of the weft thread; its combed teeth drove the thread to its place on the cloth.

Woven linen cloth was ready to be made into garments, but fine woolens required fulling (shrinking and thickening) to provide body. Woolens, when they left looms, were uneven and unsightly in appearance, but hand fulling was such an onerous and time-consuming task that early colonists fulled only those woolen pieces worn on special occasions such as Sabbath worship or holidays. Fulling removed grease (a by-product of spinning and weaving), shrunk the cloth, and consolidated fibers. The fabric was fulled by soaking it in warm, soapy water and either beating it while wet with sticks on a firm surface, or kicking the damp cloth around with bare feet for several hours. Household fulling improved the fabric's character somewhat, but fulling mills did a superior job.

Known in Europe since the Middle Ages, fulling mills were found in New England, where demand for woolen goods was greatest, from the 1640s. Mills did not appear in the mid-Atlantic and southern colonies until the beginning of the eighteenth century. Soon after their introduction, they became almost as widespread in these regions as grist- and sawmills. Fulling mills, like other mills, encouraged settlement in sparsely populated areas by lightening settlers' labors, and by providing an attraction to emigrating European people. But unlike saw- and gristmills, fulling mills were not essential to either colonial survival or easing regular daily burdens. Their rapid proliferation indicated that New England rural society in the mid–seventeenth century, and mid-Atlantic and southern rural communities some sixty years later, desired and could sustain enterprises not directly linked to their immediate survival.

Rural communities as well as cities and towns seemed to produce wealth enough to support skilled craftsmen called fullers to operate fulling mills. These artisans placed soapy, wet bolts of cloth, tied end to end to make the process continuous, in a beating trough, where it was squeezed through waterwheel-powered rollers. These rollers supplied the fulling force in early mills, but later ones added trip-hammers * to pound and intermesh fibers more completely. Fullers then proceeded to finish the fabric by stretching the wet cloth on frames and attaching it with tenterhooks, sharp hooked nails, to rid it of wrinkles and to maintain its shape. When it dried, they teased it, using either wire-studded paddles or the seed pods of a plant known as fuller's teasel. Teasing raised the nap, but not evenly; fullers then evened it by laying the fabric on a wooden cylinder and shearing the cloth with long-bladed shears.

Colonial Leather Making   Leather's durability and flexibility were prized for harnesses, aprons, breeches, saddles, and especially shoes and boots, and its manufacture was well established in New World settlements by the mid–seventeenth century. Unlike textile industries, however, leather making threw up terrible smells, which led to demands that tanneries be situated only on settlement fringes, far from dense population concentrations.

Preparation of skins and hides for tanning took time. After receiving hides or skins, tanners cut off horns, tails, and ears, and soaked the remaining parts for a day in running water. They then

---

* *trip-hammer:* a mill-driven hammer connected by an elliptical-shaped cam shaft. The long part of the cam raised the hammer, which fell of its own weight.

immersed them in vats of lime to swell the skins and loosen hair. Hides usually remained in this solution for several months. Next they were ready for beaming, during which tanners placed hides hair-side-up on an elevated log (or beam), and proceeded to scrape off hair with curved two-handled knives. They then flipped the skins over and scraped away the remaining flesh and fat. Hides were washed again, then rubbed and smoothed by stone. One final washing and they were ready for tanning.

As tanners readied skins, they also gathered and prepared their source of tannin—hemlock bark in the North and oak bark in the South—which they ground into coarse powder in an animal-powered bark mill. Such a mill had a narrow, circular trough filled with bark to be crushed. A rotating vertical post was placed at the center of the trough. A pole joined the animal and the post, which served as the axle for a wooden or stone wheel with corrugated edges. The animal walked around the trough and moved the wheel, which in turn pulverized the bark.

Tanning itself was at least as time-consuming as skin preparation. In the initial phase, tanners dumped hides and skins into a large vat and added a weak solution of bark and water. They gradually strengthened the infusion over several months, and regularly redistributed skins to insure uniform tanning. It took from four to six months to tan calfskin or buckskin, but cowhides required a second, more drastic treatment. They were removed from the vat and allowed to dry. Tanners then filled another vat with one-inch layers of ground bark alternated with layers of hide, and flooded the vat with water. The hides were repeatedly agitated, and remained in the vat for as much as a year. The leather was then washed and hung to dry. Workmen completed the tanning by taking heavy clubs and pounding the leather to compact it.

Shoe and boot manufacture consumed much of the tanners' leather, and was predominantly a local enterprise. The relative expense of shoes and boots, coupled with the demands of colonial labors, made fit and durability crucial. Customers often took such an active interest in their footwear that they purchased leather for shoes and boots directly from tanyards and supplied it to shoemakers. Artisans reciprocated by measuring a foot in several places and whittling a last, or model, to reflect the foot's unique characteristics and contours. The last provided a guide for making the sole of the shoe or boot. Thick leather was cut to conform to the last's shape, beaten to increase flexibility, and smoothed. Shoemakers also used a model for the uppers, but did not build it to individual specifications. They kept several wooden patterns on hand

Each colonial shoemaker made a complete shoe.

and chose by visual inspection the appropriate form. They then fashioned the uppers in two pieces: the vamp, which covered the toe and instep and ended in a wide tongue; and the counter, which covered the heels and sides and ended in straps for either lacing or buckling.

Shoemakers next stretched the uppers over the last, allowing some material to hang over its edge, and inserted temporary tacks. Unlike modern shoes on which uppers are tucked under and sewn to soles, colonial uppers were turned out. To hold the leather in the proper position for stitching, craftsmen glued the uppers to the sole. They then cut a narrow channel in the sole to keep the stitches from contact with the ground, and punched regularly spaced holes through the leather to facilitate sewing. Using waxen hemp or linen as thread, and hog bristles as needles, shoemakers joined the two parts together. They then added the heel by nailing to the sole several heel-shaped thicknesses of leather. Finally, they removed the tacks and last, pasted in linen linings, and swabbed on tallow or lampblack and wax to protect their creation.

Large-Scale Enterprise

Most seventeenth-century technical activities were small enough that they could be established even in sparsely inhabited rural settlements. Local persons generally met local needs. Iron manufacture and shipbuilding deviated from that pattern, however. The largest of colonial enterprises, both were pursued in limited areas, involved a range of craftsmen, and served distant communities. Both took advantage of raw materials abundant in the New World but scarce in England, and as such reduced colonial and English imports. Using technologies to transform colonial ores and trees into iron and ships, these enterprises enabled England and its colonies to retain wealth that otherwise would have had to go to other countries for the purchase of these crucial commodities; they were manifestations of the mercantile philosophy in action.

Colonial Iron Manufacture   Iron manufacture was perhaps the most demanding colonial enterprise. It required several types of skilled labor, copious amounts of wood, readily available supplies of flux, and plentiful sources of ore. These requirements determined where iron-production facilities could be easily located. Colonists found it far easier to transport finished iron bar than bulky raw materials. They therefore erected ironworks in the wilderness near resources. Unfriendly Indians and fires, however, often plagued these wilderness enclaves.

Workers required support while in residence at ironworks—a considerable expense. Instead of relying on supplies from surrounding settlements, colonists organized these wilderness production facilities to be self-sufficient entities. In that sense ironworks resembled the agricultural plantations of the southern colonists, and the patroon manors of New Netherlands. Sizable land grants from the Crown, proprietors, or colonial legislatures usually accompanied permission to engage in iron production. Such a grant offered both the raw materials necessary to produce large quantities of iron and the means to maintain the work force. On these lands artisans, laborers, indentured servants, and slaves built houses, offices, barns, mills, stores, ovens, roads, and watercourses. They also cleared portions of the area for orchards, grain fields, and grazing animals. These men and women also remained at the site after initial construction and provided support services for large-scale production.

Iron production was among the oldest of colonial technologies. Its long history stemmed from both its extensive New World application and European encouragement. Britain's perennial shortage of wood and the Crown's dependence on high-priced Swedish iron placed a premium on establishing colonial manufacture. As early as 1619, the Virginia Company of London sent 150 men to North America to erect three ironworks. John Winthrop, Jr., son of the Massachusetts Bay governor, toured England in the 1640s and attracted new investors for colonial ironworks, which led to the

*Colliers cover a mound of wood to prepare charcoal for
smelting iron.*

creation of no fewer than eight separate works in New England
prior to 1680.

Gathering and preparing three essential ingredients constituted
the first step in iron manufacturing. Workers mined ore from bogs,
outcroppings, and shallow deposits; felled trees and converted wood
into charcoal; and collected flux, usually limestone.

High-grade ores might be ready for smelting, but often ores required roasting or washing to remove impurities. Charcoal production was more problematical. Colliers, artisans skilled at making
charcoal, needed to exercise great care to burn wood slowly, evenly,
and only partially. Too rapid or complete combustion not only ru-
ined the wood for iron-making purposes but also raised the specter
of fire. Colliers controlled the rate and extent of combustion by
limiting the oxygen supply. They buried wood in earth-covered pits
or poured dirt mounds over cone-shaped wooden stacks. In both
furnace configurations, colliers tried to eliminate air pockets by
piling wood precisely—often as much as 30 cords—and filling natural gaps with smaller pieces. At the center of these charring furnaces was a chimney. It was filled with wood chips, ignited, and
then covered, the fire radiating from the chimney to the stacks

underground. Draft holes were sometimes cut into the pile to ensure uniform charring.

The entire charcoal-making process took more than a week. Colliers stood ready at all times. They used probing rods to detect uneven burning, and warily protected against open flames, gas explosions, and strong winds. The last several days were spent cooling the charcoal. They removed the outer covering of the stack slowly, a portion at a time, and kept the cooling material in small quantities away from the great mass. Only when they were sure it was cold did they carry the finished charcoal to the coal house.

The three raw materials went to a bloomery or to a blast furnace. The bloomery was a modest contraption, used primarily to furnish small quantities of iron for local consumption. It consisted only of a fireplace, bellows (usually pumped by an apprentice), tongs, and a hammer. The ore, charcoal, and flux were placed in the fireplace and heated with the aid of bellows. A spongy mass of iron—a bloom—resulted. It was lifted by tongs, reheated, consolidated by hammer, heated again, and pounded again to drive off more impurities. This wroughting process continued until refining was complete.

The blast furnace was a more magnificent structure. Standing roughly twenty-five feet tall, shaped like a pyramid with its capstone removed, it was suitable for large-scale iron production and devoured enormous quantities of raw materials. Thick stones mortared together formed its outer walls, which were lined with bricks or clay. Internally, the furnace had a narrow throat to accept the ore, charcoal, and flux. This passage gradually widened to a chamber with a diameter of ten feet. About three yards above the ground it tapered quickly to a diameter of roughly two feet; this sloping area functioned as a platform to support the three ingredients. This circular platform drained into the crucible, the reservoir for molten iron and liquid slag, or waste. Attached through a porthole were huge twenty-foot-long leather bellows, inflated by means of cams on giant overshot waterwheels. Stone counterweights enabled the bellows to discharge their powerful blasts, which helped generate the extreme temperatures necessary for iron-smelting.

Most activity revolved around the crucible. Usually once or twice daily, workers drew the slag off the top, opened a clay plug to tap the molten iron, and let it run into a series of sand molds to make bars. They also used iron for casting pots, kettles, and the like, ladling the liquid metal into clay molds. The draining of molten iron did not signal the end of blast furnace operations, however;

the cycle continued uninterrupted for as long as forty weeks. At that time, ironworkers repaired the furnace's inner lining and leather bellows, and gathered the raw materials for next season's smelting.

Only a fraction of the iron went directly to produce finished goods. It lacked the tensile strength required for high-stress applications and needed further refining. Its conversion into wrought iron occurred in two stages at a forge. In the finery, where the first part of the process took place, workmen heated the iron to melting in a bellows-assisted charcoal fire. They stirred the molten metal several times in the presence of air to burn off excess carbon, then allowed it to solidify. They next placed the red-hot metal on a stout iron anvil and beat it repeatedly with a massive trip-hammer, which was raised by cams on a waterwheel shaft and fell by gravity. The force of the hammer's blows squeezed out many of the remaining impurities. The resulting iron mass resembled a flattened dumb-bell. It now went to the chafery, a cooler, non-bellows-aided fire, for the final purification step. Attention there focused on the bar's ends. Both were heated and subjected to the great hammer. When the consolidated iron cooled, it left the chafery as bars of wrought iron.

These bars reigned as the principal form of commercial iron. Some were exported to Europe, but most remained in the colonies. In blacksmiths' shops they became axes, horseshoes, plows, wires, tools, and blades. Wheelwrights used the material to fashion chains, fittings, and iron tires necessary to outfit long-lasting wagons. For a few important products, however, the bars underwent further processing in special facilities. Most common were waterwheel-powered rolling and slitting mills. In rolling mills, heated iron bars were pressed several times through a pair of heavy rollers rotating in opposite directions. The resulting metal sheets could be used as plate iron or sent to slitting mills, where sharp-edged discs cut the iron into uniform strips suitable for manufacture of nails.

Colonial Shipbuilding   Iron also played a significant role in colonial shipbuilding. Shipsmiths used iron for each ship's several anchors, for iron braces to hold the rudder to the ship, and for iron bolts to attach the keel to the rest of the ship's frame and the rigging to the hull. But wood was the most important shipbuilding material. Colonists felled oaks by broadax, then dressed them into timbers with an adz or sawed them into planks at sawmills. They selected locustwood to supply wooden dowels (treenails) for

A master shipsmith, journeymen, and apprentices forge an
anchor.

fastening together timbers and planks. Two other shipbuilding staples, sailcloth and cordage,* were usually imported from England.

Massachusetts' possession of abundant quantities of wood and
iron contributed to its rapid ascent as the New World's shipbuilding center. So, too, did its concentration of peoples intimately acquainted with commerce—Pilgrims and Puritans—and the colony's
short growing season and rocky soil, which made agriculture difficult. By the late 1630s, the towns of Boston and Salem did a big
shipbuilding business. Their craftsmen constructed ships for English and colonial merchants primarily to carry timber and salted
fish, but by the late seventeenth century they were transporting
turpentine and rum distilled in New England. Turpentine went to
England, while rum fueled the odious "triangular trade." New England merchants traded rum in Africa for slaves, sold the slaves in
the West Indies and southern colonies, and carried back molasses
from which to make more rum. They also used Massachusetts-built
ships to trade corn, wheat, cattle, fish, and horses for molasses in
the West Indies, and rum to fishermen in the icy North Atlantic.

The shipbuilding trade relied on bulky raw materials, but unlike

---

* cordage: ropes and cords for rigging a ship.

iron production it was a predominantly urban industry. Colonists found it simpler to transport iron and wood from forests than completed ships. To ensure ease of launch and low overhead, shipyards generally were located in the lowlands and coves near rivers, on inexpensive land away from dense settlement. There worked a diverse force. Some yards were manned by only a shipwright and a few day laborers, while others employed shipwrights, shipsmiths, sawyers, carpenters, joiners, and caulkers. In a highly structured shipyard, shipwrights supervised construction; shipsmiths did iron work; sawyers worked sawmills; carpenters secured planking; joiners smoothed planks, built deck rails, and finished cabins; and caulkers filled in seams with oakum, tar-covered unraveled hemp.

The shipbuilding process began with the ship's eventual owner. He provided rough specifications for the ship to the shipwright, who drew up detailed plans—perhaps even made a scale model—submitted them for the owner's approval, and superintended construction. Workers laid the keel first, then fastened stem and stern posts (the front and rear posts, respectively) to its ends. They then built the ribs, each of which was composed of several smaller timbers—the floor anchor, floor, three or more futtocks, and two top timbers, one long and one short. Only by dividing the ribs into sections could shipbuilders find lumber of the desired curvature. Each rib was two timbers wide, with the overlapping pieces pinned together to provide the strength necessary to bear the anticipated load. The floor anchor was the first rib timber attached. The keel lay beneath and bisected it. Joined to the floor anchor successively were the first and third futtocks and the shorter top timber. The rib's other half was composed of the floor, which was nailed to the floor anchor and the first futtock; the second futtock, pinned to the first and third futtocks; and the longer top timber, attached to the first futtock and the other top timber. After they completed the framing, ship carpenters placed the keelson on top of the floor anchors, and bolted it to them and the keel, which stiffened and fortified the frame. They then planked the ship's interior and exterior, pausing at deck level to construct the deck. The deck beams were supported on both sides by wooden braces (knees) fastened to each rib, and in the middle by wooden columns (stanchions) attached to the keelson. The planking resumed when the deck stood in place.

All that remained was to finish the craft. Joiners dressed the rough surfaces with planes. Caulkers used wheels, irons, and mallets to pack every crevice with oakum. Sheathing the part of the ship that would rest underwater was the final operation. Shipwrights coated

*A large shipyard located on the fringe of New York City,
ca. 1717*

those planks with a mixture of tar and hair, and sometimes covered them with fir boards.

BEYOND SUBSISTENCE: THE EIGHTEENTH CENTURY

The colonial economy moved from subsistence to surplus in the
eighteenth century. Several generations of immigrants had successfully applied and modified European technologies and practices to
suit New World conditions and needs. As the English had hoped
when they founded the colonies, the continent's plentiful material
resources enabled colonists to produce surpluses of several commodities in short supply in the Old World and to begin to export
them. After about 1700, coastal towns such as New York, Philadelphia, Boston, Providence, and Charlestown emerged as prominent commercial centers and collected raw materials and finished
goods from rural settlements for shipment to Europe or to other
colonies. The flourishing export trade heightened demand for ships,
and North American forests furnished readily available supplies of
lumber. Shipbuilding spread to every colony; Pennsylvania, New
Jersey, Rhode Island, and Delaware contested Massachusetts' claim
of colonial shipbuilding supremacy. Expansion of iron production
exceeded that of shipbuilding. Each colony made some iron, but
the New England colonies, Virginia, Maryland, and particularly
Pennsylvania were the heaviest producers. By the third quarter of
the eighteenth century, colonial installations accounted for a full
one-seventh of the world's annual iron output.

Colonial agricultural surpluses helped support the populations necessary for increased iron production and shipbuilding, and for the expanding mercantile centers. Agricultural productivity was sufficiently great that colonists could also export some products. New York and Philadelphia became foci of the beer export trade. New York post mills (primitive windmills) and Philadelphia water-mills powered flour mills as colonists began to tap overseas urban markets, which demanded a more highly refined product. Colonists developed methods to clean grain, bolt (sift by sieve) meal to rid it of impurities, and cool and pack the flour, which they shipped to the West Indies and Europe. By the mid-eighteenth century, Baltimore, Wilmington, and Richmond also exported substantial quantities of flour. Shoemaking also developed a made-for-export component as New England shoemakers, with as many as forty artisans working in a single place, made entire shoes and boots in a handful of standard sizes and shipped them to southern plantations and the West Indies.

Signs of New Wealth

In accordance with mercantile theory, colonial merchants handled these exports and accumulated the largest portion of the wealth. New techniques, processes, trades, and institutions catered to these wealthy coastal traders. A move to more expensive brick structures—houses, churches, warehouses, and counting houses—in East Coast municipalities indicated the colonies' economic success. The clay used to make bricks was prepared in a pugmill, a pegged wooden post inside a large wooden box. An animal hitched to the post walked in a circle and turned the post as laborers poured water and shoveled clay into the box. When the mill had worked the mixture into a paste, it was drawn out through a plug in the box's bottom and placed in brick molds, which had been dampened and dipped in sand so clay would not adhere. Brickworkers forced clay by hand into the mold's every crevice, scraped flat brick tops, and left bricks in the sun to dry. They then heated the bricks in wood-fueled kilns for about a week. Bricks were now ready for bricklayers, who joined them with sand and lime mortar.

Whitesmiths, silversmiths, coppersmiths, and pewterers rose to serve the expanding merchant class. Whitesmiths specialized in finishing imported goods by filing, polishing, and assembling iron parts—labors essential to commercial enterprise.

Only a few silversmiths worked in New York or Boston in the late seventeenth century, but they could be found in all major colonial cities in the eighteenth century. No colonial artisan rivaled the silversmith's prestige. He handled the most expensive materials and possessed direct connections to prosperous colonial merchants. His product—primarily silver plate—reflected his exalted status and testified to his customers' prominence. Silver plate stood as one of the surest ways to store wealth at a time before neighborhood banks. Unlike the silver English and Spanish coins from which it was made, silver plate was readily identifiable. Often formed to individual specifications, it always carried the silversmith's distinctive markings and consequently could be traced and retrieved. Customers generally secured the silver for their silver plate. They saved coins, took them to smiths, and discussed the type of plates desired. Silversmiths complied with these requests by melting the money in a small furnace, adding a bit of copper to form a stronger alloy, and casting the alloy as an ingot. They hammered ingots to the appropriate thickness by hand on heavy anvils, shaped them, and used screw presses or dies to adorn them. Engraving was done by hand. Not all customers chose silver plate, however. Some sought more intricate products, such as silver teapots. Silversmiths made these devices by shaping or casting parts separately and then soldering them together.

Colonial coppersmithing also came of age in the early eighteenth century, and prospered in northern cities. The metal's ability to conduct heat efficiently and to resist corrosion contributed to its attractiveness. Yet copper in colonial America was dear, and smiths were never numerous. Virtually all copper worked by coppersmiths was imported as sheets or obtained by recycling old copper goods. Copper was used, not admired. Coppersmiths employed it to fashion pots and kettles for the home, and brewery and distillery apparati for the alcohol trade. They shaped it in much the same manner as silver or melted it in a foundry with lead or tin. They also mixed it with zinc to make brass for maritime and scientific instruments.

Pewterers were even scarcer than coppersmiths. Prior to 1750, there were probably fewer than twenty in the colonies. Like coppersmiths, pewterers congregated in northern cities. There they mixed and heated tin with copper, antimony, or lead, and cast the soft material as plates, candlesticks, and coffee pots. And like coppersmiths, they melted old pewter and recast it.

Emergence of popular entertainment also signified that a substantial segment of colonial society was no longer consumed by the question of subsistence. East Coast merchants sponsored subscription libraries, organizations to investigate natural history or philosophy, theaters, and especially inns, taverns, and coffee-houses. Beer remained the home beverage, but eighteenth-century colonists increasingly left the home to drink applejack, peach brandy, rum, coffee, and tea. More than ever before, tippling had become a year-round public social event. The popularity of these drinking houses fostered communication and provided informal settings in which to discuss commercial ventures. But ardent spirits played an even greater role in colonial life. Colonial elections were often in-fluenced by a candidate's free distribution of drink. For example, when George Washington stood for election in 1758 to the Virginia House of Burgesses, he spent thirty-seven pounds sterling, seven shillings on his campaign, "of which over 34 pounds was for brandy, rum, cider, strong beer and wine."[3]

But perhaps as important as the wealth gained from exports was the idea of exporting itself. That idea carried consideration beyond the locality as colonial producers and merchants sought markets for surpluses. It yielded a new awareness of other localities—in the colonies and England—as both potential importers and potential competitors. In this framework, attempts of the Crown to regulate exports would be seen as potent challenges to the long-standing mercantile partnership between England and its colonies.

TECHNOLOGY AND THE REVOLUTION, 1763–1783

Printing as a Catalyst for the Revolution

The export-generated interest in extralocal affairs produced local markets for information about other settlements and areas. The mid-eighteenth-century rise of colonial presses and newspapers fed that demand. By 1763 colonial master printers operated some forty presses, published at least a dozen newspapers, and issued thou-sands of pamphlets. This communications boom was furthered by the termination of colonial printing's dependence on European raw materials. Benjamin Franklin stood as a crucial figure in this event. He provided financial support for several journeymen printers to establish their own print shops and, more important, reigned from the mid-1730s as by far the largest supplier of white linen rags, the essential ingredient in colonial paper. Rags that Franklin collected went to stamping or Hollander mills, where they were converted into thin fibrous pulp. This pulp was transferred into vats in which papermakers dipped rectangular-framed, fine wire meshes. The captured fibers became sheets of paper, which were placed between felts, squeezed of excess water in screw presses—pressing also forced fibers to adhere to one another—hung to dry, and smoothed with stones. Indeed, Franklin's rag-collecting endeavors were so massive that he maintained they led to establishment of no less than eighteen papermills.

Franklin helped provide colonial printers with a ready, dependable supply of rough-quality inexpensive paper. He proved equally adept at ensuring they had ink. Printers' ink was oil-based, made from linseed oil boiled with rosin until it became viscous. Its color came from lampblack, the fine soot resulting from the incomplete combustion of carbonaceous substances. Although Franklin made some oil and mixed some ink, he specialized in lampblack. He purchased his first lampblack house in 1733. By 1756 he had several others and assumed in the ink trade a position as fully important as he had in the paper trade.

Franklin's extensive involvement in the printing trade suggested the significance of communication to the mid-eighteenth-century colonial mind. Colonists' resistance to English threats to restrain their ability to communicate further reflected that significance. Three acts of Parliament appeared particularly menacing. The Sugar Act of 1764, which levied taxes on sugar, coffee, molasses, and wine, posed a potent challenge to the continuance of public drinking establishments. Indeed, James Otis, a member of the Massachusetts Assembly, recognized as the "great incendiary of New England," maintained that the Sugar Act "set the people a'thinking in six months more than they had done in their whole lives before."[4] The Stamp Act of 1765 taxed colonial newspapers, pamphlets, almanacs, and playing cards. The Townshend Acts of 1767 endangered the colonial papermaking industry and tea houses. Colonists railed against these acts and often protested in the streets. And at Otis's behest, they sent representatives to Massachusetts to formulate a united response to the Stamp Act.

The vehemence with which colonists greeted these acts opened a new chapter in the imperial relationship. To be sure, England had imposed commercial regulations on the colonies almost from their inception to restrict economic competition between elements of the English-colonial commonwealth. Taxes, prohibitions on manufactures, and the rerouting of commodities served to retain the commercial viability, prosperity, and supremacy of England and *its* merchants and manufacturers. For example, the Crown had stipulated that colonial sugar, tobacco, rice, copper, furs, naval stores, and cotton could not be sold in Europe unless they first passed through England. It also maintained that with the exception of salt, slaves, and wine the colonies could import no foreign products that had not been transferred and taxed in England. Similarly, England had prohibited exports of colonial wool, woolen cloth, and hats—commodities that England itself produced in abundance—and outlawed colonial iron-finishing establishments, such as slitting mills, plating forges, and steel furnaces, to end competition for the manufacture of finished iron products within the commonwealth. To ensure that the colonies would not circumvent these regulations and erect enterprises that would squander wealth, it had also banned all colonial joint-stock companies unless expressly approved by Parliament.

These and other edicts had hurt colonial industries. Occasionally those particularly aggrieved sought redress and petitioned colonial governors or assemblies, or proprietors, Parliament, and the King. Those complaints had been relatively few, their tenor mild. Colonial reaction to these earlier acts had been extraordinary in its general acceptance of the terms and its acquiescence to the measures. No doubt the colonists' ability to circumvent these regulations—smuggling was prevalent—had contributed to their docility. So, too, had the comparably small quantity of colonial exports. But the matter went much deeper. At its heart lay mercantilist social theory, in which colonists accepted inequalities as the "normal" state of affairs, their persistence tolerable so long as the whole seemed likely to persevere. Only when the whole—usually referred to as the "public good, order or safety"—appeared threatened did drastic action seem appropriate.

The Parliamentary acts of the 1760s seemed to colonists to create that situation. They argued that England's attempts to regulate communication and various industries, coupled with introduction of massive new taxes, was an effort to transform the colonies from full, if unequal, partners into serfs. The Crown recognized the problem differently. It asserted that colonists had not borne their fair share of the Empire's costs and that the time had come to correct that imbalance; the colonists flourished while the English were burdened with increasing sums for their colonies' defense (for example, during the French and Indian War).

These varying perceptions of the well-being of the commonwealth produced an impasse in which each subsequent dispute merely exacerbated tensions. England continued to crack down on colonial communication—it temporarily dissolved the New York and Virginia Assemblies and the Massachusetts General Court—to regulate aspects of colonial commerce and manufacturing, and to levy heavy taxes and enforce their collection. Colonists persistently objected to Crown policies and, as they had with the Stamp Act, sought to gain reversal by refusing to purchase English goods. The decision to boycott English products fortified many extant colonial industries and spawned new ones. Indeed, establishment of the New World's first type foundries stood as an attempt to ensure the flow of communications in the face of nonimportation agreements.

The economic relationship between England and the colonies, which was predicated on its ability to further both—the commonwealth as a whole—could not long withstand a situation in which its constituents perceived only disadvantages. Economic skirmishing soon gave way to a full-blown conflagration, and the colonies declared their independence from Britain on July 4, 1776. Ironically, each newly emancipated republic initially pursued its own course against their common foe. To be sure, the Continental Congress and army existed, but the states contributed to them secondarily, almost as an afterthought. Each state sought to ensure its own defense and to encourage within its borders production of a full range of war supplies. Most offered inducements—bounties, land, freedom from taxation—for the manufacture of gunpowder, cannons, and guns; and all exempted from military service gunsmiths, cannon-makers, shot- and shell-casters, papermakers, and gunpowder manufacturers. Only with the Articles of Confederation's passage and ratification did the states concede that in the war their interests were virtually indistinguishable, and only then did they begin actively to provide for the common defense. Congress's creation of armories and manufactories for ordnance at Carlisle, Pennsylvania; Springfield, Massachusetts; West Point, New York; and New London, Virginia, stood as a product of that concession.

Manufacture of Arms

Of the necessary war ordnance, colonists were most familiar with firearm manufacture. Gunsmithing had been an established colonial trade, and its practitioners quickly geared up for the conflict. Two types of firearms predominated: the smooth-bore flintlock and the Pennsylvania long rifle. They had similar firing mechanisms and operated in the same fashion. Pulling the trigger caused the hammer's flint to strike a serrated metal plate, which yielded a spark and touched off the powder in the priming pan. This ignited the main charge. The dense smoke generated by the rapid combustion of gunpowder increased dramatically the pressure behind the ball, which together with a sealing patch had been jammed down the barrel's bore with a heavy iron ramrod—sometimes assisted by a small mallet—and the ball shot from the barrel.

As suited its name, the Pennsylvania rifle's barrel was longer than the smooth bore—five feet rather than $3\frac{1}{2}$ or 4—and its bore was rifled with spiral grooves. The increased barrel length and rifling produced a weapon of greater range and accuracy. Despite design and performance differences, however, barrels of the two guns were made in much the same way. Smiths heated long wrought-iron bars to white heat a section at a time, and hammered them around mandrels. After completing this forging and welding process, artisans made apertures uniform with a hand-operated boring device. Another boring machine was used to add the Pennsylvania rifle's characteristic grooves.

Some firearm manufactories attained significant size. A typical large enterprise might include three barrel forges, a similar number of lock forges, a casting shop for brass fittings, forges for bayonets and the like, watermills for grinding and polishing gun parts, and an assortment of files for finishing the locks' various parts. It would employ more than a hundred men and produce several thousand weapons a year.

Colonists also had extensive experience producing gunpowder, but only two of its ingredients, charcoal and sulfur, were plentiful in America. The third, saltpeter, was in short supply. Most was purchased from France, although Americans made approximately one-eighth of the required material in saltpeter houses. Heaps of animal and vegetable refuse, mixed with limestone and wood-ashes and moistened from time to time with urine, were left to decompose. The noxious matter was then leeched with water, which was collected and evaporated. The crystalline residue, saltpeter, next went to a gunpowder mill where a hundred parts by weight were usually mixed with fifteen parts of sulfur and eighteen of charcoal. Water was added to form a paste-like concoction. A water-powered stamp mill replicated the action of mortar and pestle, grinding the coarse materials as finely as possible to ensure even combustion. This latter process took as long as twenty-four hours. The powder was then air dried, run through a fine sieve, and loaded into rotating wooden kegs, a final guarantee against incomplete mixing.

In contrast to firearm and gunpowder manufacture, Americans were relatively unacquainted with cannons. These important devices served the military as artillery for sieges and field battles, as

*Using a handful of mobile artillery pieces, the Rebels
surprise the British at the Battle of Princeton on
January 3, 1777.*

weapons for harbor and coastline defense, and as the primary means
for conducting naval engagements. At the war's beginning, the colonists controlled only a handful, which were small and located near
cities or mounted on ships to discourage pirates. Nor was there an
established American cannon-making industry. Most cannons were
cast from brass, which was an amalgam of copper and zinc, and
copper was dear in the New World. Unable to produce brass cannons, and compelled by the war to develop an alternative, the states
encouraged iron manufactories to cast cannons. These iron armaments only found use as defensive weapons because they were difficult to move; they required thicker walls than their brass
counterparts and weighed roughly three times as much. But Americans did not have to do without brass cannons completely. They
seized British armaments and (especially after 1777) purchased
French cannons. These two sources proved sufficient to provide
Americans with a supply of mobile artillery.

THIRTEEN COMMERCIAL NATIONS, 1776–1789

Ironically, the Articles of Confederation extended only to pursuing the war. In other spheres there was little cooperation among the states. The Declaration of Independence had voided colonial governments, and colonists conceived of it as transforming the English New World settlements from part of the Empire into thirteen sovereign nations; each colony went from a section of the whole to thirteen different wholes. Each framed a governing document and became its own commonwealth. All new state constitutions proved similar in that each embraced mercantile theory; new state governments were granted extensive powers to regulate commerce and manufacturing within state borders.

Interstate Commerce

In true mercantilist fashion, each state discouraged importation and encouraged exportation. The nascent states applied the policy to each other as well as Europe; in commerce and manufacturing they stood as separate, competing entities. Each sought to free itself of the need for imports and to stimulate exports by establishing, protecting, and fostering domestic production, known as home manufactures. State governments exempted lands from taxation; lent entrepreneurs money; purchased interests in ventures; passed laws prohibiting the export of manufacturing tools; granted land, monopolies, and bounties; and enacted tariffs to defend home manufactures. For example, North Carolina, Connecticut, and New Hampshire exempted rod and nail works from taxes. Pennsylvania lent £300 to erect a steel furnace, while Massachusetts loaned Aaron Burr £200 at no interest to build carding and spinning machines. Pennsylvania fined "ill designing persons" who exported tools and machines used in manufacturing as well as those guilty of "seducing" manufacturers to leave the state. It also bought one hundred shares (valued at £1,000) in a failing cotton mill. North Carolina was willing to grant 5,000 acres of land to anyone establishing a successful ironworks, while New Jersey gave a bounty of six shillings per hundredweight for wool, flax or hemp raised in the state. The bounty in New York was eight shillings for the same quantity of hemp, while Connecticut excused forty shillings from hemp growers' taxes for each crop acre. Rhode Island put a twenty-five-percent tariff on ready-made garments, canes, and watches produced out-of-state.

State governments concentrated on decreasing imports, but efforts to step-up exportation and to bring additional wealth into a state received only a modicum of governmental assistance. Enterprising traders, especially merchants, were generally left to their own devices, and sometimes their ambition to increase exports resulted in revision of established practice. Such was the case with the rise of the "putting-out system" in late eighteenth-century Massachusetts footwear and cloth manufacture. It stemmed from the desire to increase the volume of production—to enhance exports—and consequently undercut the traditional role of artisans as financiers *and* producers. Essential to extralocal trade and the wealthiest Americans, merchants alone had money enough to bankroll ventures of this magnitude. They used their funds to purchase large quantities of leather, wool, and linen, then hired shoemakers, spinners, and weavers for a fee to make shoes, boots, and cloth. These artisans made these raw materials into finished products in their own homes or home-shops, then returned the products to the merchants, who marketed the goods in other states and overseas. That this new practice engendered little comment testified to the persistence of mercantilistic notions even after independence; the state, not England and its colonies, had become the commonwealth.

Although the states remained economic adversaries, the exigencies of the war had led them to conclude the Articles of Confederation as a mutual defense pact. This agreement looked both east to Europe and west past the Appalachian Mountains to the continent's interior; approval of the Articles had been delayed until states had ceded to the Confederation in 1781 their often-conflicting claims to western lands north of the Ohio River (some states had claimed land clear to the Pacific Ocean). The presence of unfriendly Indians, coupled with the vacuum that would be left in the West if the English were driven from North America—a vacuum that some other European nation might seek to fill—made the orderly development, consolidation, and control of the interior a top priority. The Confederation also recognized the West's economic significance; its sale would help pay war debts. Since 1776, land speculators had clamored to buy large chunks of the West and to resell them as small parcels to homesteaders at a considerable profit.

Linking the East and West   For either reason, defense or economics, it appeared crucial to link the West to the East. That meant establishing not only an administrative relationship—realized in the Land Ordinance of 1785 and the Northwest Ordinance of 1787—but also commercial ties, which required adequate transportation.

Heavy, sturdy Conestoga wagons led the settlement of the trans-Appalachian West. Courtesy of Chicago Historical Society

To be sure, Conestoga wagons carried some freight between East and West. Developed by Pennsylvania's Palatine Germans about 1750, the wagons were large and had high wheels for easy running and a floor that was curved upward at its ends to prevent cargo from shifting on steep grades. But climbing the Appalachians with these vehicles was problematical and time-consuming. Other entrepreneurs attempted to create improved forms of transport. George Washington was the most illustrious.

Soon after his retirement from the Continental army, Washington actively promoted the Potomac River as a means to join East and West. From a source several hundred miles inland near the Ohio river, the Potomac ran through a gap in the Appalachians, past Washington's Mount Vernon estate, and into the Chesapeake Bay. (Indeed, the river's attractiveness for linking the West commercially and politically to the East would later lead Washington to designate its southernmost point on the tidewater the site of the new nation's capital.) Events of the mid-1780s convinced Washington of the river's potential. Even at that early date he considered circumventing the river's waterfall and rapids no problem; short canals, such as those common in England, could be dug around them. But Washington remained puzzled about how to get vessels upstream against the stiff current. James Rumsey, a millwright familiar with the blacksmith's art, offered him in 1784 what seemed an answer. Rumsey had built a model paddlewheel-powered pole boat—essentially a floating undershot watermill geared to poles that pushed on the river bed. He demonstrated his invention to Washington, and with the general's encouragement and support, received late that winter ten-year monopolies on the use of his device in Virginia and Maryland, the two states that bordered the Potomac.

Washington was not alone in applauding Rumsey's invention. Thomas Jefferson and John Marshall also endorsed it enthusiastically. But Rumsey was not content to build a full-scale pole boat, because possibilities of steam intrigued him. He conceived of using steam pressure to force water through a narrow cylinder at the stern to propel a boat. While Rumsey's plan to employ steam to power a craft was unusual, moving of water by steam engine was not. All steam engines in America (and there were but a handful), as well as the vast majority in Europe, raised water in some capacity— either to clear mines, to provide cities with drinking water, or to power waterwheels during hot weather when streams dried up. These engines generally operated on principles different from that proposed by Rumsey. They were atmospheric engines. Their power stemmed from the atmosphere's weight forcing down a piston in a cylinder in which a partial vacuum had been created by the swift condensation of steam; Rumsey's device was to use steam at high pressure as its motive power.

Before Rumsey had made any headway, he gained a competitor, John Fitch. Apprenticed as a clockmaker, Fitch had established a brass foundry and worked as a silversmith. He was familiar with the Newcomen engine—an atmospheric steam engine in which steam was cooled in the cylinder—and proposed to connect this engine to an endless belt and to attach to the belt a series of paddles; steam power would row the boat upstream. Several prominent individuals, including Franklin and the ubiquitous Washington, offered Fitch encouragement.

Washington acted before Fitch or Rumsey had passed the planning stages. In early 1785, he helped secure Virginia and Maryland charters establishing the Potomac Company to make the river navigable as far west as possible, and to build a wagon road from that terminus to the Ohio River. Washington was named president of the company. No doubt he expected either inventor to produce his craft soon, yet he made provisions in case they failed. He intended to fasten iron chains to rocks and trees on the river banks so boatmen could pull their vessels upstream and hired Rumsey to superintend construction, and purchased slaves as well as indentured Irishmen to do the heavy labor.

Rumsey worked on his pole boat as he oversaw Washington's company. He gave a moderately successful demonstration of his device in summer, 1786, but then shifted to steam. Design of an alternative boiler was among his contributions. Rumsey deemed the massive spherical boilers usually associated with steam engines impractical for steamboats because they took up too much space, were too heavy, and required great amounts of fuel to keep the water heated. A small pipe-shaped boiler, containing only enough water to produce the steam necessary to fill the cylinder, seemed to overcome these objections. Rumsey found, however, that the highly pressured, superheated steam in these vessels tended to melt solder connections.

This stumbling block frustrated Rumsey; while he considered alternatives, Fitch gathered support from James Madison, Patrick Henry, and other notables. He was given a New Jersey monopoly for steamboat operation. Fitch attached his paddles to cranks and tried to power his boat with a horizontal, double-acting Newcomen engine. He finally settled on a vertical, single-acting Watt engine— an atmospheric engine with a separate condensor for cooling steam— with modest success. As Rumsey's pole boat had, Fitch's heavy mechanical vessel could travel upstream, but only at about three miles per hour, far too slow to justify the extra expense associated with the device. Construction of a truly functional steamboat remained a product of the next century. Nonetheless, Fitch was justifiably proud of his creation and demonstrated it in Philadelphia during the Constitutional Convention.

FRAMING AND IMPLEMENTING A COMMERCIAL PACT, 1787–1800

Few members of the Convention witnessed Fitch's demonstration. They were engrossed in hammering out a new relationship among the states. Competing trade policies of the thirteen sovereign states had caused problems; the convention call specifically urged modifications in the Articles "to take into consideration the trade and commerce of the United States." Even before the call, each state's emphasis on home manufactures had seemed to some wasteful, even menacing. In 1785 the Massachusetts legislature passed an act

placing a protective tariff on all state products. The Commonwealth conceived of this act as a defensive measure, to avoid injury
by an influx of "foreign" goods, and it urged other states to enact
similar laws until Congress gained power to regulate trade. Delaware and South Carolina agreed, but the Pennsylvania Assembly
was more outspoken, fearing that unless Congress soon regulated
trade, the Confederation would dissolve. James Madison explained
the situation graphically with particular reference to the plight of
New Jersey and North Carolina. "New Jersey," he wrote, "placed
between Philadelphia and New York, is likened to a cask tapped at
both ends; and North Carolina, between Virginia and South Carolina, to a patient bleeding at both arms. Most of our political evils,"
he concluded, "may be traced to our commercial ones."[5]

The Constitution: A Commercial Document

The new document, the Constitution of the United States of
America, provided the general government the "power to lay and
collect . . . duties, imposts and excises" so long as they "shall be
uniform throughout the United States." It also granted Congress
authority "to regulate commerce with foreign nations, and among
the several states, and with the Indian tribes." Both the Constitution's proponents and opponents recognized its commerce clauses
as liberating, not restrictive. Experience had proved that the states
together constituted an economic whole—they comprised a commonwealth—and that pursuit of separate, overlapping commercial
and manufacturing strategies merely created waste and redun-
dancy; each state's ability to corral wealth suffered from the competition. Even opponents of the Constitution felt it mandatory for
the states to act as one in economic matters. In classic mercantile
fashion, they argued that regulation of commerce and manufacturing comprised the sole justification for a federal government and
the only reason for a union. Aware that "all the states have local
advantages," these men and women urged them "to supply each
other's wants" and maintained that Congress should be empowered to ensure that the program was carried out. "A diversity of
produce, wants and interests, produces commerce," they asserted,
"and commerce, where there is a common, equal and moderate
authority to preside, produces friendship."[6]

Such a broad consensus about the importance of managing commerce and manufacturing overcame objections about the potential
abuses of power, and the states ratified the Constitution in 1789.

No matter how later generations would interpret it, the Constitution established a government that differed from the Confederation in little more than its power to regulate trade. Congress moved quickly to implement its program. Among its first acts were the placement of customs duties on all imports and a tonnage duty on all shipping. These measures were designed not only to raise revenue but also to encourage domestic manufacture and to decrease importation. Enactments of Congress in 1790 were more straightforward and pointed directly to the role accorded technology in advancing the new nation's mercantile interests. It created a patent board to rule on priority of inventions and to grant true inventors monopolistic rights for specified periods; established a United States Mint; and sought to develop a standard of weights and measures.

Action on patents had come at President Washington's request. Like his compatriots, Washington believed that government guarantees of exclusivity were essential to stimulating public disclosure of new devices and processes. Without that assurance, inventors would conceal their inventions, a maneuver that would impede the public interest by hampering American manufacturing and commerce. The new nation's secretaries of state and war and its attorney general—Thomas Jefferson, Henry Knox, and Edmund Randolph respectively—constituted the patent board, but Jefferson was the principal responsible officer. He scrupulously examined every application and rigorously tested every device. His criteria proved so demanding that the board issued only three patents during its first year.

The establishment of the mint and official weights and measures also fell to Jefferson. He put David Rittenhouse, whose orreries (mechanical planetariums) had gained him the sobriquet "The American Newton," in charge of the mint and, with Rittenhouse's assistance, penned a report arguing for the use of the decimal system in all coins, weights, and measures.

Hamilton's Call for Federal Assistance

That Congress should regulate commerce and manufacturing went without question. What this meant practically—what were the just and proper limits of federal involvement—sparked great debate. Alexander Hamilton, the secretary of the treasury, emerged as the most notable advocate of large-scale, direct federal intervention. He outlined his proposal in his "Report on Manufactures" and submitted it to Congress in December 1791. The report contained three

*Late–eighteenth-century agricultural tools commonly used
on American farms*

main sections: a survey of the state of American manufactures; a defense of their utility for furthering the nation as a whole; and suggestions of ways that the federal government might encourage them. Much of Hamilton's analysis compared manufactures to agriculture. He argued that both remained critical to the public weal, but that manufacturing held some wealth-accruing advantages. Hamilton maintained that manufacturing required each artificer to master only a single task, which increased skill; allowed women and children to participate, which expanded the number of producers; relied on extensive use of machinery, which reduced the drudgery associated with heavy labor; permitted children to choose from a larger store of occupations, which increased happiness; fostered recognition of talent and enhanced productivity; encouraged immigration, which brought America new wealth-transforming skills; and created a larger and stable demand for the nation's agricultural products, which ensured farmers a consistent, fair price for their goods.

Hamilton's arguments for federal aid proved equally compelling. American entrepreneurs needed assistance to combat what he saw as their countrymen's natural resistance to change, as well as the active support of their own manufacturing by foreign governments. Most of Hamilton's recommendations about the manner in which the federal government should encourage domestic manufactures were in line with established mercantile theory and reminiscent of the states' policies under the Articles. Protective tariffs, exclusion of designated goods, bans on the exportation of certain processes and equipment, bounties to stimulate production, and premiums and exemptions from taxes to reward production found his favor. But he also developed several innovative measures to enhance America's international commercial picture. He wanted Congress to extend patent law to cover those who introduced inventions of foreign origin into America; mandate inspection of certain exportable products to ensure their quality (this would build the new country's reputation and allow certified producers to expand their share of the export market); create a single, nationwide paper money; and finance a massive system of internal improvements such as roads, bridges, canals, and harbors. His final proposition was even more imaginative: he urged Congress to establish and finance a board to promote American manufactures by defraying immigration expenses of artificers in specified fields, paying rewards to inventors of designated processes or devices, and soliciting for and securing machinery not available in America.

This last proposal may have smacked of industrial espionage, but all of Hamilton's suggestions were intended to garner and preserve wealth for the new country. He claimed that import duties would yield the necessary revenue, and he looked to the National Bank as a source of financial support. This suggestion, like much of his report, horrified many Americans, including Jefferson and Madison, who generally were not opposed to manufacturing nor machinery. Jefferson, for instance, served on the patent board, applauded establishment of water-powered grist mills, and operated carriage, furniture, and nail manufactories on his Monticello plantation. He also invented several devices, such as a copying machine and automatic door-openers, and designed numerous others, such as a horizontal windmill, an improved plow, a fulling machine, and a spinning jenny. Although an avid supporter of mercantile theory at this point in his life, Jefferson detested the huge European manufactories and the teeming industrial cities they bred, likening their contributions to virtuous government to sores on the body politic. Fear that Hamilton's scheme would reproduce the corrupt European system in America fueled Jefferson's hostility.

Dispute between the two camps rested on the question of whether the policy would benefit the whole nation. Hamilton had offered his justification. His antagonists considered his program flawed because it favored manufacturing interests over agricultural and commercial. Initiation of federally sponsored inducements to establish and encourage American manufacture, including tariffs and import prohibitions, would surely be met by a corresponding wave of European protectionism—a move that would damage American agriculture and commerce. To his antagonists Hamilton's program seemed a product of faction, of the desire of the few to further themselves at the whole's expense. Only private investors—speculators—would benefit; others would lose. Government would be providing an opportunity for "the few to increase the inequality of property, by an immodest, and especially unwritten, accumulation of riches." In short, those opposed to Hamilton contended that his program did "not appeal to the understanding and to the general interest of the community."[7]

Private Development of Manufacturing

Although the anti-Hamiltonians offered no formal alternative to the treasury secretary's report, they prevailed in Congress and effectively removed the federal government from an active role in fostering manufacturing. Initiative rested on private citizens' shoulders, and even before Hamilton submitted his report, Americans had formed several local associations to promote manufacturing. The Pennsylvania Society for the Encouragement of Manufacturing and the Useful Arts (established in Philadelphia in 1787) was the most wide-ranging, but similar organizations were created in Boston (1786), New York (1788), Baltimore (1788), Wilmington (1789), Burlington (1791), Morristown (1792), and Newark (1793). Although their members might engage in manufacturing, these groups themselves generally sought only to inspire it. They held meetings to discuss methods of achieving that end, gathered and exchanged information about new manufacturing techniques, and published their deliberations. They also rewarded inventors and provided financial aid to those claiming special knowledge about manufacturing processes.

Only the Pennsylvania society was not content to limit its role in that fashion. Under stewardship of Tench Coxe, who would later become Hamilton's assistant secretary and help develop the manufacturing report, it raised through public subscriptions a fund to establish cotton textile manufactories. Coxe understood that the ultimate success of American textile manufacture depended on its ability to compete with the sophisticated British; it required the new late-eighteenth-century inventions—Richard Arkwright's water frame and carding machine, James Hargreaves's spinning jenny,

Samuel Crompton's mule*—as well as men experienced in their operation. To these ends, Coxe hired an agent and, "for the good of the United States of America," sent him to England both to secure "models and patterns" of textile machinery and to seduce mechanics to emigrate to America.[8] The British discovered the scheme, however, and the society's agent fled empty-handed.

Though aid from abroad was not forthcoming, the society remained undeterred. It employed several hundred women to spin yarn, and by August 1788, its weavers worked twenty-six looms. By the end of its first year of production, the manufactory had made more than 7,000 yards of cloth. A fire in its plant ended the venture in 1790, but its example apparently encouraged others. Associations to manufacture textiles were organized in Beverly, Massachusetts (1787); Hartford (1788); Baltimore (1789); and New York (1789). Each aggressively recruited European workmen; the New York organization gave Samuel Slater, who brought knowledge of the new European textile machines, his first position in America. But efforts of these manufacturing entities pale when compared to the grand design of proponents of the Society for Establishing Useful Manufactures. Chartered in New Jersey just prior to the delivery of Hamilton's report, the society was the brainchild of the treasury secretary and Coxe. Its ambition was to create "the National Manufactory," a model that would demonstrate the utility and feasibility of American industrial development. Funded primarily by a consortium of Hamilton's well-to-do friends and located at what is now Paterson, New Jersey, the national manufactory echoed in its conception colonial plantations and foreshadowed late-nineteenth-century factory towns. Its stockholders envisioned it as the home of an extraordinary number of diverse establishments, including mills to make paper, pasteboard, sailcloth, stockings, ribbons, blankets, carpets, textiles, women's shoes, earthenware, brass and iron wire, and beer. The society would own the entire complex, but it would sell workers the plots to erect dwellings. Each manufactory would be under supervision of a European emigré skilled in production of that product. For example, Thomas Marshall, who claimed to have directed installation of Arkwright's cotton mills, gained control of textiles. The society also pledged to secure from abroad "such machines and implements as cannot be had here in sufficient perfection."[9]

---

* *mule:* a machine for simultaneously spinning thread and winding it onto tubes called caps.

As befitted a site aspiring to become the "capital scene of manufactures,"[10] the society hired Pierre L'Enfant, architect of the new federal capital, to design the physical plant. Among L'Enfant's contributions were construction of a great aqueduct to carry Passaic River water to power the mills' waterwheels, and a canal to enable the society to transport its varied products to the river. Almost from the start, however, the society realized that expenses were proving steeper than anticipated and it scaled back expectations, albeit only modestly. The national manufactory remained the largest American manufacturing venture of the eighteenth century. Prior to 1796, the society had built facilities to make cotton goods, brass and iron wire, and stockings, and machines to spin hemp, flax, and wool. It also established a drying and bleach works, sawmill, gristmills, and waterworks and constructed houses for fifty workers. Some production facilities were quite large. For instance, the cotton mill was built of stone and stood four stories high, ninety feet long, and forty feet wide, while its stone carding and roping house was two stories tall, sixty-four feet long, and thirty-six feet wide.

The society's recruitment of European factory operatives was likewise somewhat successful. In addition to those European emigrés who found their own way to America, the society had attracted several French wire-makers, roughly a dozen Scottish stocking workers, and 200 Manchester cotton mill employees who fled to Paterson from Ireland. Though impressive, the influx of Europeans did not begin to fill the society's need for skilled workers. Its operations also were beset by a continuous lack of funds. Not only had the enterprise been underfinanced initially, but its investors had become increasingly restive when a return failed to materialize. By 1796, many had withdrawn from the society, and its manufacturing activities stagnated.

The case of the national manufactory demonstrated American manufacture's difficulties. Competition from Europe, a dearth of skilled workers trained in the newest techniques, ignorance about manufacturing apparatus, a lack of financing, debates over the merits of large-scale manufacturing, and modest governmental assistance combined to place America at a severe disadvantage. Nonetheless, the nation had emerged as a serious economic rival to Europe less than 200 years after the first permanent English settlement. North American inhabitants had combined New World abundances and Old World technologies to go beyond subsistence, to enter the export trade, and by the late eighteenth century, to begin to frame larger (albeit still modest) commercial and manufacturing objectives. These successive transformations each had occurred within the context of a rigid mercantilistic philosophy in which society as a whole was to accrue wealth while its members reaped benefits differentially. The Constitution brought the beginnings of a modicum of political equality, but the notion that similar equality should occur in the socioeconomic sphere waited until the first decades of the next century.

NOTES

1 Donald S. Lutz, *Popular Consent and Popular Control: Whig Political Theory in the Early State Constitutions* (Baton Rouge: Louisiana State University Press, 1980), p. 157.

2 Ian R. Tyrrell, *Sobering Up: From Temperance to Prohibition in Antebellum America, 1800–1860* (Westport: Greenwood Press, 1979), p. 16.

3 Tyrrell, *Sobering Up*, p. 18.

4 James Otis, *The Rights of the British Colonies Asserted and Proved* (Boston, 1764), p. 54, reprinted in Edmund S. and Helen M. Morgan, *The Stamp Act Crisis: Prologue to Revolution*, 2nd edition (New York: Macmillan, 1962), p. 52.

5 Madison is quoted in Catherine Drinker Bowen, *Miracle at Philadelphia* (New York: Bantam Books, 1966), p. 9.

6 Quoted in Herbert J. Storing, *What the Anti-Federalists Were For* (Chicago: University of Chicago Press, 1981), pp. 24–25. Quotation originally appeared in a letter to the *Massachusetts Gazette*, January 1788.

7 James Madison wrote a series of brief papers in 1792 for the *National Gazette* in opposition to Hamilton's plan. For a good summary, which includes the quoted material, see Richard Hofstadter, *The Idea of a Party System: The Rise of Legitimate Opposition in the United States, 1780–1840* (Berkeley: University of California Press, 1969), pp. 81–82.

8 "Contract Between Tench Coxe and Andrew Mitchell, August 9, 1787," printed in Jacob E. Cooke, *Tench Coxe and the Early Republic* (Chapel Hill: University of North Carolina Press, 1978), p. 107.

9 "Prospectus—Society for Establishing Useful Manufactures," reprinted in Arthur Harrison Cole, ed., *Industrial and Commercial Correspondence of Alexander Hamilton* (New York: Augustus M. Kelley, 1968), p. 194.

10 *National Gazette*, July 14, 1792, reprinted in Joseph Stancliffe Davis, *Essays in the Early History of American Corporations* (Cambridge: Harvard University Press, 1917), vol. 1, p. 424.

FOR FURTHER READING

Baron, Stanley. *Brewed in America: A History of Beer and Ale in the United States* (Boston: Little, Brown, 1962). Effectively demonstrates the importance of beer in colonial life.

Bishop, J. Leander. *A History of American Manufactures from 1608 to 1860*, 3 vols. (Philadelphia: Edward Young, 1868). A year-by-year catalogue of technological development.

Bridenbough, Carl. *The Colonial Craftsman* (New York: New York University Press, 1950). An older analysis of the role of craftsmen in the colonial economy.

Cole, Arthur Harrison. *The American Wool Manufacture* (Cambridge: Harvard University Press, 1926), vol. 1. An economic study of American wool cloth production.

Flexner, James Thomas. *Steamboats Come True* (Boston: Little, Brown, 1944). A spritely account of the steamboat race.

Goldenberg, Joseph A. *Shipbuilding in Colonial America* (Charlottesville: University Press of Virginia, 1976). The best study of the subject available.

Hartley, E. N. *Ironworks on the Saugus* (Norman: University of Oklahoma Press, 1957). A thoroughly detailed study of an early colonial ironworks.

Haynes, Williams. *American Chemical Industry* (New York: D. Van Nostrand, 1954), vol. 1. A compendium of facts with little analysis.

Hazard, Blanche Evans. *The Organization of the Boot and Shoe Industry in Massachusetts Before 1875* (Cambridge: Harvard University Press, 1921). Similar in thrust to Cole's wool volume.

Hunter, Louis C. *Waterpower* (Charlottesville: University Press of Virginia, 1979). A landmark of scholarship in the history of technology.

Hurt, R. Douglas. *American Farm Tools: From Hand-Power to Steam-Power* (Manhattan, KS: Sunflower University Press, 1982). A convenient study of agricultural machines.

Lewis, W. David. *Iron and Steel in America* (Greenville, DE: Eleutherian Mills-Hagley Foundation, 1976). The best short volume on the subject.

Mulholland, James A. *A History of Metals in Colonial America* (University: University of Alabama Press, 1981). Skewed to the urban scene, a nonetheless valuable contribution.

Storck, John and William Dorwin Teague. *Flour For Man's Bread: A History of Milling* (Minneapolis: University of Minnesota Press, 1952). An older book with some useful material.

Teaford, Jon C. *The Municipal Revolution In America: Origins of Modern Urban Government, 1650–1825* (Chicago: University of Chicago Press, 1975). A short volume laying out local government's powers and objectives.

Tunis, Edwin. *Colonial Craftsmen and the Beginnings of American Industry* (Cleveland: World Publishing, 1965). Surprisingly thorough treatment of many obscure colonial technologies.

Van Gelder, Arthur Pine and Hugo Schlatter. *History of the Explosives*

Industry in America (New York: Columbia University Press, 1927).
Similar to Haynes, a compendium.

Welsh, Peter C. *Tanning in the United States to 1850: A Brief History*
(Washington: U.S. National Museum, 1964), Bulletin 242. Discusses
tanning processes and materials in great depth.

Wroth, Lawrence C. *The Colonial Printer* (Portland, ME: Southworth-An-
thoensen Press, 1938). Best book on printing during the colonial period.
Accentuates artistic factors.

2
Young America and Individual Opportunity: 1800 to the 1830s

BEYOND MERCANTILISM

Americans found mercantilism an acceptable form of socioeconomic organization in a world beset by limits, but in the late 1790s a growing number began to question whether wealth was in fact limited. Oliver Evans and Thomas Ellicott, two noted millwrights, were among the earliest Americans to hint otherwise. Published in 1795, their *The Young Mill-Wright and Miller's Guide* was ostensibly a text on American water-milling practices. But its concentration on flour milling and other large, complex milling operations not normally associated with contemporary American enterprises suggested a vision of America quite different from that formerly posed. It implied that through technology the nation could create and sustain virtually limitless growth. America's abundant resources, in conjunction with technical processes, seemed capable of producing unprecedented plenty, which belied the pessimistic assessment that guided the previous two centuries. Rather than the creation of a single or even several national manufactories, Evans's and Ellicott's volume appeared to hold out the promise of the establishment of as many wealth-producing manufactories as Americans desired. Implicit in their formulation was the idea that technology in America would foster almost inconceivable material prosperity.

These views came to characterize early nineteenth-century America. After about 1800 Americans surveyed their abundant lands and natural resources, increasingly threw off the yoke of mercantilism, and emphasized progress, development, and growth. Technology would be instrumental in fulfilling these new social goals. Watermills increased from fewer than 10,000 in 1800 to well over 50,000 by 1830. The number and size of young America's manufactories increased markedly. The country's population more than tripled, creating a ready demand for additional goods. Scores of new cities were established as citizens moved west. Settlements were joined by new roads, canals, bridges, and railways, most of which were built with governmental assistance. Steamboats carried goods on waterways, and steam locomotives began to bring them over land. Congress established the Army Corps of Engineers and the United States Military Academy to provide the nation with a ready supply of military engineers. Mechanics' institutes blossomed in Philadelphia, Cincinnati, Boston, Baltimore, New York, and elsewhere.

Extending Europe

America seemed poised to open a new chapter of world civilization, potentially the greatest. Its defenders championed it as a free land with free institutions in which Old World hopes and aspirations would be played out, unencumbered and unfettered by European traditions and trappings. They explicitly looked to Europe for inspiration, but conceived of themselves as Europeans freed of their chains, as Old World men and women in a New World setting.

Americans tempered their reverence for Europe with democratic notions, heralding their country as "the great experiment in democracy." The new nation's citizens repudiated outright some European customs, such as a hereditary aristocracy, as inconsistent with a democratic republic. They debated the suitability of other customs, such as secret, restricted organizations. But infatuation with and commitment to democratic notions ran far deeper than mere custom. To early–nineteenth-century Americans, the idea of privilege, special rights granted only to some citizens, seemed an

anathema in a nation marked by plentiful resources. They identified privilege as a vestige of an authoritarian political system (in a
sense, the product of mercantilistic ideas), and defined it as an artificial barrier that hampered individual opportunity—including the
awarding of unjustified competitive advantage by the government.
They sought to eliminate privilege. It bears noting that the assault
on privilege was an effort to protect and provide the opportunity
for success. It was not a guarantee that all individuals would become successful. Indeed, it was the opportunity for success for all
that constituted America's promise.

The Ascendancy of the Individual

Early–nineteenth-century Americans rarely agreed on who received privilege, or what constituted privilege, and many disputes
turned on these two points. In that sense, privilege in the new century's first decades was reminiscent of faction in the eighteenth
century. But crucially different premises underlaid the two concepts. As in the case of the Hamilton–Jefferson imbroglio, eighteenth-century faction was deprecated because it meant a group
was pursuing its own agenda at the expense of the whole; it was a
crime against society, the whole. Privilege in the nineteenth century was attacked because its recipients gained advantages unavailable to others. Pursuit of one's own ends was not the problem—
competition was welcomed—but rather, unfair or unwarranted assistance; privilege was a crime because it hindered other individuals' opportunity to compete. Put simply, those of the eighteenth
century focused on the whole while their nineteenth-century counterparts concentrated on the parts.

The individual—the part—was the fundamental unit of early–
nineteenth-century American society. This social calculation was
manifested in the organization, conduct, and explanation of technological enterprises. It served as appraiser of propriety and adju-
dicator of efficacy. Emergence of a new kind of business entity, the
corporation, which with family partnerships became the predominant form of large-scale manufacturing and commerce, reflected the
new emphasis. Unlike the previous centuries' joint-stock companies, plantations, and town governments (each of which regulated
a whole spectrum of activities, created total environments, and in
effect replicated society in miniature), early–nineteenth-century
corporations functioned as individuals. They assumed in law the
status of individuals, and generally were permitted to tackle only
one or a handful of closely related tasks. Their mandates were neatly

circumscribed by state legislatures precisely because of their potential to impinge on the opportunity of others if left unchecked. Prohibitions against engaging in banking—congregating and controlling wealth unnecessarily—were almost always included in corporate charters. Corporations were created because of some obstacle to individual opportunity and received only the authority necessary to overcome that barrier; the grant of additional extraneous powers constituted a privilege, and would likely result in erection of hurdles for others.

BUILDING THE NATION'S INFRASTRUCTURE

A building boom swept early–nineteenth-century America; its citizens engaged in an almost maniacal spate of turnpike, canal, bridge, and railroad construction. New routes linked the West more securely to the East, connected seaboard cities with those inland, and generally facilitated travel, commerce, and communications. This boom stemmed from recognition that individual opportunity and economic development went hand in hand, and that both required creation of an "infrastructure"—roads, bridges, canals, and so on—to join producers and markets.

The federal government played a crucial role in this infrastructural revolution. As part of its constitutionally mandated responsibility to provide for the national defense, it improved harbors and seacoast fortifications, built the National Road, and provided numerous topographical surveys. The Army Corps of Engineers, stationed at West Point and constituting a military academy, planned and executed these works. Created by the federal government in 1802, the Corps recruited French engineers to teach at West Point and to direct many projects. These professors used French texts (French was taught at the academy as the military language), maps, and instruments. The dependence of the Corps on the French persisted after its earliest years. It continued to solicit French engineers for American projects, and in 1815 conducted a survey of French fortifications and engineering schools—a move that led to West Point's being remodelled after the acclaimed Ecole Polytechnique. Not until the late 1820s did the Corps begin to free itself from French influence.

From its inception, several prominent Americans sought to enlarge the functions of the Corps. As early as 1806, Jefferson called for a constitutional amendment allowing the Corps to plan and direct nonmilitary improvements, an approach seconded by Treasury Secretary Albert Gallatin's 1808 report on roads and canals.

Madison, during his administration, also urged passage of an amendment permitting the Corps to undertake those internal improvements that required "a national jurisdiction and national means" to bind "more closely together the various parts of our extended confederacy." John C. Calhoun went a step further. His bonus bill, passed by Congress in 1817, would have appropriated $1.5 million to the states to fund road and canal construction. Madison vetoed the measure as unconstitutional without an amendment.

Madison's veto message was that Congress lacked authority to appropriate money for canal and road planning in cases other than the national defense. His sentiments were outdated; the Corps was already providing localities direct and indirect assistance. Its experienced officers surveyed numerous turnpikes and canals for states and cities, and several resigned their commissions or took leaves-of-absence to direct local projects. President Monroe acknowledged the new reality in his 1822 veto of a measure to collect tolls on the National Road when he argued that Congress could fund internal improvement plans, but lacked jurisdiction over their construction and operation. The General Survey Act of 1824, which authorized the president to employ the Corps and others to conduct road and canal surveys considered of national importance, institutionalized that view. During John Quincy Adams's tenure in office, the Corps surveyed thirty-four canal routes, eighteen roads, forty-four river and harbor projects, and six railroads. Nor did Andrew Jackson's election end federal involvement in internal improvements; although he vetoed a congressional road appropriation as constituting unfair privilege, he also signed an 1836 measure to distribute all but $5 million of the federal surplus to plan and finance internal improvements in each state.

The Panic of 1837, fueled in part by rampant land speculation, negated the act's significance. Funding of non-military internal improvements remained where it always had been, with states and cities. Indeed, state and city governments had participated in the vast majority of construction projects, although they rarely undertook ventures alone. They had acted instead through mixed corporations, which joined public and private capital. Chartered by states—most often at some entrepreneur's request—each mixed corporation was authorized to construct a single project that linked two points, and was required to follow certain guidelines in return for public support. In most cases, the charter prescribed the project's exact specifications—for instance, the type and depth of surfacing, maximum grade, and drainage requirements for roads— and a state or local board was created solely to oversee its planning and implementation. Public funding usually came as land grants or as pledges to subscribe to specified amounts of stock, but sometimes it merely consisted of the right to charge tolls or of an award of exclusivity.

These transportation projects had economic implications far exceeding the money invested. States and cities kept tight rein on their development precisely for that reason. New routes raised the value of adjoining lands, fostered the creation of new cities and towns, and promised prosperity to nearby established settlements by enabling them to increase the scope of their economic influence. But new transportation pathways also placed distant cities and towns at a competitive disadvantage. Many died or declined as their citizens moved closer to the new routes, while others demanded the same advantages for themselves and called on the state to undertake new projects in their vicinity.

These political considerations often increased construction difficulty and expense; straight routes with few grades or hollows were of course the cheapest and simplest to build, but they rarely satisfied everyone. The ease with which land could be cleared also remained a factor, as was the proximity of suitable building materials. Rarely did these technical questions take precedence, however, and few projects paid their investors dividends. In fact, construction costs frequently surpassed the corporation's ability to pay. But few of these projects were ever abandoned. New investors, an increase in state or local support, or total state or municipal ownership carried the work to completion, and citizens reaped the benefits from improved transportation.

Building Turnpikes

Early–nineteenth-century Americans continued colonial precedents and sought to improve travel overland and on the nation's waterways. Most colonial roadways had been essentially unimproved pathways where only the volume of traffic kept the underbrush from growing over, while river transport upstream was limited to pole-boats and rowed vessels. Adaptation of a type of roadway common in Europe, the turnpike, constituted a major commercial innovation. By the mid-1810s, Americans had built several hundred-mile or longer spans of these for-profit tollways. A decade later turnpikes joined eastern cities to the tramontane West.

Early–nineteenth-century American commercial vehicles had no standardized form, so turnpikes needed to accommodate different kinds of traffic traveling at unequal speeds. A wide turnpike with plenty of room for passing was the American solution. Unlike European turnpikes, American tollways traversed sparsely settled areas and generally were built on inexpensive land. Those conditions permitted Americans to construct broader roads with only modest cost increases, which were then passed on to users. Often as wide as twenty-eight feet, turnpikes were slightly raised in the center and sloped towards the sides to encourage water runoff, and generally abutted by drainage ditches. Surfacing material and depth varied considerably. Earth, clay, or timber composed the most primitive surfaces, but broken granite, slate, or limestone paved heavily trafficked ways. These broken-stone surfaces usually ran between nine and eighteen inches deep. Milestones marked the distance from the road's point of origin, and guideposts indicated the termini of intersecting ways.

Strength was a prerequisite for those who built turnpikes, but specialized knowledge was not. Only surveyors, who laid out the routes, possessed even a cursory technical competence; they were acquainted with geometry and trigonometry. The turnpike corporation's contractors were selected because of their local prominence, not their abilities; only their reputations guaranteed that specifications would be followed. Each contractor was responsible for organizing and superintending construction of a road section and employed local farmers to do the actual work. Using shovels, rakes, and hoes to move earth, and two-pound hammers to break rocks, the farmers/laborers cleared the path of trees, stumps, and large stones, then dug the drainage ditches, piling dirt from the ditches in the center to form a raised road bed. The bed was then tamped down and sculpted in preparation for surfacing.

Builders of the earliest American stone turnpikes followed the then-current European practice. Stones of five-to-eight inches in diameter constituted the surface's bottom layers, and those with diameters of between one-and-one-half and two-and-one-half inches composed its top. Stone dust, gravel, or sand was spread over the surface to enhance binding. But European road-surfacing ideas changed in the 1810s, and Americans began to adopt the new view later in the decade. The beds, not the surfaces, of these new macadam turnpikes were to bear the traffic's weight. Named after Scottish road-builder and inspector John McAdam, these roads proved less costly because of their thinner, readily maintained surfaces. Keeping the bed dry was crucial. Proponents of macadamizing argued that the earlier surface's large and different-sized stones permitted water to percolate to the bed. They advocated the use of

stones two inches in diameter throughout, cleansed of all mois-
ture-trapping or gap-creating material, to allow them "to unite their
own angles into a firm, compact, impenetrable body."[1] The initial
layer was rammed into the bed, and each successive layer was
compacted by traffic before the next was laid. Regular filling of ruts
was usually the only maintenance necessary.

Building Bridges

American turnpike-builders faced a recurring problem rarely confronted by their European counterparts; the New World's numerous rivers and streams required construction of many bridges to
transport turnpike traffic. Bridge construction was expensive, so
Americans made those structures to last. States generally created
different corporations to build bridges, but occasionally authorized
turnpike companies to build their own. Separate tolls were usually
charged for bridge travel.

American bridge-building was virtually identical to classic Ro-
man practice; like turnpike construction, it required no special skills.
Laborers created a watertight enclosure by driving wooden piles
through river mud to reach bedrock, removed water and silt, and
built masonry arches resting securely on the river bottom. These
arches supported the span, which generally was made of stone,
but timbers occasionally formed the span's roadway. Wooden
superstructures often covered these timbers to protect them from
the elements.

Improved River Travel—Building Steamboats

Improved river travel did not entail large construction expenditures
but rather a new prime-mover. It was a question of inventors and
inventions, not laborers and materials. Fitch and Rumsey had attempted to adapt steam power to exploit the river trade, but neither was alive by the time steamboats became operational. Rumsey
had died suddenly in 1792; a despondent Fitch committed suicide
a few years later. Evans experimented with a high-pressure steamboat in 1802, but ultimately placed the engine in a sawmill. John
Stevens, a Hoboken mechanic, was interested in steamboats somewhat earlier but failed to build a working vessel until late 1807. By
that time, Robert Fulton's *North River* (often incorrectly called the
Clermont) was steaming on the Hudson River between New York
City and Albany.

A jeweler's apprentice and a cameo painter, Fulton had gone to England in 1788 to study with the noted artist Benjamin West. There he learned of Fitch's work and met Rumsey, but apparently neither man made an immediate impression on him. Canals captivated Fulton, however, and during the 1790s he offered his unsolicited views on the virtues of canals to George Washington, Napoleon, and Parliament. He also proved something of a mercenary, building for the French an underwater bomb-attaching vessel similar to that originally constructed by David Bushnell in 1775. Fulton's submarine failed to destroy the English Navy, but his initiative brought him to Robert Livingston's attention. Although he was then ambassador to France—Livingston held a New York steamboat monopoly, and had employed Nicholas Roosevelt, a New Jersey steam-engine-maker, to design a craft (that proved to be unsuccessful)—Livingston told Fulton of his steamboat experiments, and decided to finance the young man in that endeavor.

Fulton began work quickly. By 1802 he had determined that paddlewheels delivered steam power adequately and that a long, narrow boat with stem and stern cut sharply at sixty degrees would best slice through water. He also concluded that the boat should have an atmospheric engine, and directed England's Boulton and Watt to build the engine and ship it to America. England had learned of Fulton's underwater vessel, however, and attached the engine until he designed another submarine to attack the French fleet. Although Fulton again failed to produce a successful submarine, the Crown relented and released the engine in November 1806. In August 1807, Fulton gave Livingston's friends their first public demonstration, and the next month began commercial runs.

Livingston and Fulton made a $16,000 profit in 1808, and they hurried to capitalize on the western trade. The Louisiana Purchase of 1803 had opened the port of New Orleans to Americans, and the Mississippi, Missouri, and Ohio Rivers seemed to provide a natural means to tap and settle the nation's interior. With the exception of the Orleans territory, however, western states and territories favored competition for their trade and refused to grant Livingston and Fulton monopolistic privileges. Two technical problems also plagued the New Yorkers. First, western rivers' swift currents often destroyed Fulton's angular steamboats. A return to conventional hulls resolved that problem. The second problem proved more difficult; his steamboats lacked power to move effectively upstream. Although Fulton tried larger atmospheric engines, he died in 1815 without solving the problem. His boats operated on western rivers

U.S. Patent drawing of the North River's assembled engine and boiler. Fulton's North River was the first successful steamboat. (top)
The North River was powered by an atmospheric engine. Notice the careful gearing from the engine to the paddlewheel. (bottom)

after his death, but they met stiff competition from, and were soon superseded by, high-pressure steamboats. The first was introduced in 1816 by Henry M. Shreve, a steamboat captain disappointed by his atmospheric engine's performance. In less than a decade, more than two-thirds of the western steamboats used high-pressure steam engines, which were much lighter and as much as ten times more powerful than atmospheric engines. By 1830 some two hundred steamboats were in operation in the West.

Although Shreve was the first to adapt high-pressure engines to western steamboats, these engines had first appeared in the region in about 1806. Evans had shipped numerous high-pressure engines to power western saw- and gristmills, and in 1811 opened an engine-building manufactory in Pittsburgh. Cincinnati and Louisville entrepreneurs created competing establishments almost immediately. These three Ohio River cities quickly took advantage of their geography, and by the mid-1820s built about two-thirds of American steamboat engines.

Building Canals

From the 1820s other forms of transportation competed with steamboats and turnpikes, as well as served localities the boats and roads could not reach. Canals were perhaps the most spectacular early–nineteenth-century internal improvements. America built its first canals in New England in the 1780s. Patterned after the English canals of the 1760s, these first few artificial waterways traversed short, straight, level distances. The first canals, as well as those of the nineteenth century, were bordered by towpaths from which horse or mule teams drew special flat-bottomed canalboats.

The nation's earliest canals were contemporaneous to Fitch's and Rumsey's steamboat labors, but the steamboat's success did not lessen the country's interest in canals. In fact, the American canal era flourished after steamboats had already started to course the nation's rivers. More impressive and extensive canals over rougher terrain had become popular in the late 1810s; canal-building mania swept the nation for two decades. While Americans had built fewer than one hundred canal miles prior to 1815, during the next twenty-five years more than $125 million was spent on nearly thirty-one hundred canal miles. New York, Pennsylvania, and Ohio accounted for more than half of that expenditure.

States and corporations retained ultimate canal-building authority, but responsibility for technical matters and daily operations usually rested with each canal's superintendent. His recommendations were generally accepted on technical questions, such as the canal's width and depth, number of locks necessary to keep channels level, type of lock mechanism, and whether locks should be made from wood or stone. He was expected to ride herd on surveyors, rodmen, chainmen, and axmen, as well as other workers. These demands led corporations and states to devote considerable time to selecting superintendents. English or French canal-builders were usually their first choices, but often these men refused to come to America. Americans with European canal-building experience or who had already headed an American canal project were solicited, as were those currently working in the upper echelon of an ongoing canal venture. Corporate raiding was common. Experience was equated with, and was the test of, competence.

New York's decision to build the Erie Canal joining Albany to Buffalo on Lake Erie sparked the canal boom. Constructed between 1817 and 1825, but proposed over a decade earlier as a means for New York to counter New Orleans' growing commercial influence in the West, it dwarfed all preceding American canals. Its channel was 363 miles long and four feet deep, with bottom and surface widths of twenty-eight and forty feet respectively. The canal contained eighty-four locks, the largest of which was 110 feet by 18

Horse-drawn canal boats make their way on the Erie Canal, the largest early–nineteenth-century American canal.

Excavating the Erie was a project of unprecedented scope.
Workers often used hoists to clear the channel.

feet, for a total lift of 689 feet. Its towpaths stood about three feet above the channel's surface and were fourteen feet wide. Although the towage rate averaged less than two miles per hour, and the canal cost more than $7 million to build, it reduced freight charges to one-twentieth the overland price, a factor that encouraged use and allowed the canal's original investors to recoup their monies within seven years.

It was the Erie Canal's economic and technological potential, as well as its subsequent success, that fueled canal construction nationwide. Seaboard cities such as Philadelphia and Baltimore, already threatened by the rise of New Orleans and the western river trade, saw the canal as a further threat to their economic viability: it promised to provide New York City (connected to Albany by the Hudson River) with a tremendous competitive advantage in the western trade. Goods would flow to and from New York City or New Orleans, while Baltimore and Philadelphia would be relegated to secondary cities with limited markets. Other cities urged their legislatures to grant them the same advantages as New York City;

they wanted competing canals. Western states and cities looked to the Erie Canal not as a competitor but as an opportunity to expand their eastern trade, and constructed canals linking them to Lake Erie. The various Ohio canals, Indiana's Wabash Canal, and the Illinois Canal were testimony to that thrust.

These western canals often were built by men who had worked on the Erie Canal. The magnitude and difficulty of the Erie Canal's construction seemed the supreme American test of engineering mettle, particularly in the fifteen years after 1817. During that period, former Erie employees directed far more nonmilitary construction projects than West Point graduates and Corps of Engineers members combined. Only when canal-building waned in the late 1830s did the Erie men lose their preeminence.

Constructing level, watertight channels was a most important and demanding aspect of canal-building. Grades or currents, consequences of poor surveying, would strain boat-pulling beasts. Indeed, the absence of grades was so critical that canal companies constructed stone aqueducts, similar to turnpike bridges, over even small streams. Variations in water level also produced difficulties. Surface drainage into channels raised water levels precipitously and played havoc with locks, while seepage, often the result of channel-burrowing animals, lowered the water level and caused boats to run aground. Canal workers lined channel walls and floors in either of two ways. Workers called puddlers tempered clay in pug mills, mixed it with sand and gravel, kneaded it further by hand, and applied it to channels in layers three inches deep, taking care to roughen surfaces to promote adhesion of subsequent layers. In the Erie Canal, and many later ones, a limestone-based hydraulic cement, which hardened underwater, replaced puddling. Laborers prepared the cement by kiln-heating the distinctive limestone, breaking it up, and mixing two parts lime in pug mills with one part sand. They then spread the pasty mixture in the empty channel, after which it was flooded.

Locks were the devices by which the various level stretches of the canal at different altitudes were joined. Locks varied considerably in shape, size, and material, but all worked according to a similar principle: boats could be raised or lowered simply by raising or lowering the water level, like a water elevator. Situated in cutouts parallel to channels, the locks' chambers were framed with timber or masonry long and wide enough to accommodate the largest boats. Their walls were tall enough to reach the upper level of the canal, and thick enough to secure the water necessary to raise

Operational view of the Erie Canal at Lockport

boats to the desired height. Heavy swinging gates sealed their ends and made them watertight. Once boats entered chambers, lock operators closed the gates and pumped water in or out to achieve the appropriate end.

Building Railways

Railways were the final early–nineteenth-century transportation initiative. The first American railways replicated those in Britain a century earlier. Like the earliest British railways, Massachusetts's Quincy Railway (1825) and Pennsylvania's Mauch Chunk Road (1827) were horse-drawn, single-purpose railways. The British railways transported coal from mines to ironworks, while the American railways carried large granite blocks from quarries. Not until 1827 did a state legislature charter a general-purpose railway. Again, the British example was crucial. George Stephenson's Stockton and Darlington Railway (1825) and the Liverpool and Manchester Road (construction began 1826) helped convince recalcitrant state legislatures to approve similar plans in America.

These initial American general-purpose railways mimicked canals and turnpikes in route, purpose, organization, and design. They were treated as common roads, open to anyone who paid tolls and used proper cars, which were usually specified in the way's charter. Their advocates boosted them as faster than canal travel and as more reliable; not subject to freezing in winter, as well as flood- and drought-resistant. States and municipalities sponsored railways to provide their citizens with commercial advantages available elsewhere to others. For example, Baltimore initiated the Baltimore and Ohio Railroad (chartered in 1827 to run over 250 miles from the Monument City to the Ohio River) both to combat New York City's dominance of trade with the West and to prevent Washington, Georgetown, and Alexandria from exploiting their Chesapeake and Ohio Canal advantages. Boston took a different approach: Its 308-mile railway from Boston harbor to the Hudson River was designed to siphon off the New York City trade; instead of erecting a competing line, Bostonians tapped into an existing one.

Questions of railway methods and materials pervaded these early years. Designers and builders experimented with rails of stone, wood, and iron-plated wood before settling on iron. Even then they debated rail shape and whether to cast them (it proved difficult to cast rails straight in lengths greater than five feet) or use wrought iron. They also considered flanges to hold wheels on the track, whether railway bridges should be made of stone or wood, how best to cover ties to provide horse teams with a smooth walking surface, and whether to construct double- or single-tracked ways, the latter with periodic cutouts to allow oncoming traffic to pass.

Building railways required a curious combination of turnpike- and canal-construction skills. Making the bed level, dry, and firm were the most important tasks. Ironically, few canal engineers superintended railway projects. The Corps of Engineers' current and former members surveyed the great majority and directed many, sometimes after a fact-finding English tour. They situated the drainage ditches, laid the way to avoid abrupt curves, and tried to keep grades to less than one-eighth-inch per yard. Following this procedure reduced costs. Curves could involve deep ground cuts and expensive embankments. Steep grades required stationary steam engines to pull cars up the rails; horses could pull heavy loads on level ground, but it was assumed that the frictionlessness of iron wheels running on iron rails prohibited them from taking their burden up significant grades. Even in the face of these problems, railways grew dramatically: from twenty-eight operating miles in 1830 to more than twenty-three hundred in 1839.

The spate of railway construction in the 1830s was fueled by a new prime mover, the steam-powered locomotive. Like railways themselves, locomotives were of British origin, first introduced to carry coal in South Wales in 1804. Powered by high-pressure steam generated from spherical boilers, which housed fireboxes and flues, these early locomotives' drive wheels slipped continuously on the rails, which rendered them unsatisfactory for transport. Only in the 1810s did the British conclude that slippage stemmed from an abundance of power, a situation readily corrected, not from contact between frictionless surfaces. Still, locomotives offered horses virtually no competition until Stephenson's Stockton and Darlington Railway, and only a modicum of it in the years immediately thereafter. The 1829 Rainhill Trials turned the tide. Sponsored by the Liverpool and Manchester Road's directors and held on a flat, one-and-a-half-mile track, the trials sought to determine locomotive transport's practicality. Entrants were asked to pull twenty tons at least ten miles per hour for sixty miles. Fifteen thousand spectators witnessed the event. Stephenson's *Rocket* exceeded the directors' conditions and demonstrated that a large, complex prime mover could be dependable.

The trials also captured the American imagination. The first locomotives in the United States arrived from Britain later that year. In the following year, Peter Cooper, a Baltimore mechanic, built America's first. Eager businessmen quickly put these engines in service, and determined almost as quickly that they failed to meet American needs. The British locomotive's driving wheels turned on axles rigidly attached to the engine's frame. Such an arrangement proved suitable for Britain—where tracks were level and routes were short, straight, and heavily traveled—but disastrous in America, where uneven ground and long, winding, lightly used routes would be the rule. Two major innovations gave American locomotives a characteristic form. Albany's John B. Jervis, who had begun as an Erie Canal rodman, designed in 1833 the swivel truck—a second frame and a double set of smaller wheels under the boiler's front end—to enable locomotives to follow curves, while Philadelphia's John Harrison added the equalizer lever in 1838 to distribute shocks evenly to all drive wheels.

Other American innovations were less revolutionary but no less important. Each was a response to peculiarly American conditions. Cow-catchers appeared in 1833 to remove obstructions from tracks and prevent derailments caused by freely roaming animals. Long-distance travel through thinly populated environs on irregular schedules also gave rise to railroad bells, which by 1835 were required in several states. These same lengthy routes led to frequent night motoring and to headlights, to which parabolic reflectors were soon added. Smokestacks also received attention. British locomotives burned coal cleanly, but American wood-burning engines produced numerous embers and large sparks, which injured passengers and set fires along routes. As early as 1832, Americans employed crude wire-mesh devices as primitive spark arresters.

Nor was that all. In 1836, William Norris, owner of a Philadelphia engine-building establishment, conducted a public demonstration of locomotive practicality as significant for America as the Rainhill Trials. He demonstrated that a locomotive could run easily up a grade as steep as 363 feet per mile; grades were no longer debilitating. Without the need for the expense of precision leveling and stationary steam engines to pull locomotives and cars over even modest grades, the locomotive's potential now appeared unlimited.

CHANGES IN MANUFACTURING AND LABOR

New transportation routes and forms of transit provided an infrastructure for commerce, served as immigration pathways, and helped determine settlement patterns. European immigrants streamed into the ports of New York, New Orleans, Boston, and other coastal cities, but generally moved beyond their port of first call. Emigrant Friends' Societies in virtually every municipality in the nation's interior encouraged newly arrived Europeans to settle away from the coasts. The overwhelming majority of immigrants took up residence in proximity to a site blessed by improved transportation. These men and women became producers and purchasers of goods. America's abundance of natural resources promised, when modified by technical or technological endeavors, to yield unprecedented growth, and material prosperity seemed within reach of anyone whose opportunity remained unchecked.

Immigration, emphasis on technologically engineered production, and a desire to increase consumption heightened demand for material goods and quickened America's manufacturing pace in the early nineteenth century. Small manufacturing ventures remained the rule, and these establishments generally perpetuated shop-based master–apprentice relationships, but with master–apprentice connections even less formalized than in the colonial era. Manufactories of this type flourished in northeastern towns and cities from Boston to Baltimore, and also in the newer cities in the West. These urban manufactories generally served those outlying areas with easy access to the city, which engendered competition for rural and village markets situated between two or more manufacturing centers as each of the latter sought to extend its influence and penetrate the others' hinterlands.

Manufacturers borrowed heavily from European technology. Immigrants and industrial espionage brought European practices and techniques, especially machinery, to America, but Americans used waterwheels, not steam engines, to power the great preponderance of manufacturing machines. Lack of cheap labor led some American industries, such as clockmaking, to rely even more heavily on machinery than their counterparts in Europe, but this early machine-based production usually resulted in shoddy goods as these devices failed to duplicate the skills of artisans. Yet it would be a mistake to underestimate the scope and diversity of early–nineteenth-century manufacturing interests. For example, Cincinnatians had created before 1820 foundries for casting copper, brass, and iron; a manufactory for textile machinery and cotton gins; a steamboat works; a manufactory for white and red lead; and a chemical works. By 1830, the city housed more than three times as many chemical and steamboat works and foundries and had added several machine shops, gunpowder and paper mills, steam engine and casting manufactories, a sugar refinery, and manufactories of tallow candles and soap.

Machine Shops as Adjuncts to Mills

At the heart of water-powered, machinery-based manufacturing were mills, millwrights, and machine shops. The explosive growth of mills testified to the proliferation of manufacturing in the early nineteenth century, while their relatively small power output— generally less than one hundred horsepower—indicated the modest size of most manufacturing establishments. Demand for millwrights far exceeded that of the colonial period, and early–nineteenth-century millwrights needed to understand the sophisticated linkages often required between mills and machines. But machinery-dependent production also required a new type of craftsman/manufacturer—one who worked in a machine shop. These machine-makers fashioned devices from wood and metal to fit individual specifications, and repaired broken machines. Because completed machinery was often unwieldy to transport, machine

A modest ironworks, about 1830, combining machine shop and foundry. Observe the foundry operations.

shops were located near manufacturing complexes and on transportation routes; they served localities. The number of machine shops increased markedly during the century's first three decades, but their form remained essentially unchanged. These single-room opera-

A huge, highly articulated ironworks, about 1840

tions employed fewer than fifteen men, were fitted with hand- or foot-powered lathes, contained workbenches with vises and grindingstones, and included tool chests filled with assorted hammers, chisels, and files. Only a few larger shops before the mid-1830s housed waterdriven lathes, screw-cutting engines, or boring, drilling, milling, and planing machines.

Metal-working machine shops depended on foundry castings, rods, and sheet metal. Establishments often called *works* combined a machine shop and foundry, and generally included a drafting room to prepare plans on paper, a pattern room to store wooden casting patterns, a blacksmith shop, and an assembly room to assemble finished articles. Works sometimes employed as many as fifty hands in machine shops, and produced large items, such as steam engines.

Use of Machinery in Textiles

Works and machine shops built and used many machines, but textile manufacture was the most significant consumer industry to embrace machinery. Carding engines, spinning jennies, water frames, mules, throstles, and power looms would become integral parts of early–nineteenth-century American textile manufacturing. While most Americans continued to sew their own clothing, a lesser number wove their own cloth. As they shifted to machine production, textile manufacturers reduced labor costs by replacing skilled workers with less-skilled or unskilled workers, or even children.

Much cloth was produced for local domestic markets, but imports into America by the heavily mechanized Europeans remained formidable. American manufacturers repeatedly protested that European authoritarian regimes possessed unfair advantages because they could keep labor costs artificially low. This situation did not prevail in labor-short America, where the free labor market, and comparably profitable endeavors such as farming, forced manufacturers to compete for workers with relatively high wages: American textile workers received wages from 30 to 50 percent higher than their European counterparts. The Jeffersonian Embargo and the War of 1812 dramatized the new nation's dependence on European textiles, and led some to conclude that American security required the country to free itself from possible European tyranny by fostering domestic textile manufacture, which needed reliable domestic demand to survive. Increasingly they turned to the national government for assistance. The term "American System," popularized by Henry Clay, came from these pleas, as did protective tariffs, as high as 45 percent, imposed on imported textiles in 1816, 1818, 1824, and 1828.

Part of the American textile manufacturers' difficulties in becoming competitive stemmed from the fact that the transformation from the use of skilled labor to the use of machinery occurred neither all at once nor at the same rate in cotton and woolen textiles. Mechanization of spinning preceded changes in weaving, and changes in the manufacture of cotton fabrics predated those in the manufacture of wool. Despite these differences, early textile mills assumed a characteristic configuration. Individual operations generally were housed in separate buildings or rooms. In cotton-spinning, for example, raw cotton first went to a picker house where machines clawed apart bales, removed impurities missed by cotton gins, and shaped it into clean batting sheets for carding. In another location, rotating wire-toothed cylinders called carding engines sorted out fibers and laid them parallel to form loosely twisted cotton ropes, which water frames roved. At their next destination rovings were drawn, twisted, and wound on bobbins or spindles by throstles (for strong, coarse yarn) or mules (for fine yarn). Both machines employed mechanical rollers or plates to do work formerly done by fingers, and both used flyers to guide yarn onto bobbins and spindles.

Manufacture of textile machinery also occurred in textile mills. Machine shops might be placed in the basements of new mills, where mechanics built mill machinery while carpenters and masons completed the mill's structures. Rarely did early mills sell machinery to competitors, but trading them was common, as was leasing inventions. The de facto prohibition of machinery sales helped restrict the industry to those firms already in operation, while trading and leasing protected inventors' rights and spread improvements among industry participants. Only with the industry's tremendous expansion in the 1820s did textile manufacturers regularly offer machines for sale. These textile machine shops then were spun off; they moved to separate quarters to accommodate increased production, and also began to make other wares such as machine tools.

But development of textile machine shops was possible only after the introduction of British textile machines invented some decades earlier. Indeed, that was the crucial step. Americans sought to pirate English textile machines almost as soon as they were created. Prior to 1790, British precautions had proven adequate: Americans failed to reproduce machinery that yielded cotton yarn suitable for weaving. Samuel Slater, a former English cotton-spinning mill manager, was the first in America to produce workable devices. He left England for the United States in 1789, and was working for a New York corporation when William Almy and Smith Brown of Pawtucket contacted him about building cotton-spinning machines. Slater agreed, but only after he had used his coveted status to secure a half-interest in the anticipated textile mill. The manufactory opened in 1790 and children seven to twelve years old ran the equipment. The machines were so productive and child labor so inexpensive that by 1895 raw cotton costs amounted to two-thirds the mill's operating expenses, and it produced more than enough yarn to keep one hundred weavers occupied.

The Scholfield brothers did somewhat later for woolens what Slater had done for cotton. Experienced wool-spinners, John and Arthur left England for New England in 1793 and were joined by James in 1802. They produced an impressive series of waterpowered woolen textile machines (each of which was already in operation in Britain). These included carding machines (1793), spinning jacks (1802), machines to loosen matted wool for carding (1806), spinning jennies (1809), and mules (1810).

But neither the Scholfield brothers' success nor that of Slater stimulated a rash of textile-mill construction. As late as 1808, the brothers were virtually alone in having established woolen manufactories, and just fifteen cotton mills operated in America, of which Slater owned eight. As mentioned earlier, not until they recognized their dependence on English cloth supplies, and the southern plantation economy's dependence on British purchases of raw cotton, did Americans strive to establish their own mills. These efforts were most pronounced in cotton spinning: by 1814, 243 cotton mills dotted the nation. Most new cotton mills copied Slater's manufactories as men trained by Slater left Pawtucket and erected their own mills or built machinery for others.

American textile manufacturers, especially cotton-spinners, faced renewed English competition after the War of 1812. Concern for the nation's security, and the introduction of power looms into America, permitted many cotton mills started during the war to survive its conclusion. Common in England before 1810, cotton power looms initially appeared in America in 1813 at Francis Cabot Lowell's Boston Manufacturing Company, a textile concern located in Waltham, Massachusetts. A wealthy entrepreneur without a technical background, Lowell had planned a textile manufactory some years earlier and in 1810 went to England, presumably for his health. While there he observed power looms and returned capable of building them, assisted by local mechanics. These waterpowered looms replicated the action of hand looms, and dominated the industry after 1820. A woolen power loom was introduced in 1816 and also met general acceptance in the 1820s. Noisier but much swifter than conventional looms, these power looms created a demand for woolen and cotton yarn while simultaneously reducing cloth manufacturers' single greatest expense—weavers' wages.

Social Experiment at Lowell   These spinning and weaving mills were minuscule when compared to another textile venture undertaken by Lowell and by other Boston merchants. They planned to erect a huge textile manufactory on the lower Merrimac River near Pawtucket Falls, a site twenty miles from Boston and connected to it by the Middlesex Canal. This manufactory's location on the river would make available to its mills 10,000 horsepower. These men faced a problem of such magnitude as to dwarf any earlier waterpowered enterprise in harnessing the forceful river system, even when it was at flood level. Dams, races, and other means of controlling the Merrimac, of unprecedented size and scope, needed to be constructed.

This venture also provided an opportunity for an experiment in what might be called social technology. Lowell had visited Robert Owen's New Lanark cotton mills in Europe. That visit had suggested to him that large manufacturing ventures could provide sizable profits for entrepreneurs, and character-building opportunities for workers. He interested his associates, some of whom had participated in the Waltham mill, in this curious democratic utopian combination. What became known as the Lowell mills were the result. Opening in 1823, six years after Lowell's death, and employing almost exclusively rural New England women between the ages of seventeen and twenty-four, the mills were built on a belief that individual success rested on mental discipline and moral management—character—and that repetition and regimentation were excellent teachers of those behaviors. Working in the Lowell Mills was to provide this education. Dormitories supervised by matrons and in which all the women lived were the scheme's linchpin. Mostly widows, the matrons ensured that mill women dressed and acted properly; they enforced a code that prohibited immodesty, impropriety, dishonesty, idleness, intemperance, profanity, dancing, and gaming. Dormitories were locked at 10 P.M. Regular church attendance was required. Violation of regulations was grounds for dismissal.

Women operate banks of power looms in large textile manufactories, such as the Lowell Mills.

Idleness was hardly a problem; women ran textile machinery at least eleven-and-a-half hours daily. Despite this commitment, Lowell women established a school, created a broad range of morally uplifting organizations, and edited a literary magazine, the *Lowell Offering*. These activities and the dormitory situation refined Lowell women. Mill work provided them dowries. After a stay of about four years, many married, moved West, and became school teachers. The constant turnover, lack of advertising, long hours, and severe regimen never resulted in a labor shortage. Rural women recognized that Lowell offered them opportunities for advantageous matches. Eligible men flocked to the place to meet, court, and marry these virtuous young women.

Prominent Europeans and Americans as different as Andrew Jackson, Edward Everett, and Davy Crockett visited these mills and applauded the experiment. Crockett summed up in 1835 the prevailing view. He discovered at Lowell "thousands, useful to others, and enjoying all the blessings of freedom, with the prospect before them of future comfort and respectability."[2] The Lowell social experiment was abandoned shortly after Crockett's visit because the mill was not profitable enough. Nevertheless, mills even larger than Lowell, but without its democratic, utopian pretensions, would become common in the mid-nineteenth century on rivers in Massa-chusetts, New Hampshire, and Maine.

Reducing Labor Costs

Extensive machine use, however, was hardly the only method early–nineteenth-century entrepreneurs adopted to trim large-scale pro-duction's labor costs. For example, machines made no inroads into shoe and boot manufacture. Nor did workers employ tools other than the traditional hammers, awls, pincers, nippers, and scrapers. Yet the industry lowered labor costs dramatically during the early nineteenth century.

Modification of the putting-out system, accompanied by use of less skilled labor, cut shoe and boot workers' wages. No entire shoe or boot was made under a single roof, or by a single individual. The rise of main shops, at which leather for uppers were cut, was the critical organizational factor. Workers carried the pieces cut in the main shops to their homes, sewed seams and bindings, stitched counters and straps, and returned completed uppers to main shops. These uppers and roughly cut soles were put out to another group, makers, who would last and then sole boots and shoes in their homes. They then deposited the goods at main shops for inspection and to be readied for sale.

Replacement of high-waged artisans with lower-paid, semi-skilled labor, much of it female, reduced costs drastically and made Amer-ican shoes and boots highly competitive in the world market. South America and the West Indies imported much of this American footware, and northeastern shoemen further expanded sales in the southern states.

These cost-cutting moves did not always go smoothly. Particu-larly in traditional craft industries such as shoe and boot manufac-ture, early–nineteenth-century American workers vehemently protested employers' attempts to replace artisans with less skilled laborers, to keep wages low, and to lengthen workdays. These workers occasionally expressed their dissatisfaction by withhold-ing their labor; journeymen turned out on masters and workers struck owners. Wage-earners justified their actions in a manner characteristic to early–nineteenth-century America: they repeat-edly portrayed employers as attempting to erect themselves as an aristocracy of wealth, and maintained that this aristocracy de-stroyed individual opportunity for improvement.

Workers living in cities were the most active protesters. Groups of artisans formed local trade associations that mimicked the European guild system. These secret societies, many of which existed by 1800, established rules for members, wage scales, and conduct codes; they attempted to regulate the labor supply, and to maintain or boost wages. Agreements not to work in shops not employing association members, or failing to follow association guidelines, were their weapons, and they sometimes shut down manufacturers refusing to bow to their demands.

This power threatened manufacturers, who sought legal redress. In a series of decisions between 1806 and 1815, courts upheld the right of laborers to associate but defined trade-association activities as "arbitrary and coercive," as an attempt "to create a monopoly or restrain the entire freedom of trade." These rulings maintained that trade-association policies hampered individual opportunity in two senses. First, judges argued that they served "to injure those who do not join their society" by making it impossible for non-members to work. Second, the courts also claimed that employers had "rights as precious as" those of associations and that combinations "to deprive their fellow citizens" of these rights constituted illegal conspiracies.[3]

Outlawing closed shops left moral suasion as the trade associations' sole legal right. Moral suasion provided little leverage, however, and association influence declined. Removal of property qualifications for voting in the early nineteenth century led workers in a different direction. Newly enfranchised laborers (virtually any white male citizen over the age of twenty-one) established workmen's political parties in the late 1820s and 1830s; political power, not economic clout, would secure equal opportunity. These parties railed against a wealthy, nonproductive aristocracy bulwarked by special privilege, contended that government favored this aristocracy, and argued that their candidates' election was the only means to change the situation. More precisely, workingmen's parties demanded termination of special privilege, especially banking monopolies; imprisonment for debt; and mechanics' lien laws in which a mechanic's tools (and by extension his livelihood) were seized until he paid all outstanding debts. They advocated equitable taxation, direct election of public officials, and free public education. The last would provide workers' children with advantages formerly reserved for the rich.

Workingmen's parties never dominated the American political scene and disappeared almost as suddenly as they began. Internecine warfare marred them as various elements battled for control, and more traditional parties co-opted many of their proposals. By 1840 they ceased to exist.

PREPARING AMERICA'S YEOMANRY

The agitation of workingmen's political parties hardly constituted a rejection of the work ethic. It dramatized instead their criticism of those who used privilege to benefit from the labor of others. Those who labored were glorified: early–nineteenth-century Americans lionized such men as Eli Whitney, Oliver Evans, Robert Fulton, and Benjamin Latrobe. Commentators considered these American inventors the archetypical American workingmen, compared them favorably to Europe's best and brightest, and celebrated their lives as well as their technical achievements. To most people, the two seemed closely related: lofty attainments reflected a high state of individual cultivation, which included temperance, morality, moderation, cleanliness, godliness, and industriousness. Both lofty attainments and a high state of individual cultivation were considered to be natural products of the American political system. These technological stalwarts' careers reaffirmed democracy's virtues; they demonstrated what individuals, lacking Europe's manifest institutional advantages but freed of social and legal restraints, could accomplish. Their successes were natural, not the result of affiliation with secret organizations, special training unavailable to others, or hereditary advantage. Their triumphs seemed analogous to those of the Founding Fathers and Andrew Jackson. That Whitney, Evans, Latrobe, and Fulton worked with their hands seemed another indication of their freedom from aristocratic pretensions. Celebration of these technologists was celebration of young America's promise, the promise of opportunity.

A closer examination of the lives of Whitney, Evans, Latrobe, and Fulton would have revealed less nobility and independence than contemporaries maintained. That is beside the point. The issue is not whether those lives merited praise, but that early–nineteenth-century Americans found so few lives to celebrate. This suggests that America's early promise remained at that time largely unfulfilled, an issue of great immediacy after about 1820. Americans had difficulty in pointing exclusively to the future for that promise as they approached the Declaration of Independence's fiftieth anniversary. The nation had had a half-century to prove itself; it had ceased to be democracy's great experiment and had become an established fact. And when judged by the relative paucity of individuals worth celebrating, America seemed more than a little disappointing.

*Whitney's cotton gin was acclaimed by Americans for its
simplicity and effectiveness, and as a demonstration of
ingenuity.*

To be sure, Americans recognized that they had begun to exploit
the nation's abundance and to provide a modicum of material prosperity. But a properly functioning democratic republic in early–
nineteenth-century American terms should have resulted in more
than increased wealth and a handful of celebratory figures. For example, the Boston cleric William Ellery Channing despaired at how
few learned men America had produced and how Americans generally were unfamiliar with belles-lettres. James Fenimore Cooper
detected an alarming tendency among Americans to abandon the
principles of democracy, and to imitate European aristocracy and
expect the privileges congruent with that imitation. That it had
not measured up to these expectations implied that the nation also

was failing to generate a virtuous yeomanry, the base upon which democracy presumably rested. The category, virtuous yeomanry, was to include nearly everyone if the promise of democratic America was to be achieved. A virtuous yeoman would be self-reliant, content, healthy, educated, not necessarily wealthy but certainly not destitute, and possess a character beyond reproach. That character would allow him or her to resist material and aristocratic pretensions—the pretensions of Europe—and to perpetuate democratic sentiments and interests.

This apparent failure was a most menacing concern. It struck at the very heart of the idea of America. In contemporary parlance it signalled a lack of individual opportunity, a lack of democracy. In other words, those artificial barriers that prohibited more from achieving the notoriety of an Evans, Whitney, Latrobe, or Fulton also seemed to prevent a far greater number of individuals from developing the character on which America's present and future depended. This fear spawned several attempts to remove obstacles through the creation of increased access, including self-culture campaigns. Americans produced a massive amount of literature to supply the crucial information, and formed a dizzying variety of new character-building associations. These organizations not only provided additional opportunities for individuals to associate but also furnished knowledge and fostered appropriate habits among members. Establishment of mechanics' institutes during the mid-1820s exemplified that thrust.

Mechanics' Institutes

Although the British had erected mechanics' institutes somewhat earlier, they scarcely resembled their transatlantic counterparts. Americans identified as mechanics almost all urban workingmen except merchants and bankers. Knowledge to all seeking it served as the institutes' motto; it reflected a commitment to a large segment of the population, dissatisfaction with colleges as the privileged's bastion, and an attack on guilds as well as other restrictive societies. These mechanics' associations conceived of knowledge broadly, as not limited to workplaces. Evening public lectures on natural history and philosophical topics such as "The Wisdom and Goodness of God as Evinced in the Natural History of Water"[4] demonstrated their catholicity and were their chief educational forays. Sometimes institutes provided more formal, although hardly more technical, schooling; they established postcommon schools that taught drawing, mathematics, Greek and Latin, rhetoric, and some science to offer participants advantages similar to those able to afford college.

Mechanics' institutes stressed self-culture but they also furnished three important technically related services. Their boards of control created yearly prizes or premiums for solutions to specific technical problems and to questions of local economic interest. Institutes also housed libraries and artifactual collections. As important, they held annual exhibitions; inventors and small manufacturers displayed their productions, while larger manufacturers and merchants examined them and perhaps purchased rights to produce or market one. The exhibitions thus joined mechanics with capital.

Agricultural Societies

These latter activities served mechanics in the same ways that contemporary agricultural societies served farmers. Early–nineteenth-century agricultural organizations offered prizes for specified technical innovations as well as for improved fertilizing and cropping methods, and held contests to adjudicate rival claimants' contentions. Field trials placed implement-makers in conjunction with capital and consumers. Agricultural societies also demonstrated concern for their members' self-culture. Many presentations extolled the farmers' national position and provided guidelines for farmers to prepare themselves for it. Temperance, cleanliness, godliness, and the like figured prominently.

Technical Education for the Yeomanry

Other institutions also addressed the cultivation of the virtuous yeomanry—sometimes from an unusual angle. The Rensselaer School, founded in 1824 by New York patroon Stephen Van Rensselaer and his associate Amos Eaton, sought to create an educated yeomanry by teaching "sons and daughters of farmers and mechanics" the philosophical underpinnings of the knowledge they would use in their lives' work. Instruction in the "common purposes of life's"[5] underlying principles was not to establish new facts or procedures, but to imbue future yeomen with moral discipline, responsibility, and understanding—traits essential to life's virtuous conduct.

An improved yeomanry was the school's goal, and it cast its net widely. It diffused its influence by not teaching yeomen directly but rather training persons to teach them. Although it attracted well-to-do students, the school's educational thrust mimicked master–apprentice relationships. Eaton served as master in an extraordinary variety of areas within the more general rubric of "agriculture, domestic economy, the arts and manufactures." He tutored each student in the teaching of a particular calling, such as land-surveying. Eaton performed demonstrations of various skills, and explained the scientific principles upon which these exhibitions rested. Students practiced these demonstrations and explanations before Eaton and their peers, and were graduated when they could flawlessly execute and elucidate them in front of the school's examining board.

Few students attended Rensselaer during its early years, and fewer graduates engaged in teaching. Eaton tried to increase the school's utility in two ways. He repeatedly petitioned the New York State legislature to fund one student per county to prepare there. He also applied what he preached, using his demonstration–explanation technique in public lecture courses given directly to farmers and mechanics. His series entitled "Technology," first offered in 1830, met twice-weekly in ninety-minute sessions over a period of four months, and provided learning opportunities to its three hundred participants.

Eaton's technology lectures closely approximated the form of Josiah Holbrook's lyceums. Holbrook reportedly witnessed the burgeoning lyceum movement among English mechanics, thought America fertile ground for such a democratic institution, and adapted it in 1826 to the United States. These "associations for the purpose of mutual education" quickly proved popular; by 1834, Americans operated roughly 3,000. Like Eaton's lectures, lyceum addresses were usually open to the public and depended on scientific apparatuses—Holbrook manufactured these lyceum apparatuses—to show the principles on which commonplace occurrences were based. But the substance of Eaton's series was more nearly akin to the lectures that Harvard's Jacob Bigelow had given in the Boston area since the mid-1810s. The self-proclaimed "leading exponent of technology," Bigelow helped popularize the term in America. He divided technology into elements—separate fundamental units designating categories of human productions—and maintained that these elements were bound together as manifestations of a particular mode of thought. To Bigelow, technological advances depended on creative but formalized thinking; they were the consequences of rationality, of inductive, not empirical, thought.

Bigelow claimed that the number of American inventions indicated that this mode of reasoning had gained a New World foothold, but he worried about its lack of pervasiveness. His lectures and writings were attempts to ensure its diffusion. They served to make audiences aware of the rational mode of thought's significance; they both taught it and gave men and women opportunities to learn and apply it.

MACHINERY, PRODUCTION, AND DEMOCRACY

American concern for a virtuous yeomanry produced initiatives that aimed to ensure its creation. Americans expressed much less anxiety over the rapid introduction of machinery in industry, and undertook no moves to combat it. That was not the case in England. There groups of protesters called Luddites broke machines, and others decried machinery's effects. The noted English litterateur and student of German romanticism, Thomas Carlyle, tried in 1829 to frame the question for his generation. He attacked the "mechanical philosophy," which he maintained had become "a sign of the times." He feared for mankind's moral and aesthetic future, arguing that "mind will become subjected to the laws of matter" and that "efforts, attachments [and] opinions" will be solely "of a mechanical character." The "external and physical is now managed by machinery," and Carlyle predicted that "the internal and spiritual" would soon be controlled. "In our rage for machinery," he concluded that "we shall ourselves become machines."[6]

Timothy Walker, a young Cincinnati lawyer, challenged Carlyle's conclusions in a characteristically American way. He asserted in 1831 that "mechanical enterprise, . . . far from being unfavorable to our spiritual growth, is the one thing needful to furnish the freedom and leisure necessary for intellectual exercises." The mind depends on the body, he contended, and "if machines could be so improved and multiplied" to gratify all corporeal requirements, then all mankind could become "philosophers, poets and votaries of art. The whole time and thought of the whole human race could be given to inward culture, to spiritual advancement." Walker surveyed past civilizations to discover how they achieved the leisure necessary for intellectual activities and found that freedom "from the bondage of perpetual bodily toil" always involved tremendous moral costs. Past civilizations operated on the "barbarous principle that one nation may purchase itself leisure . . . by aggressions upon the rest"; they provided their citizens opportunities "by levying tribute upon all other nations, and keeping slaves to perform their drudgery at home." Walker claimed that such drastic measures were no longer necessary. "Force did for [the past] what machinery does for us."[7]

The Walker–Carlyle dialogue reflected European and American differences. Carlyle's critique stressed the approach's creative and aesthetic consequences, a predominantly European concern. It suggested to Americans an unseemly preoccupation with things generally reserved for the aristocracy and appeared to be a plea to perpetuate aristocratic privilege. Walker accentuated the body, arguing that machinery expanded spiritual growth opportunities by providing for corporeal needs. Indeed, the debate typified by the arguments of Carlyle and Walker was less about machines than about European and American perspectives. Americans debated machinery's implications among themselves, but restricted discussion to its relation to democracy, to individual opportunity. Daniel Webster, for instance, championed material prosperity as democracy's natural result. Others feared that fondness of machines was but a short step from materialism, a glorification of wealth and material possessions. John Adams and Lyman Beecher claimed that production caused temptation, and despaired that many Americans lacked the moral strength necessary to withstand the temptations to adopt aristocratic pretensions. Matthew Carey was more optimistic and argued that "prosperity generally expands the heart, and leaves it accessible to the suggestions of benevolence and beneficence."[8]

NOTES

1 John Loudon McAdam, "Report to the Right Honourable President and Board of Agriculture," in John Loudon McAdam, *Remarks On the Present System of Road Making,* 6th edition (London: Longman, Hurst, Rees, Orme, and Brown, 1822), p. 51.

2 Quoted in John F. Kasson, *Civilizing the Machine: Technology and Republican Values in America, 1776–1900* (New York: Viking Press, 1976), p. 81.

3 Quoted in John R. Commons, et al., *History of Labour in The United States* (New York: Macmillan, 1918), vol. 1, pp. 141, 142, and 146.

4 Quoted in Bruce Sinclair, *Philadelphia's Philosopher Mechanics: A History of the Franklin Institute, 1824–1865* (Baltimore: Johns Hopkins University Press, 1974), p. 113.

5 "Letter from Stephen Van Rensselaer to Rev. Dr. Blatchford, 11/5/1824," printed in Ethel M. McAllister, *Amos Eaton, Scientist and Educator, 1775–1842* (Philadelphia: University of Pennsylvania Press, 1941), p. 368.

6 Thomas Carlyle, "Signs of the Times," *Edinburgh Review*, vol. 49 (1829), pp. 439–59.

7 Timothy Walker, "Defence of Mechanical Philosophy," *North American Review*, vol. 33 (1831), pp. 122–36.

8 Matthew Carey, "On the Effects of Prosperity and Adversity," *Miscellaneous Essays* (Philadelphia, 1830), p. 451.

FOR FURTHER READING

Armroyd, George. *A Connected View of the Whole Internal Navigation of the United States* (1830; New York: Burt Franklin, 1971). A contemporary volume providing material on early-American transportation routes.

Bathe, Greville and Dorothy. *Oliver Evans: A Chronicle of Early American Engineering* (Philadelphia: Historical Society of Pennsylvania, 1935). The best study of Evans presently available.

Calhoun, Daniel H. *The American Civil Engineer* (Cambridge: Massachusetts Institute of Technology Press, 1960). A provocative analysis of the development of engineers in early–nineteenth-century America.

Calvert, Monte A. *The Mechanical Engineer in America, 1830–1910* (Baltimore: Johns Hopkins University Press, 1967). Charts the importance of machine shops and the master–apprentice system.

Cole, Arthur Harrison. *The American Wool Manufacture* (Cambridge: Harvard University Press, 1926), vol. 1.

Dodd, Edwin Merrick. *American Business Corporations Until 1860* (Cambridge: Harvard University Press, 1954). An early attempt to interpret the development of business corporations.

Dulles, Foster R. *Labor in America: A History*, 3rd edition (New York: Harlan Davidson, 1968). An older, mainstream account of American labor.

Durrenberger, Joseph A. *Turnpikes* (Valdosta, GA: Southern Stationery and Painting Co., 1931). Contains considerable detail about early road construction techniques.

Flexner, James Thomas. *Steamboats Come True* (Boston: Little, Brown, 1944).

Gibb, George S. *The Saco–Lowell Shops* (Cambridge: Harvard University Press, 1950). Documents the spinning off of textile machine shops.

Hamlin, Talbot. *Benjamin Henry Latrobe* (New York: Oxford University Press, 1954). Latrobe was an early civil engineer responsible for the Philadelphia waterworks.

Handlin, Oscar and Mary F. *Commonwealth. A Study of the Role of Government in the American Economy: Massachusetts, 1774–1861* (Cambridge: Harvard University Press, 1969). Pioneering study of a state's role in internal improvements.

Hartz, Louis. *Economic Policy and Democratic Thought: Pennsylvania, 1776–1860* (Cambridge: Harvard University Press, 1948). Similar in thrust to the Handlins' account, but more interpretive.

Hazard, Blanche Evans. *The Organization of the Boot and Shoe Industry in Massachusetts Before 1875* (Cambridge: Harvard University Press, 1921).

Hill, Forest G. *Roads, Rails and Waterways: The Army Engineers and Early Transportation* (Norman: University of Oklahoma Press, 1957). A solid history of the national government's involvement in internal improvements.

Hunter, Louis C. *Steamboats on Western Rivers* (Cambridge: Harvard University Press, 1949). A meticulously researched volume. Excellent on early steamboat technologies.

Meyers, Marvin. *The Jacksonian Persuasion: Politics and Belief* (Palo Alto: Stanford University Press, 1957). Details the cultural tensions of the 1820s and after.

Pursell, Carroll W., Jr. *Early Stationary Steam Engines in America* (Washington, D.C.: Smithsonian Institution, 1969). Pioneering book on the subject.

Roe, Joseph Wickham. *English and American Tool Builders* (New Haven: Yale University Press, 1916). Brief biographies of prominent mechanics.

Scheiber, Harry N. *Ohio Canal Era: A Case Study of Government and the Economy* (Athens: Ohio University Press, 1969). Accentuates political and economic questions.

Smith, Merritt Roe. *Harpers Ferry Armory and the New Technology* (Ithaca: Cornell University Press, 1977). A challenging look at early-American manufacturing techniques.

Somkin, Fred. *Unquiet Eagle: Memory and Desire in the Idea of American Freedom, 1815–1860* (Ithaca: Cornell University Press, 1967). An interesting discussion of the evolving relationship between Europe and America.

Tucker, Barbara M. *Samuel Slater and the Origins of the American Textile Industry, 1790–1860* (Ithaca: Cornell University Press, 1984). A recent social history of textile manufacture.

Wallace, Anthony F. C. *Rockdale* (New York: Alfred A. Knopf, 1978). An impressive modern study of a textile manufacturing community.

Ward, John W. *Andrew Jackson, Symbol for an Age* (New York: Oxford University Press, 1955). Effectively demonstrates how the myths about Jackson reflected American notions.

Ware, Caroline F. *The Early New England Cotton Manufacture* (Boston: Houghton Mifflin, 1931). An older volume stressing the economics of the industry.

White, John H., Jr. *A History of the American Locomotive. Its Development: 1830–1880* (Baltimore: Johns Hopkins University Press, 1968). A massive compendium of locomotive technology.

3

America as a Social Unit: the 1830s to the 1870s

THE SEARCH FOR A UNIQUELY AMERICAN CHARACTER

The concern in America in the early nineteenth century about the glorification of wealth, and the resulting effects on a democratic republic, receded in the late 1830s. Americans had begun to realize that their pursuit of individual objectives had, in fact, produced a civilization fundamentally dissimilar to that of Europe. The technology of the early nineteenth century had helped foster that new civilization: it produced the transportation and communication links that tied together more intimately the country's diverse settlements; produced several self-reliant, burgeoning industries; and began the development of a national infrastructure. Leaving their earlier concerns behind, mid-century Americans sought to consolidate and build upon the fruits of their immediate past. They saw their young nation as a unique social unit, marked by a people exhibiting, and defined by, a character unlike that of any other country.

A full and precise definition of the uniquely American character proved elusive, however. Mid-nineteenth-century residents of the United States repeatedly asked such questions as, what is an American? or, what do specific acts or lack of action say about Americans? but rarely did they achieve an absolute consensus. To be sure, everyone recognized that Americans should love their families, go to church, be compassionate toward their fellow humans, believe passionately in democracy, and be generous but thrifty as well as neat, clean, orderly, and moderate in all things. Similarly, most agreed about what an American was not. Blacks, Mormons and some other religious groups, and the poor were thought by the majority of citizens to lack the uniquely American character. Beyond those broad categorizations and platitudes, however, or whenever a particular situation emerged, disagreements, sometimes heated, usually surfaced. (The question of slavery—the treatment of people as property—stood as the most noteworthy and explosive issue.)

Yet the absence of an explicit, commonly held definition of what constituted the uniquely American character is and was beside the point. What was crucial and what focused the attention of mid-nineteenth-century citizens was that they perceived there was or ought to be a distinctly American character, and that that character was manifested in the behavior and activities of the nation's inhabitants. Mid-century Americans acted on those notions.

The Break from European Society

An explicit repudiation of Europe and a conscious attempt to distance America from the Old World accompanied recognition of the nation's distinctiveness. Americans saw little in Europe worth imitating, and rejected not only Europe's artificial and arbitrary privileges but also its manners, morals, and arts. They also called for creation of "a distinctly American school of good manners,"[1] and the development of a body of knowledge appropriate for the United States. In 1837, Ralph Waldo Emerson spoke for his generation. He claimed that "we have listened too long to the courtly muses of Europe," maintained that "our long apprenticeship to the learning of other lands draws to a close," and called for the emergence of a new type of person, the American Scholar. Emerson's American Scholar would come from American soil and reflect American civilization; it would be a new person serving a new people.[2]

Emerson's plea was emblematic of a subtle yet profound change in how Americans defined themselves. Other commentators began to take for granted America's distinctiveness, and concentrated instead on identifying the mechanisms that gave its people their common character. Rarely did anyone point to technological change. The European lyceum, ironically designated as a "curiously American invention," was most usually given the credit; to many mid-century Americans the lyceum provided a common experience, which served as "powerful social ligaments, binding together the vast body of people" to produce a unique civilization.[3]

No matter what explanation was offered for this new American distinctiveness, the notion highlighted the idea of nationwide coherence, and Americans labored in the 1840s and after to achieve and maintain this coherence. They embraced technology as a means to further unify and homogenize America. Improved transportation and communication technologies were among the most obvious ways to achieve this unification: railroad expansion, coupled with telegraphy, bound city and country, state and nation. Clipper ships joined ports, steam commuter railroads connected cities and distant suburbs, and omnibuses and horse-drawn street railways linked city sections. Mid-century technologies also permitted Americans to tap the nation's natural resources in new ways, at new locations, and on a grander scale. Waterpower remained the dominant form of motive power, but coal- and wood-fueled steampower began to make inroads, which freed enterprising manufacturers from dependence on rivers and streams. No longer restrained by the requirements of watermills, manufactories spread into previously impractical areas and soon blanketed the nation; manufacturing became a process to which few Americans could remain unexposed, whether they actively participated in it or not. Some entrepreneurs recognized that the idea of a unified, homogenized America implied the existence of a nationwide demand for similar goods—a common material experience—and geared up their establishments to milk that demand. They introduced new production techniques, established the factory as the quintessential industrial locus, and distributed their products nationwide over the new transportation ways. They also devised new promotional strategies to demonstrate to a sometimes skeptical public the intrinsic American quality of themselves, their products, and their production methods.

These efforts stemmed from the notion of American distinctiveness, as did the rejection of things European. But it would be incorrect to suggest that the Old World ceased to influence American affairs. The nation repeatedly sparred economically with Great Britain, and passed twelve additional pieces of tariff legislation to control the influx of European goods. Irish, German, and Swedish immigrants flocked to America in unprecedented numbers—more than five million between 1850 and 1870—bringing their skills and expanding the labor force. Europe remained a fertile ground for inventions, furnished much-needed railroad and industrial capital, and sympathized with the Confederacy during the Civil War. But questions of the nation's distinctive destiny, whether manifest or not, captured the American imagination. They fueled a geographic expansion and the Mexican War, a nativist movement, competition between the industrial North and agricultural South, and a persistent argument about slavery culminating in the Civil War. Ironically, these same questions led Americans to interact with and challenge Europe—even as they repudiated the Old World—as the nation strove to achieve what its citizens considered to be the country's rightful place as a leading and vital member of the world of civilizations.

Using Technology to Find America—Photography

The way that Americans manipulated and employed the European invention of photography demonstrates the persistence of a relationship between Europe and America, and provides a graphic representation of how the notion of a distinctive American character influenced technology in mid-century America. In August 1839, L. J. M. Daguerre announced the first photographic process, the daguerreotype, to the French Academy of Sciences. Polished silver-plated copper sheets, subjected to iodine vapor, constituted his film. Daguerre shielded the light-sensitive plate in a box with a lens, which was his simple camera. There was no shutter in his camera: Daguerre simply removed the lens cap to expose the plate. He developed the image by subjecting the plate to mercury vapor, and fixed (stopped development and stabilized the image) with sodium thiosulfate. Daguerre had conceived of his process as an extension of the diorama * and used it to photograph landscapes. That it required a long exposure time (a minimum of fifteen minutes at high noon in the summer sun), and that it was suitable only for reproducing inanimate objects, caused him no concern. That was not the case in America. Within months of Daguerre's announcement, several Americans, including Alexander Wolcott, a New York manufacturer of dental supplies; Samuel F. B. Morse, a chaired professor of sculpture and painting at New York University; and

---

* diorama: enormous transparent painting, which under different lighting conditions gives the appearance of three-dimensionality.

John William Draper, Morse's N.Y.U. colleague, sought to modify the process.

Unlike Daguerre, these Americans attempted to adapt photography to portraits. Although Morse failed to improve upon Daguerre's process, his contemporaries were more successful. By October 1839, Wolcott had successfully produced his first portrait, while Draper followed two months later. Draper decreased the focal length of his lens, which sharpened the picture and speeded the process. Wolcott adopted the astronomical technique of using a concave mirror to concentrate the image on the plate. These two modifications reduced daguerreotype exposure time considerably. Indoor portraits now took from forty seconds to two minutes, while the sitting time for outdoor portraits varied from a scant twenty to ninety seconds.

Portraits captivated mid-century Americans. By 1841, most principal towns and cities had daguerreotype studios or galleries, often decorated like elegant parlors. A decade later, an estimated 2,000 Americans earned livelihoods as daguerreotypists and took more than three million portraits annually. More than 100 portrait studios operated in New York City alone. That Americans had embraced daguerreotypes was striking because these portraits were expensive, generally costing between one and two dollars. Wooden cases with velvet, metal foil, and glass layers were required to protect the copper plates' fragile image, adding markedly to the price.

The daguerreotype's popularity was even more remarkable because it competed with another simpler, much less-costly technique, the calotype. Developed by Britisher William Henry Fox Talbot around 1840, calotypes quickly became well known in America. Calotypists used cameras similar to those employed by daguerreotypists, but the finished calotype did not need an expensive case. This was because calotypes used a paper film, not copper plates. A photosensitive silver salt was spread on translucent paper, exposed, developed with gallic acid, and fixed, producing a negative. Similarly treated paper was placed under the negative and exposed to the sun. The result was a positive, a replication of the initial image. Each negative could be used to yield any number of nearly identical positives. Unlike daguerreotypes, each of which was unique, calotypes could be replicated.

The calotype's most significant drawback to Americans was that light diffused through the translucent, sensitized paper and created a somewhat fuzzy picture. (Europeans viewed it as desirably "stylized.") Daguerreotypes more exactly recorded a subject, and Americans were willing to pay for a faithful representation of their appearance. Appearance seemed an outward manifestation of character; it revealed the inner self. Comparison of portraits taken some time apart enabled subjects to check character development, and to measure how closely their character approximated that of the distinctly American archetype. That American daguerreotypists often were phrenologists (examiners of skull shape to determine character), physiognomists (examiners of body conformation to determine character), or portrait painters, was no accident. Those daguerreotypists who were not, tended to locate their studios near people who were. To the overwhelming majority of mid-century Americans, daguerreotypes, portraits, and phrenologic and physiognomic analyses revealed the critical element of character—to be held up to the new American ideal.

Although daguerreotypes remained popular through the 1860s, a new photographic process began to challenge them in the mid-1850s. Reported in 1851 by Frederick Scott Archer, an English sculptor, the collodion process used guncotton dissolved in sulfuric ether—collodion—as a binder/carrier of the photoactive chemicals. It was spread on a glass plate, impregnated first with bromides and then silver nitrate to produce a photoactive surface, and inserted in a camera. Photographers developed the image with gallic acid, then fixed it. A paper positive could then be made from this glass negative, as with calotypes. At first few chose to do so because of similar limitations. Not until the late 1850s, when the diffusion problem was resolved and accurate portraits could be produced, did the inexpensive positive collodion-paper prints become common. American photographers printed the image on salted, unsensitized albumen covering the paper, not the paper itself, which did not blur images because neither light nor chemicals migrated. Before the albumen technique, collodion photographers were content to use negatives to make ambrotypes and tintypes. Both gave faithful portraits by relying on reflected light to present the impression of a positive. Ambrotypists placed black backgrounds or black varnish behind developed glass plates, while tintypists dispensed with glass and used photosensitive collodion-covered thin black-laquered or japanned iron sheets as film.

PROVIDING A COMMON MATERIAL EXPERIENCE

Setting the Stage

Daguerreotypes and collodion photography took America by storm because they satisfied a national desire to record and measure character. Several industries also were able to capitalize on the growing nationalist sentiment, and to market and distribute successfully their products nationwide. Their pitch incorporated the notion that a homogeneous, unified, and distinct America demanded a characteristically American material experience. The concept of similar products distributed nationwide would prove to be profitable because it tapped the forementioned national desire, and served to further cement the nation's people.

Demand for a common American material experience proved so compelling that consumers repeatedly welcomed nationally distributed goods even as local manufactories were hindered or destroyed. National distribution was pronounced in mid-nineteenth-century food industries. Buffalo and Rochester became national flour-milling centers, and Baltimore, then Cincinnati, and finally Chicago dominated the American hog-slaughtering trade. The demand was likewise evident in manufactured products. For example, New England emerged as the nation's textile headquarters. Cyrus McCormick and Isaac Singer successfully marketed nationwide reapers and sewing machines respectively. Yet it would be mistaken to conclude that each of those with national marketing aspirations adopted an identical approach to the question, or that identical goods were distributed throughout the nation. Neither was the case. Production techniques and sales strategies differed drastically among manufacturers, as generally did the quality—and sometimes even the style—of goods produced by a single firm and distributed nationally under a single name. Indeed, furnishing a common American material experience required industrialists only to convince purchasers of the uniformity of productions. It did not necessitate producing identical goods.

Several mid-nineteenth-century initiatives facilitated national production, marketing, and distribution. Governmental reassessment of the corporation's place within American society increased corporate numbers and size, unleashed capital by enhancing the attractiveness of corporations as investments, and accentuated the role of technical capabilities of corporate managers. The Supreme Court's Charles River Bridge decision (1837), among the earliest expressions of that corporate reassessment, was certainly not immediately recognized as a potent agent of change. The Court found corporate charters neither inviolate, perpetual, nor exclusive, and permitted states to modify or revoke charters whenever the public interest dictated.

States were mollified by the Court's ruling establishing public interest as distinct from and superior to private, and society as superior to the individual (because corporations continued to be treated as individuals in law). But the Court's affirmation of state power made states generally confident enough to relax corporate supervision, and loosen the corporate mechanism. The ability of states to revise corporate charters, coupled with the demise of mixed corporations implicit in the public/private separation, led state governments to create numerous new corporations, to become less immediately involved in corporate activities, and often to cede their oversight authority. Creation of general state laws of incorporation to replace special legislative acts as means to form corporations reflected both state optimism in their capacity to rein in any renegade corporate entity and recognition that corporations, especially manufacturing and commercial enterprises, could further the public interest. Often framed during the spate of mid-century state-constitution redrafting, general incorporation laws required only that individuals follow generalized guidelines, file incorporation articles with state secretaries, and report their dealings annually to these officials.

The states' lenient exercise of power made corporate investment attractive, which produced a proliferation of corporations, and more capital for technological endeavors. Large corporations no longer needed to comply with state-mandated sets of particularized regulations, or report continually to specially constituted state oversight boards. The absence of these restrictions made investors more directly accountable for a corporation's commercial or manufacturing success. Profit now depended on corporate initiatives, not state-specified procedures, and investors elected to place their firms under control of people well-versed in their corporations' technical activities. These technicians understood the business's complexity, which seemed to ensure investors larger returns. Conversely, freedom from day-to-day corporate cares and reporting permitted investors to back a wider assortment of enterprises, which insulated them from the era's periodic business fluctuations. Unlike the early nineteenth century, when fortunes were routinely made and lost, mid-century capitalists frequently retained their wealth because they diversified their economic activity.

Mid-century governmental activities and policies helped create and stabilize fortunes. They supported some corporate ventures—the public/private distinction seldom was equated with absolute government/corporate separation—though not as partners but as contractors. These limited-duration wares or service contracts did not impinge on investors' abilities to set corporate priorities, missions, or activities, but provided favored entities additional operating capital and guaranteed markets. Government contracts, then,

.s to gain, without sacrificing autonomy, secure
.urns on their investments, which could be used to
other projects. The effect of this policy, however, like
.ication of assets by the wealthy, was to begin to create in
.crica rigid and permanent socioeconomic inequities at the very
.noment that the people paraded their newfound national coherence.

Railroads Ascendant

Expansion of railroads and their adoption of new communications
technology, telegraphy, also enhanced national development. Railroads quickly outstripped canals as the nation's primary overland
transportation mode. In the two decades after 1840, Americans laid
track totalling 26,000 miles. During the same period almost as many
canal miles were abandoned as built. Prewar railroad construction
focused in the Northeast and Old Northwest, and many new roads
paralleled and supplanted established canal routes.

Railroads offered entrepreneurs the virtues of speed, all-weather
dependability, and flexible as well as larger cargo capacities. Huge
capital and maintenance costs for rails, locomotives, and rolling
stock were their drawbacks. Despite these expenses, Americans built
and maintained the overwhelming majority of prewar railroads with
no or little governmental assistance. The East's and Old Northwest's relatively dense population, their comparatively many manufactories, and the earlier success of turnpikes and canals made
railroads attractive investments. But the massive sums of money
necessary to construct and keep up lines required investment arrangements more substantial than ad hoc consortiums of local citizens. New York City merchants and their bankers were two groups
able to fund large enterprises. The city's fine harbor had brought
them much of the European trade—establishment of packet lines
to England and the cotton South also provided crucial assistance,
while its Erie Canal connection captured a large share of the western trade. These men had experienced the financial benefits of improved transportation and proved willing to venture significant
amounts on railroads, especially in the wake of the new corporate
policy of state governments.

Mid-century railroad capitalization pioneered new ground, but
railroad hardware underwent no similar revision. The lines' cast-
iron T-rails, steam locomotives, and rolling stock differed little from
that developed prior to 1835. Grading, curve cutting, and bed-prep-
aration methods remained basically the same. Only the mid-century railroads' scale separated them from their predecessors. That

Railroads were complex mid-nineteenth-century enterprises, instrumental in catering to national markets.

was no small matter. Conduct of national or regional transportation businesses created logistical problems. Larger railroad lines operated over unprecedented distances, employed huge, diverse staffs, and emerged as America's biggest iron consumers. They encountered a wide array of difficulties and hazards as each sought to coordinate its internal affairs. Profits depended on a line's smooth function; iron shortages, insufficient or mispositioned rolling stock, mechanical breakdowns, accidents, and the like cost investors money. They confronted these questions in three important ways. First, investors selected men with broad funds of practical, technical, railroad-related know-how to oversee company activities. These railroad generalists usually had graduated from the school of experience—although they might have had formal training at West Point or elsewhere—which enabled them to anticipate and resolve problems, assess situations promptly, and develop plans to maintain efficient operations. Second, investors demanded that railroads keep detailed records. Meticulous record-keeping produced data for rational choices; it served as the technology to enable managers to make sound decisions. Exact records also furnished investors information to adjudicate managerial effectiveness. Third, railroads embraced electrical telegraphy.

Railroads and Telegraphy   Telegraphy offered almost instanta-
neous communication over long distances; its judicious use permitted railroads to prevent situations and conditions from becoming
problems, to remedy unanticipated difficulties, and to synchronize
operations. This American telegraph–railroad nexus flowered after
1850, but electrical telegraphy was much older, a predominantly
European invention. Alexandre Volta's explanation in 1800 of electrical current as the consequence of two dissimilar metals separated by a poor conductor raised the possibility of generating a
dependable, continuous electrical flow, an idea first realized by
Humphry Davy and others, and later in the wet-cell batteries of J.
Frederic Daniell and William Grove. Similarly, Hans C. Oersted's
and Andre M. Ampere's electrical and electromagnetic discoveries
enabled Carl Gauss, Karl A. Steinheil, M. H. Jacobi, William F.
Cooke, and Charles Wheatstone to devise workable telegraphs. Indeed, in 1838, the Cooke/Wheatstone apparatus controlled some
English railroad traffic.

   These early cumbersome telegraphs expressed reception of electrical impulses through electrolysis or deflection of galvonometer
needles. These systems required multiple circuits to represent the
entire alphabet. Samuel F. B. Morse adapted the telegraph to American conditions; the country's great size made multiple-circuit
transmission prohibitively expensive. His practical high-speed devices worked effectively and cheaply over long distances; they were
single-circuit telegraphs in which long-distance transmission was
accomplished by relay. Morse had begun his telegraphic deliberations in 1832, fortified only by a lecture series on electromagne-
tism that he had attended five years earlier. An artist and portrait
painter who, like Fulton, had studied in Europe with Benjamin West,
Morse was an inveterate mechanical dabbler and had tinkered with
fire pumps and marble-cutting machines. Simplicity marked his early
telegraphic speculations. Morse initially conceived of using coded
type to transmit words. A unique series of spaces and teeth characterized each type letter. Telegraphers would arrange letters on
composing sticks and draw them past a contact. The electrical circuit would open at spaces and close at teeth, transmitting words
as dots and dashes. Morse's receiver/recorder was similarly elegant.
The incoming signal tripped an electromagnet, which activated a
pen that moved across paper advanced by clockwork. Operators then
translated recorded dots and dashes.

   Morse had considerable difficulty in translating his ideas into a
working line, but he received some valuable assistance. Leonard D.
Gale, professor of geology and mineralogy at New York University,

The key with the sounder speeded and simplified telegraphy.

alerted him in 1836 to Joseph Henry's 1831 paper, which indicated that long-distance electromagnetic transmission required high voltages (many battery cells connected in series) and powerful electromagnets (using many turns of wire). Their empirical understanding of a crucial aspect of what would become Ohm's Law enabled Gale and Morse to transmit a signal ten miles. Morse further extended the telegraph's range with a relay (a distant electromagnet), which, upon receipt of the original signal, automatically engaged and disengaged an independent battery to replicate the message. He then acquired another partner, Alfred Vail, an N.Y.U. graduate with substantial mechanical skills who built instruments from Morse's designs.

In 1838, the group publicly solicited financial support and gave demonstrations at Vail's father's Speedwell Ironworks in Morristown, New Jersey, at N.Y.U., and at the Franklin Institute. They also exhibited the device to President Van Buren and his cabinet, and the House Committee on Commerce, but not until 1843 did Congress grant Morse $30,000 to string an experimental line. By that time, Morse had replaced composing sticks and type with his famous key. Substitution of sounders for recorders came several years later.

The experimental line was completed a year later, followed the Baltimore & Ohio Railroad's tracks, and connected Baltimore and

Washington. Locating the nation's first telegraph line along a railroad proved prophetic. Even before the wire was strung, some Americans were championing telegraphy's potential to create a "most speedy intercourse . . . between the most distant parts of the country" and predicted that "space will be, to all practical purposes of information, completely annihilated."[4] It would become yet another means to achieve and maintain national coherence.

Business recognized telegraphy's advantages from the start. The postal service and military quickly adopted it, and newspapers and traders soon followed. By September 1846, American telegraph companies had strung hundreds of miles of wire, and had joined New York to Boston and to Harrisburg, Pennsylvania, and Washington, D.C., to Mobile. The Ohio River cities were added the next year and New Orleans a year later.

Establishing telegraphic connections to New Orleans corresponded to the beginning of a five-year period (1848–1853) of particularly rapid railroad expansion during which small, independent lines were consolidated to form the four great railroad corporations operating east–west trunk lines: the Baltimore & Ohio, Erie, New York Central, and Pennsylvania railroads. These trunk lines each travelled without disruption between the eastern seaboard and western river cities. Independent lines before consolidation lacked uniformity; each had used tracks of different gauges (or distances between the rails). Trains could run only where gauges for locomotive, rolling stock, and track matched. Absence of uniformity had meant that each time a line with track of a different gauge was encountered, goods had to be unloaded from one train and reloaded on another. Formation of trunk lines, each with a uniform but different gauge, saved time, money, and equipment costs, and rendered national distribution of products more attractive.

Not until 1852 did a trunk line—the Erie—employ telegraphy extensively. But once that precedent had been set, a railroad–telegraphy nexus became a foregone conclusion; telegraphs and railroads developed hand in hand for the remainder of the decade. Railroad managers recognized the telegraph's significance to efficient railroad operations, and telegraphers realized the convenience of using their largest customers' rights of way. Between 1852 and 1860, Americans built 23,000 miles of telegraph lines, and railroads stretched to Burlington, Iowa, and St. Joseph, Missouri.

Telegraph companies confronted economic questions in the mid-1850s not unlike those that had fostered the formation of railroad trunk lines several years earlier. To be sure, telegraph lines were cheaper to erect than railroads, but telegraphic instruments, receiving stations, batteries, poles, and wires required substantial capital. Most early telegraph companies failed; competition over routes meant duplicate lines and low revenues. In 1857, the six largest remaining companies cooperated to reduce expenses by dividing the nation into six sections, using each others' lines in their respective sections and sharing profits on messages between sections. Three of these companies did not survive competition from smaller companies in their sections, and were absorbed by the remaining three. These survivors further pooled resources in 1866 to create Western Union. This giant company then turned to New York investors for additional capital to expand operations, destroy competition, and dominate the telegraphic industry nationwide. Like the railroads, Western Union was superintended by men with broad funds of practical experience.

Governmental Investment in Railroads   Telegraphic expansion continued after Western Union's formation. About 25,000 new miles were added from 1860 to 1868, and a line first joined the coasts in 1861, beating the more expensive railroads by nearly a decade. The pace of rail construction had slowed during the Civil War but reached unprecedented rates immediately upon the war's conclusion. Over 35,000 miles were laid between 1865 and 1873. Some of this railroad mileage in the East went to creating single-gauge express lines on high-volume, particularly profitable runs. Rather than compete for this business, the new trunk lines pooled expenses, rights of way, and equipment. They laid out these lines so as not to interfere with the partners' other runs, placed them under a single manager, and shared profits. Most of the remainder of the new railroad mileage—the majority—was built in the West and South. In 1869 the Central Pacific and Union Pacific railroads were joined at Promontory Point in the Utah territory to form the nation's first transcontinental railway. This event both symbolized national unification and linked the East Coast commercial center of New York City with its West Coast counterpart, San Francisco. It was commemmorated by the use of a silver sledge to drive golden spikes to connect the two lines' rails.

Tying the West and South to East by railroad was essential for establishing nationwide distribution of goods, but these railroads attracted few private investors. New York City-based capitalists remained leery of western and southern railroads because neither the South, devastated by the Civil War, nor the sparsely settled West promised the high-volume runs necessary for quick monetary returns. Citing an explicit national purpose—military defense and settlement of the nation's interior—the federal government frequently interceded when private sources refused to furnish the requisite capital. As it had in a handful of cases prior to the Civil War (such as the Illinois Central, linking Chicago and Cairo), Congress granted to the states public lands to turn over to railroad companies as they completed laying sections of track. Congress itself handled these transactions in the territories. These free lands usually abutted lines, which converted them into prime real estate. Railroads would sell these parcels at premium prices, then use revenues to build further.

These measures were sometimes insufficient, especially in the arid western mountains where relatively few people chose to purchase land, and Congress more directly shouldered the financial burden by lending railroads money. Such was the case with the Union Pacific and Central Pacific. In addition to granting them large tracts of the public domain, Congress loaned the two lines a combined sum of more than $50 million. This loan more closely approximated a gift, albeit one justified by the idea that a transcontinental railroad was a matter of compelling national interest. The companies had thirty years to repay the debt, which was secured only by mortgages on railroad property.

Railroads and the Iron Industry   The expansion of railroads and introduction of telegraphy to form nationwide transportation and communication linkages depended on a ready supply of iron for rails, wires, locomotives, and bridge-girders. So, too, did development of other industries; nails, suspension-bridge cables, steam engines, stoves, lineshafting, and structural elements such as construction beams devoured large quantities of the metal. The explosive growth of the American iron industry in the decades prior to the Civil War was caused by these demands, and the nation produced an impressive 821,000 tons in 1860. Americans relentlessly promoted their product, claiming that the country's ores gave American iron a distinctive "character" that was "superior . . . in its strength and purity" to European irons.[5] Yet patriotic blustering and statistics paint a somewhat misleading picture; although America probably was the world's second largest producer of iron, American iron demand exceeded American supply, and the country remained an iron-importing nation. The country's production lagged far behind the British total of 3.8 million tons in 1860, which hampered American industrial development through higher prices and by periodic shortages.

Dependence on foreign iron had long galled Americans, but the nation's abundant forests, which made it seem unlikely that another fuel could supplant or compete with charcoal, prevented the adoption of the European coal-based smelting practice. As late as 1830, the large majority of American iron was produced in a manner similar to colonial iron: it was smelted by charcoal in forests close to ore deposits. But the situation changed rapidly soon afterward. By 1860, over half was smelted using anthracite coal in or near eastern Pennsylvanian cities.

Americans had known of northeastern Pennsylvania's large anthracite deposits from the 1810s, and had roasted the coal in large iron retorts to generate gas for illuminating cities late in the decade. But difficulties igniting the hard, almost-pure carbon substance, coupled with American skepticism about coal's value for smelting (Europeans dealing with bituminous, or high-sulfur coal, had found that this impurity, unless removed, yielded a brittle inferior iron), and the nation's ubiquitous forests precluded the use of coal in iron manufacture. The nation's initial period of railway construction placed sudden, severe strains on the eastern charcoal-iron industry, however, and new avenues were explored to lessen pressure. Americans turned to anthracite in the late 1820s, but only to reheat wrought or cast iron for final product manufacture. This was a low-temperature process of brief duration in which the iron was not subject to possible contamination. Hoping to develop a substantial market for coal, Pennsylvania coal-mine operators seized upon this precedent, modified Henry Cort's puddling technique, and introduced anthracite in the more challenging conversion of pig iron to wrought iron. Pig iron was melted in an anthracite-fired reverberatory furnace * (success in igniting the coal and controlling furnace temperatures were the most serious problems with this process), which kept the metal from direct contact with coal and its impurities. An iron bar stirred the molten metal, exposing its carbon to the air and causing some of it to burn off. The iron was taken from the furnace, allowed to stiffen somewhat, and squeezed through rollers to drive off additional carbon.

---

reverberatory furnace: a furnace in which the heat is deflected down from the roof. Often used to prevent contamination of metals by fuels.

Use of anthracite coal in the primary smelting phase of iron production began after 1840. Eastern mine operators, confident that their anthracite contained virtually no iron-ruining sulfur, again looked to Europe for inspiration. They adopted the hot-air-blast method, first patented in Britain in 1828 as a means to ignite hard Scottish coal, heating the air before a steam-powered bellows blew it into the furnace.

Hot-air blasts solved the anthracite ignition problem, and since anthracite, unlike the friable charcoal, could withstand enormous weight, smelters availed themselves of the possibilities of economies of scale to build huge new furnaces. Further savings were generated because these new furnaces could operate year-round; steam engines freed smelters from dependence on often-frozen rivers to power their bellows. Further, use of anthracite liberated them from highly paid colliers and slow, costly charcoal production. By 1849, sixty anthracite furnaces were located in eastern Pennsylvania, employing an average of eighty persons and capitalized at an average $83,000. Just four years later, forty new furnaces had sprung up.

Anthracite iron quickly proved competitive with and superseded charcoal iron in the East, but made few inroads west of the Appalachian Mountains. Transmontane shipment of anthracite, heavy iron goods, or ore were impractical. Western ironmasters were left to their own devices. Like their eastern counterparts, westerners had had extensive experience with coal. A great coal seam ran near Pittsburgh, but this coal was high-sulfur bituminous coal, not anthracite. Pittsburgh ironmen had employed this material in rolling mills quite early, shipped finished goods down the Ohio River, and created the leading western iron city. But Pittsburgh iron had been smelted in country-situated, charcoal-powered blast furnaces, and refined in similarly fueled and located forges. Not until the late 1840s—about a decade later than the East—did reverberatory furnaces, in this case bituminous coal-driven ones, secure a foothold in the city. By 1850, their domination was nearly complete. Eighty-five percent of Pittsburgh rolled-iron was made from city-refined iron (using bituminous coal) smelted in country furnaces (which had used charcoal).

Pittsburgh coal and coal technologies travelled the Ohio with finished iron products and took root at Wheeling, Cincinnati, and Ashland, Kentucky. These nascent ironworks, like Pittsburgh itself, depended exclusively on charcoal-smelted iron until just prior to the Civil War. Bituminous coal-fueled smelting furnaces emerged then and gained significance in the postwar decades.

Rendering coal into coke (heating coal in an oxygen-deficient atmosphere) rid it of much of its sulfur and produced a purer, easily ignited, hotter-burning material. Coking had been carried on in Britain since the early eighteenth century, and this imported technology ultimately transformed Pittsburgh into America's iron-smelting capital.

Railroads and the Steel Industry   Bituminous coal-smelted iron became as important to the nation's railways in the 1860s as bituminous coal-refined iron had been to the Old Northwest's railroads from the 1840s. But in the mid-1850s, several railroad owners and rail manufacturers had begun to express concern about the rate at which huge locomotives wore out the finest iron rails—sometimes in less than two years—and investigated steel as an alternative.

Until then, virtually all American steel had been blister steel. In this method, thin wrought-iron strips were sealed with charcoal dust in clay vessels and baked at high temperatures for up to ten days. This long process cemented carbon from the charcoal dust to the iron's surface to form a thin steel layer, which blistered under the great heat. This expensive, superior-strength material was used primarily for springs and fine cutlery, and railroad men understood that blister steel's cost and performance would not suit their needs. Instead, a new English steel-manufacturing technology captured their attention.

Henry Bessemer, an English inventor, had determined that air blown through molten pig iron would ignite the carbon in the iron and fuel the metal's refinement. He recognized, moreover, that if combustion could be controlled and terminated before all carbon was consumed, the superior form of iron known as steel would result. Bessemer designed a pear-shaped converter to regulate the reaction, and received a United States patent for his process and converter in 1856.

Yet Bessemer's patent did not result in a rapid shift to steel in the railroad industry. The patent was challenged almost immediately by William Kelly, a Kentucky ironmaster who maintained that he earlier had attempted to refine pig iron through combustion of its carbon. Although Kelly had been unable to develop a commercial process, the patent office was persuaded by his documentation, reversed itself, and awarded him patent rights to the idea; Bessemer retained his converter patent. The situation was muddled further by a United States patent issued to England's Robert Mushet, who had been frustrated by his inability to stop carbon combustion at precisely the proper point. He circumvented that difficulty by decarbonizing the iron completely, then introducing spiegeleisen, a compound of iron, manganese, and carbon, to impart an exact amount of carbon to form nearly uniform steel.

Large-scale commercial steel production depended on these three patents. No American organization controlled them all until after the Civil War. As late as 1865, Americans produced a paltry 15,000 tons of the material, nearly all of it blister steel. That was soon to change. In the early 1860s, a Wyandotte, Michigan, group had purchased the Kelly and Mushet patents, while entrepreneurs from Albany and Troy, New York, had obtained Bessemer's rights. Although both consortiums built steel-manufacturing facilities, they chose in 1866 to avoid possible protracted legal battles and pooled their patents to form the Pneumatic Steel Company. The pool granted licenses to use steel-production patents and limited the output of each licensee to ensure a high price for the material. Even these restrictions did not dampen American steel's meteoric rise. In 1876, the nation's producers made 470,000 tons of steel, a thirty-one-fold increase in eleven years. Almost all of this Bessemer steel went to manufacture rails.

Furnishing Uniform Goods

Attempts to provide a common American material experience were concurrent with efforts to establish nationwide transportation and communications connections. Both efforts were manifestations of the idea of America as a distinct social unit, a civilization unlike those of Europe. In manufacturing, this notion encouraged manufacturers to expand production capabilities and to accentuate regularized production; money was to be made by producing nearly uniform goods for a nearly uniform people. Mid-century production for the masses required neither that finished products be identical nor that their parts be interchangeable. Uniformity—similarity or certainty that products were similar (even when they were not)—generally satisfied customers. Achievement of uniformity could be accomplished in several ways. Establishment of brand names (a company's reputation as guarantor of similarity) and identification of product type (familiarity of nomenclature as guarantor of similarity) were common means to achieve uniformity, as was machine-based production itself.

The New American Factory   A characteristically American locus of large-scale production, the factory, came to be identified during this period. Generally located on cities' fringes near railroad terminals, mid-century factories took on a particular form, and differed in both form and function from artisans' shops and, according to contemporaries, from European manufacturing establishments. These large industrial facilities contained batteries of machines, foundries, or furnaces; employed large numbers of operators; and were driven by central power sources (either waterwheels or steam engines). Power was transmitted from prime movers to upright iron shafts (shafting was wooden in some of the earliest American establishments), then throughout buildings by iron horizontal line-shafts. Located near ceilings, line shafts were supported by lubricated journals, which preserved motive power. In early-American factories, wooden and cast-iron gears and pulleys carried power from line shafts to individual machines, but this millwork could not withstand the stress of the higher shaft speeds of the late 1830s and after. The English developed a rigid system of precision, toothed, wrought-iron gearing suitable for the task, but Americans lacked a highly refined gear-cutting industry. They pioneered instead the use of leather belts. Joined in endless loops and stretched over pulleys or drums fastened to line shafts, belts ran around pulleys on machine drive-shafts. Belts were quieter, performed more smoothly, cost less than metal gears, and were readily installed and repaired. Load dictated the arrangement of machines in factories; those requiring the greatest amount of power were placed at the beginnings of lines.

American belting provided a flexible power train when compared to European gear-based practice—a fact that did not escape American notice. Use of cone pulleys permitted some change of speed ratios at individual machines, and clutches allowed Americans to engage and disengage machines independently. In European manufacturing plants, the engagement of prime movers usually produced continual operation of all individual machines, each at only one speed. But this relative (and quite minimal) American flexibility, while championed by American millwrights as a concrete advance over European methods and a tribute to American ingenuity, also had its costs. Millwrights fretted about changes in belt tension and slippage, as well as lubrication problems. The first two caused uneven and undependable machine performance—and therefore less-uniform products—while inadequate lubrication resulted in power loss, excessive wear, breakage, overheating, and fires. Indeed, American belting was so complicated and so troublesome that in the thirty years prior to the mid-1870s, at least eight different

American millwrights published volumes of rules to govern power-train construction and care.

Several economic factors might have given mid-century Americans pause in their desire to establish factories. Factory power-trains were very expensive to build and maintain, and securing production machines and erecting appropriate edifices required great capital investment. Manufacturers made a profit on their investment only as machines replaced skilled workers, reduced labor costs, and increased production. That these economic calculations were rarely formally considered before entrepreneurs made decisions to begin factory construction indicates that factory establishment was a matter not simply of economics, but also, for want of a better term, of style. In mid-century America, machine-based production was equated with uniform production. In shoe and boot manufacture, for example, adoption of steam-powered pegging and sewing machines, and binding and punch presses, promised production of more nearly uniform footware.

The New American Factory Worker   But mid-century factories manufactured more than uniform goods. The routine of factory life, which accompanied uniform production, also yielded American factory workers who were more uniform than their predecessors. A common work experience was "manufactured" by employee-regulating work codes, which stressed responsibility, morality, and cleanliness. Those most adept at demonstrating their adaptation to these codes rose from the ranks to assume supervisory positions, working as foremen on factory floors.

Establishing a Brand Name—I. M. Singer   Uniform American factory workers were a product of the new factories, but uniform products were often the result of something more. Character could certainly figure in; it formed the basis of evaluation in mid-century America. In manufacturing, character served as a guide to a firm's dependability. It provided consumers with security by emphasizing predictability, precision, and time-tested production techniques— all of which echoed uniformity. But character was an inward trait, not always known, so Americans usually equated it with the more public reputation (or appearance, as in daguerreotype photography). Reputation and character were not necessarily synonomous, however. Reputations could be created and marketed. Recognition by some mid-century entrepreneurs that they could create and advertise reputations for themselves, their products, and their companies catapulted their corporations into the first rank.

I. M. Singer and Company was perhaps the most notable example of the marketing of a reputation. Singer himself never maintained that he had invented the sewing machine, only that he had
perfected it—a claim he reiterated whenever he modified its design. His participation in the pooling of rival claimants' patents in
1856 indicates the degree of Singer's indebtedness: to Walter Hunt
for his lock stitch, to Elias Howe for his grooved, eye-pointed needle
and second-thread shuttle, and to Allen B. Wilson for his four-mo-
tion-cloth-feeding mechanism, among others. Presser-foot and needle-
cam-bar patents were Singer's most noteworthy contributions to
the combination.

But the 1856 agreement merely ended patent infringement suits
among litigants; it did not mark the beginnings of sewing-machine
production. As early as 1850, several manufacturers had made machines. All promoted their product, but only Singer showed in these
early years a true genius for promoting himself. Identifying himself
as a "self-made" man in "a country of self-made men," Singer
wrapped himself in patriotic garb and barnstormed the nation with
his machines. He gave demonstrations at carnivals and fairs and
sometimes even rented halls. He regularly advertised in the penny
press, circulated pamphlets, hired agents, and in 1855 published his
own periodical, the *I. M. Singer and Co. Gazette,* which he distributed free of charge.

Singer and Company proved even more inventive after the 1856
patent accord. As competitors concentrated on developing a superior product manufactured by more elegant processes, Singer focused on spreading the company name. For instance, Wheeler and
Wilson Manufacturing Company produced interchangeable-part
sewing machines by 1863, and Willcox and Gibbs had adopted that
production technique even earlier. But Singer still persisted, in
manufacturing machines requiring fitters, even though he sold his
less-precisely manufactured product at prices comparable to others.
Singer machine owners could not readily replace broken parts with
those supplied by the company, but had to pay artisans to fashion
and install substitute parts or purchase new machines. Advertising
compensated for Singer's technical deficiencies. The company
claimed to use "'precision in the application of mechanical means"
to make the "universal sewing machine"; Singer attempted to create a reputation for precision production rather than engage in it.

Singer's marketing strategy was to forge a reputation for itself as
archetypically American, and as the sewing-machine company that
demonstrated most nearly those characteristics thought to define

*An elegant atmosphere characterized Singer showrooms
such as this one in New York City.*

the distinctly American civilization. In particular, the company advertised itself as generous, benevolent, reverent, and family ori-
ented. In 1856, for example, it held a "Grand Invitation Ball" for
Singer machine owners and operators as well as company employees. More than 3,000 attended. It later sold machines to church
pastors at half-price, leased machines to individuals with options
to buy, and established a trade-in policy in which owners traded
Singer their machines, whatever make (the company destroyed these
"inferior" articles), for a $50 credit toward purchase of new Singers.
   The American family received the company's most sustained attention. Trumpeting its products as "the most important labor saving machines of the century,"[6] Singer marketed the idea that

ownership of Singer machines ought to characterize the nation's families. The establishment of sewing emporiums in large and medium-sized towns was essential to the plan. Kept clean, proper, and cheery to encourage women to enter, Singer emporiums furnished refreshments and sometimes presents. Agents personally demonstrated machines to each visitor, commented on their superiority to hand-sewing, and maintained that they reflected the highest-quality workmanship. That Singers lightened a mother's load rested at the heart of the agent's pitch: time saved sewing freed women for other familial activities and duties; conversely, women without Singers were implicitly culpable of neglecting more crucial familial responsibilities. Singer's emphasis on the family culminated in 1873 with the introduction of its "New Family" model. That machine dominated the market. Singer produced about 500,000 a year by 1880. The brand name had become synomomous with sewing.

Establishing a Common Nomenclature—The Pork Industry Brand names suggested uniformity, but so, too, could an established nomenclature. A familiar nomenclature gave nonidentical or even dissimilar goods the appearance of similarity, and promoted confidence in consumers. It permitted Americans to share a common material experience and fulfilled the public's desire for uniformity, even as they consumed different products. Mid-century expressions of this phenomenon were legion, particularly in those industries relying on hand rather than machine labor. Midwestern hog-slaughtering and packing operations provide an especially apt example. Distribution of their goods required that local butchers and wholesalers nationwide be familiar with a common nomenclature.

Prior to about 1840, hog slaughtering was a local activity, and packing generally was unnecessary. After that date, however, the infusion of capital from New York, New Orleans, and Europe enabled midwestern pork merchants to finance expansion. They initially conveyed the pork on rivers, but by the 1860s had shifted to railroads. National pork-slaughterers and packers divided their product into general categories: fancy hams, common hams, bacon, shoulders, and four grades of bulk pork—clear, mess, prime, and cargo. As early as 1840, they slaughtered hogs by regularized procedure. Animals were led to slaughterhouse roofs, where they were confined in small pens. They were knocked unconscious or killed by a blow to the head with a two-pointed hammer, then dragged inside to sticking rooms where their throats were cut and they were hung to bleed. When bleeding ceased, carcasses were scalded in caldrons of boiling water and tossed on scraping tables for hair and

THE GAMBRELS.   DISEMBOWELING AND WASHING.

TRIMMING-TABLES.

ear removal. They were then hung by gambrels * in front of gutters, who removed the entrails and sent the carcasses to basement cooling-rooms. The next day cutters severed feet and heads, and divided remains first into hams, shoulders, and sides. Those parts were then butchered more finely, and the meat was distributed in graded tubs "with the exactness of machinery."[7] It was finally smoked, pickled, or salted, then packed.

Regularized slaughtering yielded a more-uniform product. Different hog breeds complicated matters, but packers in larger slaughtering centers such as Cincinnati appointed a single individual to visit area abattoirs to confirm that butchered meat conformed to packers' categories. Indeed, the packers' reputations and ultimately their markets depended on each barrel's accurate characterization. The introduction of more sophisticated slaughtering tools and techniques in the 1860s, such as animal-shackling and cutter teams, steam-powered saws, lifting machines, and scraping apparatuses, failed to placate packers. They continued to protect their reputations by area-wide slaughterhouse inspections.

Regularized hog-slaughtering produced not only more uniform meat cuts but also more nearly similar wastes, which helped foster and regularize a host of slaughterhouse by-product industries. In the late 1830s, for instance, two Cincinnati entrepreneurs, James Gamble and William Procter, capitalized on these wastes to form Procter and Gamble, which produced oil for lamps and machinery, "a fine article of clarified Pig's Foot Oil—Neat's foot oil," as well as a variety of soaps.[8] Other manufacturers tanned hides; converted hair, hooves, offal, and blood into prussiate of potash, which textile workers used to make Prussian-blue dye; transformed offal, hooves, horns, and bones into glue; and used blood, offal, and bones for fertilizers. Pig heads, feet, backbones, ribs, and fat were rendered into lard (through use of superheated steam at pressures of five to seven atmospheres for several hours), which formed the basis of the lard oil, tallow, and stearin industries. Hog bristles became

---

gambrel: a frame shaped like a horse's hind leg for hanging carcasses.

The labor-intensive hog slaughtering and packing industry was an apt example of a common-nomenclature-derived uniformity.

brushes. By-products were big business; as early as 1851, four Cincinnati hog-bristle processing firms employed nearly 200 workers.

Providing Interchangeable Goods—Armory Practice   Perhaps the
least characteristic mid-century way to provide a common material
experience was through production of truly identical items. The
manufacture of interchangeable parts had been a dream of eighteenth-century French military men, and had been embraced by
the American military early in the nineteenth century as part of
its fascination with things French. This dream was realized in mid-
century American armories. Well before American military men
could manufacture interchangeable parts for their armaments, they
had attained "the grand object of uniformity."[9] This technique using similar, but not interchangeable, parts reduced expensive and
time-consuming fitting and filing. Prior to 1840, armory men had
adopted drop- and die-forging to make roughly similar pieces, which
were filed and fitted when soft, checked by gauges, tempered or
hardened, and finally fitted, filed, and gauged again. But achieving
uniformity was much less complicated and costly than interchangeability. As developed in the army's Springfield Armory, interchangeability generally depended on drop- or die-forging, but it
also demanded special-purpose milling and/or planing machines,
rationally designed jigs and fixtures, precision gauges, and employees capable of and willing to apply the most exact standards. Interchangeability found its initial grand-scale application in the fully
interchangeable Model 1841 percussion rifle and the Model 1842
percussion musket, both produced by the armory in 1847.

High armory practice—interchangeability—required an exceedingly large capital investment and tight quality control. That it
originated from a governmental agency—the army—which did not
need to be profitable or accountable should not be a surprise. Most
private arms manufacturers, including Samuel Colt in his famous
Hartford arms factory, did not use armory practice. Nor did this
technique find much employment in other mid-century production
spheres or outside New England. A handful of companies, especially machine-tool firms such as the Providence Tool Company,
adopted high armory practice, but they proved most exceptional.
Its lack of application did not stem from ignorance. Armory workers left government employ for private enterprise in droves and
carried their knowledge with them. But the precision-based technology did not easily transfer to the private sector, although some
armory-designed hardware did; fitting and filing were increasingly

Assembling Colt firearms was a time-consuming task.
Notice the large number of fitters and filers.

mechanized. For the overwhelming majority of manufacturers, interchangeability did not make sound economic sense. It would raise production costs prohibitively and render companies uncompetitive. Uniformity, not interchangeability, ruled the day.

UNIFORM AGRICULTURE

A similar emphasis on uniformity pervaded mid-nineteenth-century agriculture, but the manner in which it was introduced and achieved was quite different from the manner used in manufacturing. Americans viewed factories as institutions new to mid-century America, and consequently felt free to define and orient these manufacturing facilities without regard to the past; they had only to create an institution consonant with mid-century notions. That latitude was lacking with agriculture. Farming had been a traditional enterprise, and any modification had to overcome centuries of inertia following examinations of the state of the art. Numerous commentators, many of them farmers, did survey American agriculture, albeit informally and on a small scale. They were virtually unanimous, however, in concluding that farmers frequently disregarded the condition of facilities and fields. They called on husbandrymen to adopt new practices. To these observers, rundown or haphazardly conducted farms signalled rundown farmers; appearance of fields and facilities were considered to be (like the reputations of companies or photographic portraits) outward reflections of farmers' characters.

The discovery that some farmers apparently lacked the ideal American character, so fundamental a facet of mid-century American thought, engendered concern far greater than the number of cases actually cited. It suggested there existed among the class of farmers an individualism, independence, or ignorance so distasteful and contrary to mid-century assumptions that it demanded immediate action. Contemporaries advocated a new kind of farming—a machine-based agriculture, which mimicked machine-based manufacturing—to regularize farming and get agriculturists up to snuff. But reorienting established practice is rarely simple. That American agriculture already possessed an institutional structure—its societies and newspapers—limited options, but also identified mechanisms for change. Mid-century agricultural societies sponsored thousands of farm-implement competitions, and agricultural publications were awash with advertisements for similar tools. These plows, drills, cultivators, and harvesters generally required animal or steam power and were constructed of iron or steel rather than wood. They eased farmers' burdens and enhanced users' reputations by facilitating row cropping. Neat, clean farms with rowed fields and soils prepared to a proper, nearly uniform texture were identified as the domain of the "improved farmer." The mid-century diffusion of uniform agricultural implements encouraged regularized-row agriculture (or, in the case of grains, regularized stands) and created more nearly uniform production. Use of these tools defined the distinctly American farmer.

Economic justifications also led farmers to adopt these implements. They enabled agriculturists to secure and harvest greater yields from their lands, and to cultivate additional fields, while national and international marketing possibilities promised handsome returns for whatever quantities they could produce. That prospect, coupled with the new tools, fortified eastern farmers and inspired others to tap the prairie lands of Illinois, Iowa, Minnesota, Nebraska, and Kansas. Located far from sizable cities, farmers there worked the rich black land and sent their huge bounties to the East.

Equipment for a Uniform Agriculture

Mid-century manufacturers bombarded farmers with a dizzying variety of agricultural implements from which to choose. All equipment manufacturers issued the same claim that their product would best improve farming and farmers. While some implements certainly worked more proficiently than others, the relative merits of each for the physical act of farming were hardly ever clear. Farmers often selected their farm equipment the way that they, and Americans generally, chose other manufactured products (such as sewing machines): they equated how well a piece would perform with who manufactured it and how it was supposedly manufactured.

The McCormick Reaper   Such was the case with reapers. The reaper boom began in the 1830s when Cyrus H. McCormick and Obed Hussey independently patented workable devices. Their reapers were joined by those of John H. Manny, William F. Ketchum, and several others in the next decade; by 1850, more than thirty American companies manufactured different reapers. But McCormick's firm consistently led reaper sales. In 1876 it produced 14,000 machines, a success that testified to the importance of marketing reputations.

McCormick certainly never produced the finest article. His reapers mowed poorly, operated unsatisfactorily in tangled fields, and broke down frequently. Similarly, McCormick production techniques were anything but uniform. In the 1830s, McCormick made all reapers in the family's Virginia blacksmith shop. During the next decade he allowed companies across America to manufacture reapers under the McCormick name; no two firms produced the same McCormick reaper. Beginning about 1850, he discontinued that practice and opened a steam-powered factory in Chicago, but it did not signal a major change in production techniques. The Chicago plant only assembled reapers; McCormick provided several contractors with specifications and depended on them to supply parts. The Chicago facility enabled McCormick to produce roughly similar reapers each year, but no two years' models were alike; McCormick redesigned the machine annually. Even in Chicago, the company used no jigs, fixtures, or precision gauges. It did not drop-forge fittings, but cast them. It relied entirely on skilled craftsmen—blacksmiths, machinists, and woodworkers—not special purpose machinery. Fitters and filers far outnumbered other employees.

McCormick nevertheless established his reputation for precision manufacture of a superior product in much the same way as would

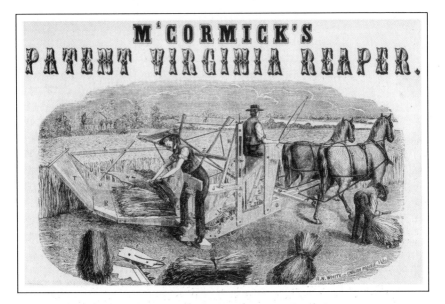

McCormick's reaper exemplified the marketing of reputations.

Singer and countless others. In the late 1830s and early 1840s, McCormick himself gave demonstrations at farms, participated in formal field trials, and secured testimonials from satisfied users. He also claimed that he had created the "perfect reaper" even as he altered its design, and that his product was "manufactured in the best manner."[10] He collected endorsements from agricultural societies and newspapermen, and warranted his machine's performance because "I have a reputation to maintain."[11]

McCormick stepped up his campaign beginning in 1843. He hired several agricultural newspaper editors to serve as agents and display his reapers locally. He engaged in a bit of "honest graft" by selling reapers to officers of agricultural societies and distinguished farmers—particularly those holding judgeships—at reduced prices. At the same time, he championed the motto "one price to all, and satisfaction guaranteed."[12] He also spoke to farmers in a different way than other manufacturers. McCormick published an agricultural newspaper, the *Farmers' Advance,* which was distributed free to more than 100,000 farmers. His advertisements there and elsewhere showed reaper-owning farmers working their machines dressed in top hats and waistcoats. McCormick sold machines on

The factory emerged in mid-nineteenth-century America.
Above is McCormick's factory soon after the firm moved to
Chicago.

credit to permit those strapped for cash to purchase them, and so to achieve the gentlemanly state to which his advertisements suggested they were entitled. He also gave his annual product names, such as the *Reliable* and the *Advance*, to signify quality and progress.

But perhaps McCormick's policy of establishing local agencies was his most effective strategy. Begun in the early 1840s and in full force by 1850, McCormick agents received sales commissions and had exclusive company contracts forbidding them from working with other reaper concerns. Unlike Singer operatives, McCormick employees had no citified emporiums. Their display rooms were local courthouse and county-seat lawns, or their own barns and wheat fields. These men traveled throughout their territories, often taking sample machines. In addition to canvassing their territories, their responsibilities included delivering reapers, instructing purchasers in reaper final-assembly and operation, and stocking spare parts—each McCormick reaper came to agents with a full complement of replacement parts. These agents needed sufficient mechanical skill to do repair work and render field service. Indeed, agents in most cases were called on to fulfill the McCormick warranty's terms, a warranty of dependability based on the company's allegedly superior production techniques.

The John Deere plow helped break the western grasslands.
This early version had a steel share and a cast-iron
moldboard.

New Plows   The McCormick name was synonomous with reapers, but three different plows shared the mid-century limelight. Cast-
iron Eagle plows led the more generalized onslaught of cast-iron
plows with wrought-iron shares in both the North and South during the late 1830s and 1840s. At least 25,000 of these plows (with
elongated, radically curved moldboards to facilitate plowing on rough,
heavily stubbled ground) were purchased yearly in the two decades
prior to 1860. But Eagle plows were unsuitable for western grasslands. The region's heavy, sticky soil clung to the cast iron and
required farmers to stop periodically to clean their moldboards. The
Prairie Breaker, a huge plow with a 125-pound wrought-iron strap-
plated moldboard, a 15-foot beam, and a wrought-iron share, was
the first important western plow. Reigning through the 1840s, the
Prairie Breaker's enormous weight required as many as seven yoke
of oxen to pull it, but cut furrows only two or three inches deep.

Neither the Prairie Breaker's power requirements nor the shal-
lowness of its furrows satisfied farmers, who expressed interest in
a plow made by John Deere in Illinois. Deere popularized the agricultural use of steel when he built a steel-shared plow with a highly
polished, cling-resistant cast-iron moldboard that needed only half
the Prairie Breaker's draft. Deere offered his plow for sale in 1841.
While farmers were immediately impressed with its performance

and durability, its cost gave them pause. Deere's price was double that of the expensive Prairie Breaker, because he hammered his shares from steel ingots imported from Germany. In 1846, Deere began to melt ingots and cast shares, and by the mid-1850s also had casted steel moldboards. Casting reduced the price somewhat, and western farmers began to purchase Deere's plows in increasing numbers.

Other Farm Implements   Few other farm implements or equipment manufacturers achieved the market share of the forementioned plows, or the notoriety of McCormick. Nonetheless, harrows, drills, and cultivators underwent changes in design, materials, and construction even more significant than those of plows and reapers. These three implements emerged as the staples of uniform row agriculture, and manufacturers began to design their products to further that approach to farming.

For example, square- and A-framed wooden harrows with fixed, regularly spaced iron teeth appeared in the early 1840s and persisted virtually unchanged for two decades. Later modifications dealt with materials, not design; sturdier steel teeth were substituted for iron, and iron frames replaced wood. Apparatuses for altering the angle of teeth were also introduced, but the basic design, ideal for uniform row agriculture, was not changed. Not until after the Civil War were harrows redesigned, but their new form did not mark a de-emphasis of row agriculture. In these new flexible-disk harrows, revolving cast-steel disks, each a foot in diameter, did the work of teeth. Farmers could adjust distances between disks, which cut several inches deep, and often used disk harrows for plowing.

Iron also replaced wood in seed drills, but the increased usage of these sowing machines by farmers in the decades after 1840 was itself critical. Enabling farmers to deposit specific seed quantities at regular intervals, drills superseded broadcast sowing for grains; they reduced waste, produced crops with more nearly uniform stands, and enhanced mechanized harvesting. They also supplanted dibble sticks and hoes for planting corn, peas, and beans, freeing farmers from the constant bending associated with those hand implements and facilitating a more nearly perfect row agriculture. Adjustable, precise mechanisms allowing farmers to use drilling techniques on uneven ground were marketed in the 1860s.

Mechanical cultivators rapidly replaced hoes as weeding implements after 1840. Farmers had initially adapted shovel plows, and harrows with the appropriate teeth removed, to the task. But by the mid-1850s several manufacturers were advertising straddle-row cultivators, which weeded both sides of a row at once. Essentially iron-shovel plows connected by arching bars that permitted cultivator bodies to pass above rows of plants, straddle-row cultivators were pulled by horses, but required farmers to walk behind and adjust the shovel depth. Sulky cultivators of the 1860s allowed farmers to ride when weeding, since sulkies were situated directly over shovels, which were engaged by a series of foot-operated levers.

Mid-century manufacturers also designed two steam-powered agricultural implements. The success of one, the steam thresher, testified to the desire to mechanize agriculture, as well as to the relationship between national and international markets and farm size. Steam threshers appeared about 1850 as replacements for horse-powered devices that threshed grain, separated straw, and winnowed. Principles for both machines were the same. Fluted threshing-drums beat grain from their heads, while conveyor belts or vibrating canvases carried straw, grain, and chaff away from the drums. Grain and chaff then were separated from straw by vibration as they dropped through spaces between slats on endless belts, or through holes in agitating pans. Fans winnowed the material as it fell. By 1870, conveyor-belt elevators had been added to lift straw to stacks, and steam threshers had become grain-farm staples.

But it was the device that failed to secure a mid-century market, the steam plow, that most captivated farmers. Indeed, the Illinois Agricultural Society was so taken by the possibilities that it offered a princely $5,000 premium in 1858 for the best steam plow. Steam plows epitomized uniformity in agriculture. Farmers could use them to prepare a nearly uniform soil to plant regularized rows or stands. Americans applied for hundreds of steam-plow patents in the two decades after 1850. Although some were of exotic design—using disk plows or replicating spading—most fell into two classes. Many were modeled after a steam plow introduced in Britain about 1810, improved in the mid-1850s, and common in that country by the decade's end. Steam engines pulled cables attached to gang plows back and forth across fields; horses moved the engines after each pass to position them on unplowed land. These devices proved too expensive for small American farms, and inadequate for large ones because the cable apparatus did not work effectively over vast distances. Nor did Americans import the British product. British steam plows were used on only two American farms in 1867 even though Congress had removed all steam-plow import duties earlier in the decade.

Self-propelled steam plows were the second type. These ponderous vehicles with large boilers and cylinders carried their own fuel

Threshers were essential machines on large grain farms.
Attached by belt pulleys, they were run by steam engines.
(top)
Steam-traction engines sometimes weighed as much as
twenty-five tons. (bottom)

and water and moved by channeling some engine power to drive wheels. Self-propulsion required great power. The steam plow's enormous weight caused it to sink in soft soils; dynamometer tests estimated that it took thirty-five times as much power to propel engines in loose dirt as it did on rails. Inventors tried to increase traction by enlarging wheel surface area, but that approach also added weight. High capital, fuel, and maintenance costs also contributed to the steam plow's lack of utility. The first such device was not even marketed until 1873. It sold poorly.

THE CREATION OF INTERNATIONAL MARKETS

America's initial interest in British steam plows even while it expressed contempt for things European, and its subsequent attempts to develop improved steam devices on American soil, symbolized the nation's relationship to Old World civilizations. The country certainly had not withdrawn into itself. Fortified by its citizens' conception of their country as having a distinctive culture, the nation pursued its destiny aggressively within the world of nations. The Mexican War, repeated disputes with the British, Seward's Folly, and other foreign-policy forays indicate that Americans deemed their nation fit to lead, not follow. With America's natural abundance and distinctive civilization, it seemed the country's duty to assert itself in world affairs.

The expansion of international commerce appeared natural within this chauvinistic context. A supposedly superior America would inevitably gain significant new markets, which would secure great wealth for Americans and improvements for the nation's trading partners. Spurred on by this explanation of why international commerce would prove profitable (and finding that it exceeded expectations), Americans tripled their exports in the two decades after 1840, both increasing trade with Europe and establishing markets in Russia, China, India, Latin America, and Japan.

Mid-Century Shipping

During the early burst of American international trading enthusiasm in the late 1830s, most of the nation's products were carried overseas on American built and designed clipper ships. These were long, sleek-hulled ships with concave bows, convex sides, and rounded sterns. They were smaller than earlier packet ships, but the speed generated by their tall masts and enormous canvas spreads more than compensated for reduced cargo space; some clippers covered more than 400 miles a day. These ships regularly sailed around Cape Horn to Asia as well as operated on packet runs to Liverpool and California. But the labor-intensive clipper's heyday was short-lived.

By the mid-1850s, oceangoing steamships proved superior in speed and cargo capacity to the clipper. Pioneered by the British in the mid-1830s, these wooden, paddlewheel-driven ships made packet runs as early as 1840. American steam vessels challenged the British in the mid-1840s, bolstered by federal transatlantic mail subsidies, but never matched their European counterparts. They fell

Clipper ships joined the East and West coasts and America to Europe and Asia.

further behind in the 1850s when the British introduced swift iron-hulled steamships powered by screw propellers. By 1860, these ships averaged less than ten days crossing the Atlantic. The clipper record was thirteen and a half days.

The International Industrial Exhibition

American clippers lost out to British steamships, but Americans capitalized in the 1850s on a new British-inspired institution, the international industrial exhibition, to penetrate further into world markets. As did local exhibitions in the early nineteenth century, worldwide industrial tournaments placed manufacturers in touch with potential markets and enabled interested parties to compare relative merits of products.

The first international exhibition was proposed by Prince Albert, president of England's Royal Society of Arts and Queen Victoria's husband, to "afford a true test of the point of development at which the whole of mankind has arrived."[13] The Great Exhibition was held in London in May of 1851. The British erected a remarkable building to house the event: a single huge structure made from a modular iron frame covered with glass, named the Crystal Palace. Despite Prince Albert's lofty sentiments, his exposition was clearly big business; he no doubt expected the Exhibition to create new markets and appreciation for the highly regarded British. But other nations as well embraced the opportunity to have their products testify to their own level of development, and to open new trade vistas. The Exhibition's organizers heightened its competitive aspect by organizing the exhibits by nation, not product.

American manufacturers quickly recognized the twin significance of the Exhibition, and the federal government established a commission to act for America. Headed by President Millard Fillmore, the commission included Joseph Henry; Thomas Ewbank, Commissioner of Patents; and other national leaders. Their work paid off, and America did not embarrass itself in London. It displayed about 500 exhibits—less than three percent of the total—and received roughly three percent of the 170 council awards, the Exhibition's highest honor. Another 102 American products, or about 3.5 percent, received one of the 2,918 prize medals. American agricultural implements, firearms, and safes attracted the most notice; the McCormick reaper, Colt revolver, Robbins and Lawrence rifle, and Day and Newell safe were singled out for public acclaim. McCormick's reaper also won a council award. Other American council-award winners were Charles Goodyear's vulcanized rubber products; Gail Borden's nutritious dried-meat biscuit; David Dick's antifriction punch press; and William C. Bond's astronomical clock.

America's performance speeded the introduction of its goods into world markets, and enabled its citizens to establish European distribution centers for American goods. The country's exhibition showing reaffirmed that the nation had become, at least with respect to manufacturing, a full-fledged member of the world of civilizations. Its success encouraged a similar industrial contest in New York City in 1853, and the country to be party to several other international industrial competitions: Paris, 1855; London, 1862; Paris, 1867; and Vienna, 1873. America's Crystal Palace prominence also led the British to establish in 1853 a special commission to investigate American arms manufacture. Committee members coined the term "the American System" to describe high armory practice, and suggested that those manufacturing techniques generally characterized American industry. Indeed, to the British, the

American System of manufacturing came to define the distinctly American civilization, even though only a handful of American firms practiced it.

The British inquiry into American arms manufacture provides a vivid demonstration of an important by-product of international industrial exhibitions—the internationalization of technical knowledge. Exhibitions became regular mechanisms by which interested persons surveyed the state of a given art, gained ideas and insight, and met like-minded people. Americans in particular availed themselves of these forums to learn about new machinery, devices, and processes, which they took back to America and often adapted. Because of informal interchanges at these exhibitions, as well as the actual observation of products, industrialists worldwide had access as never before to a common base of information about the latest practices. This had several implications. Individuals and corporations could acquire through exhibitions the knowledge necessary to begin production and, conversely, to rebound from poor-practice-related setbacks to seize the initiative; exhibitions provided opportunities to get started, and cushioned established operators from failure. Their potential for diffusing technological knowledge worldwide also suggested that new processes, machines, and methods could spread from country to country at unprecedented rates, and that worldwide products and techniques could assume unprecedented similarities.

Transoceanic Telegraphy

Acknowledgment that America operated in a world economy placed a premium on up-to-date international information. Prior to the mid-1860s, transoceanic crossings by ship were the sole source of this intelligence, and in light of the new economic realities, they began to seem much too slow. The speed with which international intelligence arrived was paramount: information about world market conditions would reduce speculation and hoarding, permit merchants to carry reduced inventories, and allow them to tap territories most advantageously.

Several Americans in the early 1850s proposed laying underwater telegraph cables insulated with gutta percha* to transmit information instantaneously between continents, but Cyrus W. Field

---

gutta percha: a rubber-like insulating substance formed from the milky juices of trees native to Malaysia.

deserves the most credit. A retired New York wholesale paper merchant, Field gained Samuel Morse's support and put together a consortium of Anglo-American investors to run a cable on the ocean floor from Newfoundland to Ireland. The consortium demonstrated its British makeup and common sense (for the British were far better acquainted with electrical phenomena and cable construction than Americans) by consulting with Robert Stephenson, Michael Faraday, and others, and by contracting with a British firm to manufacture the cable. William Thomson, later Lord Kelvin, was a director of that company. It produced a cable about five-eighths of an inch in diameter. Seven copper strands formed its conduction core, and were packed in three thin gutta percha layers. A tarred hemp sheath and eighteen strands of iron wire surrounded this cable and served as further protection.

Consortium-owned steamers attempted to lay the cable in summer 1857, but it snapped 380 miles offshore. A similar misfortune occurred in June 1858, but a foray a month later seemed successful. On August 4, telegraph service between Europe and America began. Less than a month later the signal failed. August's joyful celebrations devolved into September's recriminations; subsequent investigations suggested that inappropriate testing and faulty storage techniques had damaged the cable's insulation. The 1858 failures and the American Civil War gave supporters pause, but the consortium resumed activities in 1865. It now faced a substantial competitor. Western Union also sought to connect continents, but from Russia's Asiatic coast. Running a line from Western Europe across Russia, underneath the Bering Strait and down through Russian America (Alaska), the Western Union venture would join existing lines at San Francisco. This threat animated the consortium; it employed a group of British electricians, including Thomson, Charles Wheatstone, William Fairbain, and C. F. Varley, to design the new cable and award the contract. The new cable was one and one-tenth inch in diameter, contained three times more copper than the earlier version, and had three times the tensile strength. Nine layers of gutta percha-based insulation bound the core, and ten stout soft-steel wires protected it. Pitch-treated hemp sheathed the steel.

Despite the new cable and extra precautions, the 1865 expedition failed. New cable was made for 1866, identical except that zinc-galvanized steel rendered the hemp sheath unnecessary. This voyage proved successful, and on July 27 the telegraph linked the continents. The consortium's crews also managed to grapple and hoist the previous year's cable, splice it to additional material, and run it beneath the ocean. The second line opened on September 1.

THE CIVIL WAR: A MASS EXPERIENCE

War conditions place strains on a nation's industrial and agricultural capacity, but they also demand redistribution of resources and encourage adoption of improved production techniques, which may improve the nation's production capacity. America's civil war was striking in how little permanent change it made in northern industry and agriculture. The case was somewhat different in the South, however. That section expanded production but it rarely innovated; it merely applied established northern practices. During the Civil War the South built armories and powder mills to manufacture weapons, erected textile factories to make uniforms, laid railroad tracks to facilitate transportation, employed horse-drawn agricultural implements to increase food production, and started to tap Alabama's abundant iron ore and coal assets to produce iron. Although most attempts proved rather disappointing, southern industry and agriculture during the war more nearly approximated the North's than it had for several decades.

The Civil War's aftermath intensified southern industrialization. Northern and English capital fueled expansion, but the war itself provided a homogenizing experience. Northern soldiers wore similar uniforms, used similar weapons, ate similar foods, and slept in similar facilities. Their southern counterparts shared among themselves similar material experiences. Civilian populations also were affected. Northerners suffered the same shortages and dislocations. Southerners, too, felt roughly the same deprivations. Both reviewed the war through Mathew Brady's photographs. The most significant technological effect of the war was the commonality it reinforced. It was the archetypical mass experience in an era distinguished by attempts to provide common experience.

NOTES

[1] Nathaniel P. Willis, *Hurry-Graphs; or Sketches of Scenery, Celebrities and Society, Taken From Life,* 2nd edition (New York: Scribners, 1851), p. 300.

[2] Ralph Waldo Emerson, "The American Scholar," reprinted in William H. Gilman, ed., *Selected Writings of Ralph Waldo Emerson* (New York: New American Library, 1965), pp. 223-40.

[3] *Cincinnati Daily Times,* November 21, 1856, p. 2.

[4] United States House of Representatives Report #753, 25th Congress, 2nd Session.

[5] Enoch Reed, "American Iron," *The Plough, The Loom and The Anvil,* vol. 5 (1852/53), p. 290.

6 Quoted in Ruth Brandon, *A Capitalist Romance: Singer and the Sewing Machine* (Philadelphia: J. B. Lippincott, 1977), pp. 111, 112, and 124.

7 Charles Cist, "The Hog and Its Products," in *Report of the Commissioner of Agriculture For the Year 1866* (Washington: Government Printing Office, 1867), p. 386.

8 *Cincinnati Daily Gazette,* June 29, 1838, p. 4.

9 Quoted in Merritt Roe Smith, "Army Ordnance and the 'American System' of Manufacturing, 1815–1861," in Smith, ed., *Military Enterprise and Technological Change* (Cambridge: MIT Press, 1985), p. 56.

10 Quoted in William T. Hutchinson, *Cyrus Hall McCormick*, 2 vols. (New York: Century, 1930), vol. 1, pp. 187 and 331.

11 Quoted in Cyrus McCormick, *The Century of the Reaper* (Boston: Houghton Mifflin, 1931), p. 46.

12 Quoted in Hutchinson, vol. 2, p. 489.

13 Quoted in *The Crystal Palace Exhibition: Illustrated Catalogue, London, 1851* (New York: Dover, 1970), p. xii.

FOR FURTHER READING

Albion, Robert G. *The Rise of New York Port, 1815–1860* (New York: Charles Scribner's Sons, 1939). An excellent summary of the economic importance of New York City in the mid-nineteenth century.

Aldrich, Darragh. *The Story of John Deere: A Saga of American Industry* (Minneapolis: C. C. Webber, 1942). An older, but useful history of the company.

Ardrey, Robert L. *American Agricultural Implements* (New York: Arno Press, 1972). A catalogue of nineteenth-century farm tools.

Broehl, Wayne G., Jr. *John Deere's Company* (New York: Doubleday, 1984). A detailed, modern history. A model company history.

Chandler, Alfred D., Jr. *The Visible Hand: The Managerial Revolution in American Business* (Cambridge: Harvard University Press, 1977). A provocative assessment of new mid- and late-nineteenth-century organizational strategies for businesses.

Cooper, Grace R. *The Sewing Machine: Its Invention and Development* (Washington: Smithsonian Institution Press, 1976). The best single volume on early sewing machine technology.

Dibner, Bern. *The Atlantic Cable* (Norwalk: Burndy Library, 1959). A readable, comprehensive treatment of the subject.

Gates, Paul W. *The Illinois Central Railroad and Its Colonization Work* (Cambridge: Harvard University Press, 1934). A pioneering study of the influence of railroads on settlement patterns.

Gernsheim, Helmut and Alison. *L. J. M. Daguerre: The History of the Diorama and the Daguerreotype* (New York: Dover, 1968). A convenient history of the daguerreotype in Europe and America.

Gibbs-Smith, C. H. *The Great Exhibition of 1851* (London, 1950). Highlights aesthetic concerns.

Gilchrist, David T. and Lewis, W. David, eds. *Economic Change in the Civil War Era* (Greenville: Eleutherian Mills–Hagley Foundation, 1965). A series of essays downplaying the economic implications of the war.

Hounshell, David A. *From the American System to Mass Production 1800–1932. The Development of Manufacturing Technology in the United States* (Baltimore: Johns Hopkins University Press, 1984). A challenging examination of American manufacturing practices.

Hurt, R. Douglas. *American Farm Tools: From Hand-Power to Steam-Power* (Manhattan: Sunflower University Press, 1982). A readable account of the development of agricultural technology to the end of the nineteenth century.

Jenkins, Reese V. *Images and Enterprise: Technology and the American Photographic Industry, 1839–1925* (Baltimore: Johns Hopkins University Press, 1975). An excellent survey of the rise of the American photochemical industry.

Kutler, Stanley. *Privilege and Creative Destruction: The Charles River Bridge Case* (Philadelphia: J. B. Lippincott, 1971). A modern discussion of this landmark case.

Lewis, W. David. *Iron and Steel in America* (Greenville: Eleutherian Mills–Hagley Foundation, 1976).

Mabee, Carleton. *The American Leonardo: A Life of Samuel F. B. Morse* (New York: Alfred A. Knopf, 1944). Still the best treatment of Morse's life.

McKelvey, Blake D. *Rochester: The Water-Power City, 1812–1854* (Cambridge: Harvard University Press, 1945). Discusses the rise of flour milling in this western city.

Persons, Stow. *The Decline of American Gentility* (New York: Columbia University Press, 1973). An incisive examination of the rise of mass culture.

Reid, James D. *The Telegraph In America* (New York: Derby Brothers, 1879).

Rolt, L. T. C. *A Short History of Machine Tools* (Cambridge: MIT Press, 1965). A handy catalogue of the development of many devices.

Ross, Earle D. *Democracy's College: The Land-Grant Movement in the Formative Stage* (Ames: Iowa State College Press, 1942). A valuable discussion of the mid-century concern over the state of American agriculture.

Schisgall, Oscar. *Eyes On Tomorrow: The Evolution of Procter and Gamble* (Chicago: J. G. Ferguson, 1981). A company history containing useful information about secondary industries dependent on hog slaughtering.

Smith, Merritt Roe. *Harpers Ferry Armory and the New Technology* (Ithaca: Cornell University Press, 1977).

Taft, Robert. *Photography and the American Scene: A Social History, 1839–1889* (New York: Macmillan, 1938). A breezy discussion of mid-century American photography.

Temin, Peter. *Iron and Steel in Nineteenth-Century America: An Economic Inquiry* (Cambridge: MIT Press, 1964). An analysis of the economic parameters of the industry.

Thompson, Robert Luther. *Wiring A Continent: The History of the Telegraphic Industry in the United States, 1832–1866* (Princeton: Princeton University Press, 1947). Especially valuable on the formation of Western Union.

Wallace, Anthony F. C. *Rockdale* (New York: Alfred A. Knopf, 1978).

Walsh, Margaret. *The Rise of the Midwestern Meat Packing Industry* (Lexington: University Press of Kentucky, 1982). Deals with the economics of meat packing. Weak on technology.

White, John H. Jr. *American Locomotives: An Engineering History, 1830–1880* (Baltimore: Johns Hopkins University Press, 1968).

P_{art}
T_{wo}

Systematizing America

THE 1870s TO THE 1920s

"System" was the touchstone of late–nineteenth- and early–twentieth-century America, and it influenced virtually every aspect of American life. Electrical and telephone systems dotted the nation's landscape and electric traction systems coursed through cities. Miners and metal workers systematically extracted and smelted ores. Manufacturers systematized factory production, agriculturists farmed with system, and organizations were formed to systematize entry into professions.

The idyllic crusade for nationwide coherence and homogeneity had collapsed in the decade following the Civil War. Unlike their mid–nineteenth-century counterparts, late–nineteenth- and early–twentieth-century Americans routinely made distinctions among peoples, places, or things. Moreover, they treated these distinctions as real and crucial, the categories as static and discrete, and the elements in categories as both representative of those categories and fixed within them. Creation of these rigid divisions produced strife; Americans disputed what constituted legitimate categories and where those categories ended. As important, they realized that their categories did not stand alone, but instead were parts of systems, a notion that accentuated ordering the diverse categories as well as determining the appropriate locus of each system. This process of ordering, of systematizing the various categories, itself caused disagreements and consternation, but all agreed that establishing a "true" hierarchy based on objective methods and criteria was critical, because only in that manner could a system operate at optimum efficiency.

During this fifty-year period, Americans in all walks of life developed allegedly nonpartisan strategies and techniques to measure what constituted a category, to standardize the elements of those categories, and to place them in their proper place, position, role, or the like. This systems thinking had profound technological implications. Potential new processes and technologies as well as longstanding ones were subjected to scrutiny and dissected into their component parts; new technologies were conceptualized according to this precept and old technologies were reconceptualized in a similar fashion. Indeed, it would not be incorrect to state that in the fifty years after about 1870 Americans made over their nation.

4

Systematizing Power, Communications, and the Power to Communicate

THE GREAT CENTENNIAL EXHIBITION: LATE–NINETEENTH-CENTURY AMERICAN NOTIONS IN MINIATURE

In the early 1870s, America prepared to celebrate the hundredth anniversary of the Declaration of Independence. An international industrial exhibition in Philadelphia was to be the focus of the commemoration, and Congress characteristically chartered a corporation to handle finances. Design of the Centennial Corporation's gaudy 24- by 20-inch stock certificates provided an inkling that Americans had abandoned their mid-nineteenth-century quest for uniformity. The certificates celebrated diversity, while carefully joining relics of American civil religion—the nation's political heritage—with the country's technological productions. A woman personifying America formed the apex of a pyramid at the certificate's center; female representations of Fame and Art sat at her feet. Independence Hall and the National Capitol were in the background. Fitch and Fulton with steamboat models stood in front of Independence Hall, while Franklin and Morse with electrical and telegraphic instruments

135

assumed a similar posture near the Capitol. Elias Howe and a ship-wright also were depicted, offering America a sewing machine and clipper ship respectively. A freedman, a soldier, and a mechanic stood to the pyramid's right; a farmer, a planter, a miner, a trapper, and an Indian were at its left. John Trumbull's painting, "The Dec-laration of Independence," was placed at the center of the pyra-mid's base. At its right was a busy manufacturing city, contrasted to a neglected windmill to exemplify progress; at its left was a lo-comotive, telegraph, steamship, and reaper, contrasted with a con-estoga wagon, pony-express rider, and sickle-wielding farmer to connote civilization's development.

The Centennial Exhibition took place between May 10 and No-vember 10, 1876, and attracted more than eight million visitors (more than one-sixth the nation's population). The exhibition itself was as precisely arranged as the objects on the stock certificate. Indeed, it was structured far differently than the Crystal Palace ex-hibition of twenty-five years earlier. Separate buildings demarcated governmental and conceptual boundaries. Each foreign nation had its own promotional pavilion, as did every American state govern-ment. The federal government also established one, a harbinger of its growing industrial interest. There was even a pavilion for women. Various private enterprises, either singularly or collectively, erected their own; the Singer Sewing Machine Company, Pacific Guano Company, American Fusee Company, and consortiums of shoe and leather manufacturers, brewers, and casket-makers each set up pa-vilions. But the exhibition's real business took place in just three buildings: the Main Building housed displays of mining, metal-lurgy, manufacturing, and educational techniques; Machinery Hall was the site of machines in motion; and Agricultural Hall stocked farm implements and food-preservation processes. Products were organized within these three buildings by nation and type; diver-sity was therefore accentuated. For example, the American exhibit in Machinery Hall included a discrete machine-tool section. The British display in Agricultural Hall placed that nation's steam plows next to one another. The world's industrialists competed at these three venues.

The Centennial Exhibition's judge-selection criteria and award procedures also distinguished it from predecessors. Its commission-ers dispensed with apportioning judgeships by nation—nations had selected prominent men, men of character, as judges—and replaced that practice by choosing men "individually for their high qualifi-cations," no matter what nationality; expertise in technological processes of manufacture became the *sine qua non*. These experts did not issue medals across production categories, or discriminate between the best and second best within categories, but instead tried to separate the outstanding from the mundane and to reward only excellence. They considered each entry's "inherent and comparative merits" and used these "properties and qualities" as the basis for granting the exhibition's sole award, the Centennial Medal.[1] Judges were not required to give any medals within a production category, nor were they limited to awarding a maximum number either within a production category or throughout the Exhibition. As important, judges produced written reports explaining their selections, including dispassionate analyses of award-winners' particulars. This process opened the experts' evaluations to the public, a factor that protected the judges' and the exhibition's integrity by eliminating potential charges of favoritism. Among those American productions cited were Goodyear rubber products; Yale locks; Edison's automatic telegraph; Midvale Steel's axles and shafts; Sharps rifles; Westinghouse's air brake; Roebling's steel suspension-bridge cables; machine tools produced by Pratt and Whitney and William Sellers and Company; Otis elevators; Gatling guns; Fitts road steam-engines; Brown and Sharpe universal milling and grinding machines; and Pullman sleeper cars. Alexander Graham Bell's telephone was also displayed, as were arc lights and several typewriters.

Neither the qualitative distinctions among products nor the Exhibition's meticulous differentiation by nation and product type undercut its essential coherence and unity. The highlight of the exhibition, a gigantic Corliss engine—actually two double-acting vertical high-pressure steam engines working in tandem—symbolized the Exhibition's cohesiveness by powering all apparatuses in Machinery Hall; it bound together all the disparate parts. Contemporaries repeatedly likened the engine to an animal heart, systematically pumping life blood through metallic arteries and leather capillaries to the Hall's other machines. The engine was produced by Rhode Island's George Corliss. It stood nearly 40 feet tall, weighed 680 tons, was driven by 20 boilers, and had cylinders 44 inches in diameter with 10-foot strokes. Load dictated engine speed, a characteristic of Corliss-developed engines; rotary valves automatically governed the quantity and rapidity of steam injected in the cylinders so that the engine adjusted itself to the power demands. Motion was stored in a huge flywheel, which made 36 revolutions per minute (RPM), weighed 56 tons, and was 30 feet in diameter and 2

*At the 1876 Centennial Exhibition, the huge Corliss engine
drove all the machinery in Machinery Hall.*

feet thick. The flywheel's cogs matched cogs on a pinion and turned
a primary underground line, which was geared to eight secondary
underground shafts. Pulleys and belts connected the secondary shafts
to eight principal line-shafts located near the ceiling and traversing
the building's length. The Hall's machines ran from belts attached
to these 650-foot-long above-ground shafts. One line turned at 140
RPM, while the others rotated at 120 RPM. Each line could transmit 180 horsepower. Types of machines requiring the greatest
amount of power were situated closest to the beginnings of lines.

THE IDEA OF SYSTEM: THE PERVASIVE
LATE–NINETEENTH-CENTURY AMERICAN NOTION

The emphasis at the Centennial Exhibition on differentiation, while simultaneously championing coherence, was a microcosm of late–nineteenth-century America generally. Many internecine tensions buffeted the country in the half-century after 1870, which belied the national spirit that had been evinced by the mid-century population. Rather than continuing to seek nationwide homogeneity and uniformity, late–nineteenth-century Americans waged innumerable struggles with each other. The gap between the richest and the poorest Americans grew precipitously. Businessmen ruthlessly attacked and swallowed up competitors. Industrialists pumped unprecedented sums of money into production machinery to replace skilled workers, even while they battled the laborers being replaced by those machines. Industrial giants fought among themselves as well. By 1909, fewer than 1 percent of American firms produced 44 percent of the nation's manufactured goods. In the West and Southwest ranchers fought with farmers over the use of barbed wire. Ethical and racial confrontations also flared incessantly. The clergy, politicians, businessmen, and professionals called for social, economic, and political reforms while they competed with each other for power. City, state, and federal governments aggressively entered these frays, resulting in a series of jurisdictional quarrels.

Americans in the years between the 1870s and the 1920s proved far more discriminating than their predecessors. They had a proclivity during these decades to divide all types of people, things, and ideas into groups. This phenomenon was reflected in the complex layout of the Centennial Exhibition, as well as in its criteria for selecting judges. Americans during this period constantly made spatial and functional distinctions, and rigorously differentiated among various types of enterprises, peoples, and tasks. For example, at one time or another, they divided the nation's population into groups according to intelligence, socioeconomic status, geographical location, religion, national origin, body type, special knowledge, training, abilities, and the like. Apparently obsessed with identifying diversity, Americans treated these categories as real, crucial, and limiting. The criteria for each category were fixed, and all its elements were treated as if they were representative of that type. Each categorization defined and bounded a specific set of cohorts.

But late-nineteenth-century Americans also recognized that their categories did not stand alone. Although each category was discrete (its unique features could be isolated and studied), and each had its particular merits and liabilities, people of this period conceived of processes, activities, and social structures as being composed of sets of these individual components, like machines built from many parts. What made the various processes, activities, and social structures what they were was the relations of its components to each other. And because each component varied in capacity and quality, each had to be assigned its proper function or position in the "machine"; relationships had to be *systematized* in a hierarchical fashion. Only in that way could processes, activities, and society operate smoothly and at peak efficiency. In this view, the country's processes, activities, and social structures were *systems* of discrete, fixed, hierarchically arranged parts.

This notion of system placed a premium on uncovering the principles that organized relations by ordering the many different parts. In the case of the Centennial Exhibition, the new criteria for judge selection and its enormous Corliss engine played that role for the award process and Machinery Hall, respectively. Rarely, however, was consensus easily achieved on actual systematizing principles. Desire for systemization ruled the day, but disputes raged about particulars, especially in social, political, and economic matters. The identification of categories of peoples, position of categories within systems, and even the loci of systems (for example, city, state, or nation in the political sphere) proved sticking points. Most groups sought to serve as the system's guiding spirit, the indispensable decision-makers. Few remained content to rest at the hierarchy's base. These leadership contests spawned the period's recurring battles.

Hierarchies were most clearly delineated within industrial enterprises, where ownership granted those possessing it great latitude to impose systems of their choosing. Division of labor, specialization of function, standardization of technique, and rationalization of operations became industrial production's touchstones, as did distinctions among managerial elements. System's efficacy also encouraged powerful industrialists to attempt to organize entire industries; vertical and horizontal integration of companies, mergers, pools, trusts, and holding companies became commonplace.

The system mania translated into increased industrial output and a flurry of capital expenditures. Indeed, industrialists took the machine as their metaphor and passionately set about exploiting the nation's natural abundances. Their efforts were successful and profitable. The value of American production multiplied twelve times in the half-century after 1865. As early as 1890 the value of the country's manufactured goods nearly equalled the total of its three closest competitors. Pig-iron production increased eightfold between 1876 and 1901, and Americans smelted more than seven million tons of steel in 1897. Patents soared from about two thousand yearly in the 1850s to more than twenty-one thousand a year in the two decades before 1900. Between 1880 and 1890, American capital investment in machinery more than doubled. It doubled again in the decade after 1900, while the value of the nation's industrial products rose 76 percent. Americans added seventy-three thousand miles of railroad track in the 1880s and another thirty thousand in the 1890s. They laid electrical and telephone lines and completed water-delivery and waste-removal systems. Paved roads and concrete bridges provided the nation a durable infrastructure.

Reverence for systems likewise held important intellectual implications. It manifested itself in efforts to rationalize the inventive process—prompting the establishment of research laboratories or, as Thomas Edison termed them, idea factories—and served as a guideline of how to invent. Inventors faced a crucial conceptual choice. They either designed components for existing or anticipated systems, or created new systems to fit their inventions.

THE ELECTRIFICATION OF AMERICA

America in 1920 was dramatically different than the nation a half-century earlier. By the 1920s, for example, virtually every American municipality had been electrified. Central power stations generated alternating current and distributed it throughout cities to light streets, homes, and offices, and to power streetcars and factories. Few in America had foreseen electricity's potential in 1870 and fewer had anticipated the creation of central generation and distribution systems.

An examination of how the nation came to adopt electricity is an apt place to begin consideration of technology in late–nineteenth- and early–twentieth-century America. Electricity's significance was considerable, and its partisans were quite explicit about how they conceived of electrifying the nation. Americans were well acquainted with battery-generated electricity through telegraphy, although the rapidity with which the nation embraced and used centrally generated electricity did not stem from an established electricity-investigating tradition. Only a handful of early Americans engaged in electrical speculations, and only Joseph Henry offered concrete contributions. Moreover, central stations relied not on battery power, which was an estimated twenty times as expensive as steam power, but on generators to provide cheap, plentiful, dependable sources of electricity. Without development of a new technology, effective generators, there could be no inexpensive large-scale electric power.

The Development of Generators

The first practical, cost-efficient electric generators—dynamos—were produced in the 1870s by European builders who incorporated principles that other Europeans had uncovered earlier in the century.

Among the earliest and most important of these discoveries was Michael Faraday's announcement in 1831 to England's Royal Society that he had generated an electric current in a wire merely by moving it through a magnetic field. His discovery of electromagnetic induction intrigued both Europeans and Americans, and sparked interest in constructing magnetoelectric machines to convert mechanical work into electrical energy. Early labors yielded commutators—metallic brushes that collected current to form unidirectional (DC) flow—and awareness that coils needed continually to intersect magnetic lines to ensure constant electrical flow. They also produced two main ways to increase power: use of more powerful magnets or cutting of magnetic lines at optimum angles. The latter was a matter of design; investigators tried to situate armatures most expeditiously or to vary wire arrangements and number of turns. In 1820 Oersted discovered that current in a conductor created a surrounding electromagnetic field, and in 1840 this discovery was finally adapted to magnetoelectric machines. The replacement of permanent magnets with battery-supplied soft-iron-cored electromagnets made these primitive generators somewhat more powerful, but attaching them in series and using current produced by one to create the electromagnetic field for the next—as transformers—boosted output further. It also made them cumbersome, a situation remedied in 1855 by diverting part of the electricity generated by a machine to enhance its own electromagnetic field. A permanent magnet or battery seemed required to create the initial magnetic lines, but about a decade later, several Europeans, and the American Moses G. Farmer, found almost simultaneously that an electromagnet's soft iron core retained sufficient residual magnetism to render batteries or permanent magnets unnecessary.

The discovery of this self-excitation principle reduced machine size and cost, while redirecting attention to armature shape and wrappings. Preventing loss of efficiency caused by overheating and obtaining maximum power per armature rotation stood as the major technical problems. In the early 1870s, the dynamos of Z. T. Gramme of Belgium and Hefner Alteneck of Germany were most celebrated. They had ring- and drum-shaped armatures, respectively. These dynamos neither drove machines nor provided light; situated in factories, they generally replaced batteries and furnished electricity for electroplating.

The Advent of Electric Lighting

Arc Lights   These working dynamos also held out prospects for developing new commercial uses for the comparatively inexpensive electricity. Charles F. Brush was one of the first Americans to investigate these possibilities. In the mid-1870s, he designed independently circuited two- and four-light, single-dynamo arc lamps.

Brush had not discovered the phenomenon of arc lighting. Humphry Davy, Faraday's mentor, had passed current across a gap between two pieces of carbon as early as 1808 and generated bright, blinding light. By the 1860s, several lighthouses had installed arc lights, run off either batteries or inefficient dynamos; preventing shipwrecks, saving lives, and protecting property justified high electricity costs.

Nor was Brush alone in adapting dynamos to arc lighting. A few private residences' exteriors and a handful of commercial establishments' interiors were arc lighted in the 1870s.

But Brush was exceptional in that he quickly realized that the brilliance of arc lights rendered them unsuitable for tapping a substantial home or business market; the future for arc lighting would be its use outdoors. He looked to arc lighting as replacement for gas street-lamps. Sixteen-candle gas lamps had lit American municipal thoroughfares since early in the nineteenth century, and more than four hundred gaslight companies were in operation in 1875. Brush recognized that arc lights needed to be cheaper and better to displace the established gas street-lights. Independently circuited lights could never fit the bill because they would require numerous dynamos and extensive quantities of wire; they would be prohibitively expensive. He set about to design an entire system based on centrally generated and transmitted electricity, to fit precisely the criteria of cheapness and superiority. Each of his system's components had to be economical and compatible. He faced a host of technical and design problems in developing his new system: producing long-lasting, regularly burning carbons to reduce carbon replacement costs; manufacturing dynamos powerful enough to light numerous arc lights connected in series (jumping each gap decreased voltage about fifty volts, while wire's high cost prohibited use of separate circuits); designing a means to adjust gaps as carbons burned away; and devising methods to perpetuate series circuits when individual lights failed. Brush proved equal to the challenge. He invented high-voltage dynamos; automatic regulators to keep current constant, not dependent on load; an automatic feeding mechanism to maintain gap distance; and automatic short circuits to circumvent burned-out lights. He made better, less expensive carbons from petroleum coke rather than gas coke, binding them with coal tar-pitch rather than tar. He also rounded his arc carbons, tapered their tips to promote longer life, copper-plated their bases to cut resistance between carbon and carbon holder (which further slowed carbon consumption), and developed a double-carbon arc lamp (essentially two lamps in one) with an automatic switch to reduce the need for maintenance.

Brush took these elements of his system and in the summer of 1879 erected the nation's first central electric station to power twenty-two electric arc lamps in San Francisco, the commercial capital of the West. Within a year he had stations in New York, Philadelphia, and Boston, and more than five thousand arc lights in operation. Brush's initial success in manipulating dynamos, circuits, and arc lamps to form outdoor electric lighting systems engendered competition and yielded improvements. Before 1882, Elihu Thomson, then of the American Electrical Company and later of Thomson-Houston Electric Company, had devised smaller, more efficient dynamos, superior current regulators, air-blast insulation to promote high-voltage commutation, and lightning arresters. Edward Weston, then of Weston Electric Light Company and later of the United States Electric Lighting Company, increased dynamo efficiency by using laminated armatures. These advances translated into larger lighting systems. A dynamo that would light 40 lamps first appeared in 1882 and was followed by 65-lamp varieties in 1884, 80 in 1890, and 125 in 1894.

Early electric-arc-lighting firms competed against each other and gas lighting, but sought to capture markets in similar ways. They operated free demonstration systems for specified periods of time, allowed experts to test their equipment, and secured depositions from policemen and others testifying how arc lighting assisted them in performing their duties. They advertised their systems' dependability and efficiency in electrical trade journals, and published their own newspapers. These companies entered into agreements with local manufacturers to make apparatus and act as sales representatives. They sold systems, not components; they tied public lighting companies to exclusive reliance on their products. They also diversified in the mid-1880s and produced an assortment of electrical equipment used for purposes other than street lighting.

Not all companies were equally adept at promoting their goods. Nor were all systems equal. By 1888, Thomson-Houston had emerged as the leader and began to buy controlling interests in its competitors. That process enabled Thomson-Houston to eliminate rivals and gain patents, talented employees, and manufacturing plants. By 1890, it controlled more than two-thirds of the nation's arc lighting systems.

Incandescent Lights   Arc lighting was hardly the sole or even most significant late–nineteenth-century form of lighting, of course. Incandescent lighting paralleled its development, and much of the credit for devising and manufacturing an incandescent lighting system must go to Thomas Edison.

Edison had acquired a reputation as an inventor par excellence before he ever produced his first incandescent lamp. He had contributed to the "printing telegraph" (stock ticker), "speaking telegraph" (telephone), and "talking telegraph recorder" (phonograph). His youthful experience as a telegraph operator introduced him to shunts, relays, and continuous-feed belts (design elements Edison used repeatedly) and provided a rudimentary understanding of electrical principles such as Ohm's law. It also introduced him to investors familiar with electrical phenomena—telegraph investors. Edison parlayed their backing and his prominence to assemble an impressive staff of technical associates, and to stock extensive facilities. But Edison was not the first to investigate incandescent lighting. Several Europeans and Americans attempted in the two decades after 1840 to develop incandescent bulbs. Using electric batteries as power sources, these men sought to produce long-lasting filaments and vacuums strong enough to prevent combustion.

The inability to evacuate cylinders adequately doomed these early efforts. Not until after 1875 was an improved mercury air pump, developed a decade earlier by the German Herman Sprengel, employed in incandescent lighting work. Joseph Swan, a British pharmacist active in the mid-century lighting quest, was the first electrician to adapt the Sprengel pump, and he was soon joined by Americans Hiram Maxim, Moses Farmer, and others.

On one level, these investigators were Edison's competition. But he pursued incandescence from a different angle, one more nearly akin to the electrical efforts of Brush, Weston, and Thomson. Edison did not explore the possibility of incandescence; he did not begin by developing a lamp. He simply assumed that incandescence was possible and moved to the twin issues of marketability and practicality. Edison understood that success depended on incandescence's ability to supplant gas lights, but unlike the arc-light contingent, which was compelled by its light's brightness to develop outdoor systems, he focused indoors. Indoor gas-lighting systems were Edison's real competitors; he sought to create a DC incandescent system that would offer as much illumination (about sixteen-candle power) as gas lights, at a lower or equivalent price. He tackled the problem by first analyzing indoor gas lighting's costs. This was his single most important calculation, and would guide his attempt to produce an incandescent lighting system.

Edison decided to pursue incandescent lighting seriously in fall 1878. A public relations blitz accompanied that decision; Edison asserted in newspapers his priority in the field—in fact, he had done virtually nothing—and assured an attentive public that results soon would be forthcoming. Edison excelled at this type of campaign; it preempted other investors' claims, garnered him additional financial support, and began to accustom his market to the anticipated new technology. Even as he puffed his venture, Edison computed that an economically feasible system would require high-resistance filament lamps and parallel wiring. The former would reduce copper wiring costs substantially, while the latter would ensure uninterrupted service. During late 1878 and 1879, Edison and his associates experimented with materials of various compositions and shapes to devise appropriate filaments, always removing gases occluded in filaments by continuing to evacuate bulbs during heating; Edison's team had found that filaments often contained oxides, the oxygen from which was liberated during the initial heating, causing the filaments to burn. They also investigated electrical generators and designed efficient, low-internal-resistance dynamos suitable for parallel lighting. Edison named these devices Jumbo generators after the great elephant brought to America by P. T. Barnum; Edison appreciated Barnum's adroit use of publicity.

Edison's crew devoted themselves to developing the system's other components in 1880 and 1881. They devised safety fuses to prevent fires, and manufactured lighting fixtures, bulbs, and sockets for

Huge electricity-generating, central-station-located dynamos were the first requirements for electric light and power systems.

household use. Edison himself envisioned a feeder–main distribution network to cut copper wire costs by about 85 percent; rather than attach large copper mains directly to generators, thin feeder lines would carry current from dynamos to short sections of mains placed at points of dense electrical distribution. Pilot projects were established at Menlo Park and London to demonstrate the system's feasibility and to work out bugs. By late 1881, Edison was ready to construct his first American commercial central station; he placed it at 257 Pearl Street, in the heart of New York City's financial community. Edison men laid underground conduits, and insulated the wires with hot asphaltum. They wired households and establishments, and installed meters based on electrolytic deposit of zinc to measure individual electric usage. Jumbo generators drove the DC system and were regulated individually by handwheels, which increased or decreased field resistance when turned to produce a relatively constant voltage. The Pearl Street station began operation on September 4, 1882, and Edison distributed electricity free for the remainder of the year to induce reluctant New Yorkers to tap into the system.

To Edison, Pearl Street was merely the first of what he assumed would be hundreds of stations located throughout America. He moved to capitalize before Pearl Street opened and helped create a system of corporations to control future developments. The Edison Electric Light Company, chartered in 1878, stood at the apex. It owned patent rights to Edison's electric lighting inventions and licensed subsidiary companies. The Edison Electric Illuminating Company of New York (1880) was its first subsidiary. The New York firm built the Pearl Street station, sold its power, and was to serve as the model for expansion in large municipalities. By 1888, Edison Electric Illuminating Companies had been established in Chicago, Brooklyn, Detroit, Boston, New Orleans, St. Paul, and Philadelphia. The Edison Company for Isolated Lighting (1881) installed incandescent lighting equipment in residences and businesses not served by central stations, and built central stations in cities with populations less than 10,000. Other Edison companies held exclusive licenses from the parent company to manufacture components: the Edison Lamp Works (1880) made incandescent lamps; the Edison Machine Works (1881) built Jumbo dynamos; the Edison Electric Tube Company (1881) constructed underground conductors; Bergmann and Company (1881) manufactured a wide range of electric lighting accessories.

Edison's strategy positioned his companies to dominate incandescent lighting, and he solidified their standing in 1883 by introducing the three-wire feeder–main distribution system. It reduced copper wire costs by nearly an additional 62 percent. In this system a series circuit joined two Jumbo dynamos, doubling their effective transmission voltage; feeders ran from each side of this circuit (they used only one quarter as much copper as two-wire feeders) to mains. A third or "balance" wire ran from the connection between dynamos to the connection between mains, which served to complete the transmission circuit and to allow lamps to operate independently.

No company could seriously challenge the Edison organization's preeminence in DC incandescent lighting; Edison's system was dependable and reasonably priced, and his companies held the important DC patents. But the Edison consortium proved unprepared to deflect competition from another quarter. George Westinghouse's Westinghouse Lighting Company devised in 1886 an alternating current incandescent lighting system. By 1892, more than 1,000 AC central stations were in operation.

THE
ALTERNATING SYSTEM.

Incandescent Electric Lighting from Central Stations made Universal, Economical, and Profitable, irrespective of distance.

The Westinghouse Electric Co.,
PITTSBURGH, PA.
Eastern Office, 17 CORTLANDT STREET, NEW YORK.

Westinghouse took great pride in and publicized the virtues of alternating current.

Like Edison, Westinghouse was self-educated. When he turned to electric lighting in 1884, Westinghouse's reputation as an inventor almost rivaled that of the Wizard of Menlo Park. He had designed a mechanism to return derailed railroad cars to tracks, his famous air brake, and several automatic switching and signaling devices. He first investigated DC lighting systems, and even began to manufacture a system by William Stanley that was inferior to Edison's, but soon focused on the results of a London lighting demonstration. Like several inventors before them, Lucien Gaulard and John Gibbs had attempted to overcome the arc light short-circuiting problem through electromagnetic induction; they generated electricity through primary induction coils joined in series, and established separate secondary coil circuits for each arc lamp. But unlike their predecessors, Gaulard and Gibbs employed high-voltage AC in their primary circuit to reduce long-distance transmission costs (it required thinner wire), and used their secondary coils as step-down transformers to lower voltage for lighting. Westinghouse and

Stanley bought Gaulard-Gibbs transformers in 1885, improved efficiency by replacing iron wire in the core with iron plate, and copper plates in the coils with copper wire, and placed the revamped transformers in parallel. They then tested the system while Stanley built well-regulated alternators for power generation. In fall 1886, the company opened its first commercial central station in Buffalo.

AC incandescent lighting's chief advantage over its DC counterpart was the decreased cost of electric transmission over distance. Difficulty insulating wires and the public's fear of the system's high voltages were its disadvantages. Heavily populated areas magnified AC's disadvantages and minimized its advantages. The situation was reversed in thinly settled regions. In that sense, Westinghouse and Edison tapped different markets. Westinghouse did not lack competition, however. As early as 1884, Thomson-Houston sought to enter the DC incandescent-lighting business, but found itself frustrated by Edison's ownership of patents and his system's excellence. It therefore moved decisively into AC lighting in 1887, and by 1892 served almost as many customers as Westinghouse.

The Growth of Electric Power

DC's headstart over AC in electric lighting carried over to other uses of electric power. The central stations of the early 1880s, as well as the improved dynamos of the preceding decade, inaugurated an era of cheap electricity, and DC electric-light distributors engaged in sustained efforts to encourage electric motor development and industrial usage. Three important factors entered into their decision. First, and perhaps most basic, DC distributors looked to industrial motor applications as a means to increase electric consumption and profits. Second, distributors had gained valuable motor design experience from dynamo construction—electric motors were essentially dynamos running in reverse—and hoped to market a new product. Third, they recognized that motor usage would enable them to balance loads somewhat—light by night, industrial power by day—and produce electricity at reduced costs. Edison again led the way. As early as 1880, he employed electric motors to drive some of his lamp factories. By 1887, American industry used nearly ten thousand DC motors, generally for pumping, hoisting, or hauling. Most motors generated less than one horsepower, which made them completely unsuitable for turning ponderous lineshafting. Manufacturers either ran them on current supplied by central stations or operated isolated stations.

Horse-drawn street railways dominated mass intraurban transit until the late 1880s.

The Development of Electric Traction These industrial motors were virtually insignificant in terms of both future developments and the first popular use of electric power—electric traction. In the later nineteenth century, proponents of electric traction capitalized on the increased population and density of cities, and the division of urban space into distinct commercial, industrial, and residential neighborhoods to push their new product.

To be sure, they were not the first to profit from a desire to flee urban congestion. Steam commuter railroads, omnibus lines, and horse-drawn street railways preceded electric traction. Each of these systems had its shortcomings, however. Omnibuses were slow and required a dozen or more horses, each of which had a working life of only four or five years. Horse droppings produced constant complaints, and the threat of epizootic diseases engendered continual fears. Locomotives threw off clouds of soot and sparks, and seemed likely to cause fires. Street railways suffered from many of the same problems of omnibuses. While they used only a half-dozen horses and were speedier, the constant pounding of hooves near rails led to depressions disruptive to the wooden-wheeled commercial trade. Despite these drawbacks, street railways had become the predominant form of intraurban mass transportation. By 1880, more than 100,000 horses pulled nearly 19,000 streetcars on about 3,000 miles of track.

American entrepreneurs had attempted to market alternatives in the 1870s. They devised compressed-steam and compressed-air locomotives. They used stationary steam engines and either inclined planes or endless chains to pull street-cars from place to place. (The cable cars of San Francisco are survivors from this period.) But after 1880, several Americans investigated streetcars powered by electric motors driven by electricity supplied by central stations. Edison was among the first. In conjunction with S. D. Field, Cyrus Field's nephew, Edison developed a three-railed elevated railway. Wire brushes picked up current from the "hot" center rail and carried it to a standard Edison generator, which, hooked up in reverse, served as the motor. The two men demonstrated their contraption at the Chicago Railway Exposition in 1883, but shortly after dissolved their partnership. Edison then abandoned electric traction.

Several technical problems plagued these pioneers of electric traction. Streetcars started and stopped often, changed speeds frequently, and operated with variable loads. These conditions strained the DC motors then available. Stationary copper brushes in contact with rotating commutators wore down quickly; arcing, sparking, and fires resulted. Inventors also needed to devise effective means to convey current to motors, as well as methods to transmit power from motor to axle. The significance of these problems often was not obvious until systems were in operation. Moreover, improved design in one sphere sometimes exacerbated another situation. For example, motor placement under cars simplified power transmission, but complicated brush adjustment or replacement, as well as commutator removal for machining.

Marginal commercial enterprises marked electric traction's first generation as inventors encountered the aforementioned difficulties, which at the time of the initial construction of lines had often been unforeseen, and labored to resolve them. They placed conductors in underground wooden conduits, which were slotted to permit plow-like devices to gather current; electrified tracks and used

*New York was among the first cities in the world to have
an elevated railway.*

the car's wheels to bring electricity to motors; and strung conductors near rails and employed brass pulleys beneath cars to pick up
electricity. They also experimented with one or more overhead
conductors and used poles to tap current as well as overrunning
wheeled travelers. They situated motors on platforms in front or
back of cars or located them under car bodies and attached them
to cars or cars and axles. Various chain drives and double-gearing
systems transmitted power to axles.

These early systems each had flaws. It remained for Frank J.
Sprague to develop a line incorporating their best features and adding new ones to make a truly practical line. Sprague came to electric traction already experienced in DC motor construction. He had
investigated motor design as he worked installing Edison's first three-
wire central station and left the company the next year to devote
himself full-time to industrial motors. During the mid-1880s he
became perhaps the foremost American supplier of large motors;
industry used more than 250 Sprague motors in 1887, some as large
as fifteen horsepower. Two years earlier, Sprague had recognized
electric traction's large market potential for his motors, and he began to experiment on a short section of elevated railway in New
York City. Although he failed to interest the city in electrifying
the way, Sprague gained valuable experience and important financial contacts. His acquaintances soon secured a contract to operate

*Americans quickly adopted electric traction. By 1893, more
than 60 percent of the nation's street railway mileage had
been electrified.*

a streetcar line in Richmond, Virginia, and in 1887 they hired Sprague
to design and build it. He quickly decided that an electrified street-
level line with single overhead conductor would be most economical. Sprague's motors were joined to conductors through univer-
sally swiveling trolleys running under the conductor and kept in
contact by strong springs. He decreased commutation wear by sub-
stituting carbon brushes for metallic ones (Sprague's commutation
system required only monthly inspections rather than the normal
daily care), and attached the motors beneath his cars. He connected
motors to car and axle to ensure continuous gear engagement on
rough ways.

Sprague's Richmond success got him several additional contracts
and a slew of competitors. Thomson-Houston entered the electric
traction business in 1888 and established lines very similar to
Sprague's operation. Westinghouse, too, was impressed. That organization had tried since 1886 to develop AC traction systems but

failed to devise motors suitable for those enterprises' peculiar demands. It turned to DC traction in 1889 and employed several of Sprague's engineers to become competitive. The Edison interests joined the bandwagon by purchasing Sprague's company in 1889.

Electric traction proved immensely popular. In 1888, 130 electric streetcars transported American urbanites; four years later, more than 8,000 provided similar service. In 1890, 16 percent of American street-railway mileage was electrified. That percentage nearly quadrupled by 1893. In 1903, it reached 98 percent.

Not everyone championed electric traction as it existed in the early 1890s, of course. Trolley wires and poles were unsightly, and rails disturbed street traffic, although elevated electric railways and underground subways were two possible solutions. Nor did entrepreneurs abandon their quest to improve systems; for example, Westinghouse continued to experiment with AC motors and in 1904 opened the first AC-powered street railway.

The Ascendancy of AC Current   Westinghouse's AC motor experiments began in 1888, and owed at least as much to the company's desire to meet DC power's general challenge as it did to a desire to develop electric traction. Lack of a comparable AC motor gave DC companies a competitive edge, which sometimes proved decisive when a city selected its electrical supplier; DC seemingly had more applications, including industrial use. But DC's advantage was more illusory than real. DC motors never gained universal adoption because DC electricity's high transmission costs made steam power cheaper for most tasks. Edison's staff and financiers understood DC's limitations; Edison himself refused to acknowledge the drawbacks, but when his interest and organizational influence waned in the late 1880s, his colleagues started to explore AC's potential. There they encountered difficulties not unlike those faced by competitors of Edison's DC products in the early 1880s; Westinghouse and Thomson-Houston controlled the relevant patents and employed the top investigators. The Edison combine could withdraw, develop its own AC operation from scratch, or merge with either of the major AC firms. It opted for the last, and in 1892 joined with Thomson-Houston to form General Electric.

Westinghouse's quest for an AC motor had led him to Nikola Tesla, a Serbian emigree who announced in 1888 that he had constructed just such a device. Three wires rather than the normal two supplied power to Tesla's motor; the third wire carried a voltage out of phase with the other voltage. The two distinct phases created a rotating magnetic field in the motor's stator* windings, which eliminated commutators and brushes. Westinghouse purchased Tesla's patents in July 1888 and brought him to Pittsburgh to perfect the engine.

Westinghouse's commitment to AC power extended to more than Tesla's motor. The adoption of the three-wire motor would require designing an entirely new AC system, a polyphase or multiple-phase system. So Westinghouse's staff continued to look for an effective AC motor to adapt to the power generation and transmission machinery it already had in place, but it also examined the polyphase system's possible parameters. It settled on a two-phase system, which could be treated as two independent AC circuits for lighting, at a frequency of thirty cycles per second—low frequencies reduced impedance† and further lowered long-distance transmission costs. The company abandoned its single-phase motor search by 1892, and unveiled its polyphase system the next year. General Electric (GE) did not lag far behind. In 1894, it sold its first polyphase system. Unlike Westinghouse, the GE system had three phases, one at a higher voltage to circumvent the Tesla patents.

The competition between GE and Westinghouse persisted into the twentieth century, but on terms different than previous struggles between electrical suppliers. Earlier customers had been forced to choose the type of electrical system (AC or DC) to employ. By 1900 various devices freed them from that decision: rotary converters could change between AC and DC current; phase converters could change AC phases; motor–generator couplers could run on one type of electricity while supplying another. Customers could select devices best suited to their particular needs; the various electrical systems themselves had been systematized, and customers could pick from a wide range of components. Employing the appropriate converters, couplers, or apparatuses, a single central station could generate electricity for electric traction, three-phase AC motors, arc lights, and anything else—an event first demonstrated by Westinghouse at Chicago's 1893 Columbian Exposition. This new situation contributed to the 1896 decision by Westinghouse and GE to pool patents for fifteen years; competition now rested on excellence in manufacturing, not ingenuity. Patent-pooling provided GE access to Tesla's patents, and the company re-

---

* *stator:* a stationary, wound coil.

† *impedance:* the apparent resistance in a circuit to the flow of alternating current.

*Electric generating plants relied on water turbines or coal-
or oil-powered steam engines to produce electricity.*

placed unbalanced AC with balanced, three-phase AC, which was
vastly superior for induction motor operation and quickly became
the electric industry's standard. Similarly, the sixty-cycle frequency introduced by Westinghouse in the early 1890s as a com-
promise between lighting and power requirements served by 1920
as the standard electric frequency.

Electrical suppliers' drive to effect economies of scale joined great
areas and large populations, and the resulting reductions in cost
both reflected and encouraged increased electric consumption.
Westinghouse's Niagara Falls project, completed in 1896, stood as
the first practical demonstration of large-scale power generation.
Steam engines gave way to water and steam turbogenerators—turbines—and transmission voltages reached 40,000 volts in 1897;
60,000 volts in 1900; and about 150,000 volts by 1910. Larger, more

economical stations utilized what had been inaccessible water power to supply low-cost electricity for hundreds of miles.

THE DEVELOPMENT OF TELEPHONES

As with electric lighting, the development of American telephony graphically demonstrates both the centrality of systems in late—nineteenth-century thought and practice, and the debt that many late-century inventors and entrepreneurs owed telegraphy. Alexander Graham Bell himself proposed telephonic central stations (even before Brush erected his San Francisco operation), and he had labored to redesign telegraphic devices, while much of the telephone's early funding came from men experienced with telegraphy. But unlike electric lighting, the telephone system provided direct competition to the telegraph and ultimately led to its demise.

Bell certainly did not set out to destroy telegraphy. He was passionately involved with speech. Professor of elocution at Boston University, the Scottish-born Bell used "Visible Speech"—mouth and tongue drawings of properly articulated sounds—to teach deaf children to talk. Tutoring provided Bell a meager income, which he hoped to supplement through inventing. With no electrical training, Bell tackled the thorny problem of sending many signals over a single telegraph wire. His familiarity with speech and sound waves led him to contemplate in 1872 the possibility of a harmonic telegraph, one that would use a series of tuning forks to send and respond to precisely oscillated currents. Work went slowly, and in 1874 he paused to develop a phonoautograph, a device to convert sounds into written markings that employed an actual human ear. Although the phonoautograph had few practical applications, the vibration of the tiny eardrum and bones intrigued Bell, and he wondered if electric impulses could vibrate a membrane to reconstitute speech. That speculation was the germ of Bell's telephone labors, and he explored it as he continued his harmonic telegraph studies.

Bell's breakthrough came in June 1875. Working on his harmonic telegraph with Thomas A. Watson, a mechanic, in Charles Williams's electrical supply shop, Bell distinctly heard a spring twang on his circuit; his telegraph's make–break circuit* had apparently

---

* *make–break circuit:* an electric circuit with only two options: either electricity is passed (make) or it is not (break). Its lack of variability renders it unsuitable for transmitting such complex vibrations as are found in human speech.

fused, and a magnetized steel spring vibrated over a magnet, which varied the intensity of electricity produced and sent sound over the wire. During the next several months, Watson and Bell attempted to refine the elements of this discovery. They constructed transmitters of gold-beater's skins,* with sheet iron glued to the centers, before settling on iron-plate diaphragms. Sound waves vibrated the diaphragms, established electromagnetic currents in copper induction coils located at ends of permanent magnets, and transmitted current through wires. These electromagnetic impulses were received by other diaphragms, which vibrated to emit sound waves.

Bell submitted a patent application on February 14, 1876. His application also included a battery-powered telephonic device. A diaphragm in a battery circuit was connected to a wire resting in acidified water. Sound waves vibrated the diaphragm, which varied wire depth and resistance, producing a variable current. The receiver was similar to Bell's electromagnetic telephone. Bell's last-second addition of the battery telephone—he scribbled its particulars in the margin—proved propitious. That afternoon Elisha Gray, who also had been investigating a harmonic telegraph, attempted to file a *caveat* with the patent office announcing his intention to construct a similar instrument.

Bell's patent was approved in March. He gave public lectures trumpeting his invention and concentrated on improving the electromagnetic version. Only 230 telephones were in use in July 1877; competition from telegraphy and the electromagnetic telephone's poor voice-quality delayed initial acceptance. The situation changed when some Western Union customers replaced telegraphs with telephones. That giant company immediately recognized telephones as a threat to its business preeminence, and it challenged the Bell patent, maintaining that Gray had priority. It also employed Edison to improve the battery telephone's transmitter (in Edison's version vibrations of the diaphragm changed pressure between fine carbon granules, which varied circuit resistance), and converted its subsidiary, Western Electric, hitherto the nation's largest electrical supplier, into a telephone manufacturer. Bell countered by filing a patent-infringement suit, purchasing Francis Blake's battery-transmitter patent (a voice-activated diaphragm varied pressure on a connection between a hard carbon block mounted on a strong spring and a platinum bead mounted on a weak spring, which changed circuit resistance), and abandoning the inadequate electromagnetic telephone.

* *Gold-beater's skin:* the membrane of the large intestine of an ox.

Contemporaries feared that a maze of telephone wires would plague cities. Central exchanges reduced that threat.

Western Union's interest in telephony ironically guaranteed the device's future. The number of phones in service increased markedly as the dispute moved through the courts. By the time the two parties reached accord in late 1879, and Western Union withdrew from the telephone business and sold its phones to Bell, the telegraph company's telephones numbered nearly 56,000. Six hundred other groups contested the Bell patent, but it stood until the early 1890s.

Western Union's agreement to exit telephony was shortsighted. At least a year earlier Bell had conceived of telephones in a new way—as linked to central exchange systems. "It is possible to connect every man's house, office or factory with a central station, so as to give him direct communication with his neighbors," he wrote in 1878.[2] Bell envisioned for each locality a system of branch wires running from dwellings and shops attached to subordinate cables buried underground and paralleling thoroughfares. These cables would be connected to an underground main cable, leading to a central office. Each call would go to the central office and be channelled to the appropriate party. All telephone lines were above ground when Bell made his statements, and only a handful of rudimentary exchanges existed. Most telephone lines ran from place to place, and contemporaries feared that expansion would result in a Byzantine maze of wires. Invention of the switchbox and, a bit later, adaptation of telegraph switchboards to telephony enabled those connected to a particular exchange to converse with each other, but those mechanisms accommodated few subscribers. Indeed, switchboard development was a crucial gain for Bell Telephone in its acquisition of Western Electric in the Western Union settlement. The addition of Western Electric gave Bell control of important switchboard patents, as well as the capacity to produce an almost inexhaustible supply of phones to lease.

Western Electric's comparatively sophisticated switchboards joined more callers and enlarged exchanges, but demand quickly exceeded their capabilities. Bell purchased individual inventors' switchboard patents, and Bell exchanges (local companies ceded their stock and often control to Bell Telephone for rights to lease Bell equipment) experimented with modified switchboard designs. Peg switchboards and jackknife switches fueled further exchange expansion. The introduction of multiple switchboards in the mid-1880s allowed exchanges to handle ten thousand phones and paved the way for a mature interexchange telephony.

Bell Telephone began to implement its founder's underground cable ideas in cities during the early 1880s. Already a mass of poles and wires, New York City was the first site of extensive underground cable use. Conduits were made of iron pipe, asphalt, concrete, or creosoted wood. Wires were wrapped in cotton, twisted into cables—each cable contained about one hundred wires—and soaked in oil to keep out moisture. Dry-core cables started to supplant these early carriers around 1890. Each wire was lightly packed in paper; air provided insulation. Cables were composed of as many as twelve hundred twisted wires, sealed in lead, and placed in tile conduits. By 1896, Bell had laid 200,000 miles of dry-core cable. It had thirty times as much in 1910.

Replacement in the mid-1890s of individual telephone batteries with large common batteries situated in central exchanges completed the transformation from individual phones to central stations. There female operators—the first appeared in the late 1880s and were known as call girls—responded to signals transmitted in underground cables that lit their switchboard lights, and then manually connected the parties. The powerful common battery both facilitated voice transmission and activated individual telephone bells.

Central exchanges tied telephones into a system, but each system was limited to the locality. Telephony's initial reliance on

From the start, Bell Telephone selected women as operators.

grounded, single iron wire circuits made long-distance transmission almost impossible. These proved extremely sensitive to telluric disturbances (disturbances deep in the earth), weather, and electromagnetic radiation such as that generated by central power stations. Interference affected even relatively short calls. The substitution of copper for iron wire, addition of a second circuit-completing wire, and insulation of wires with enamel reduced static markedly. Telephone connections were established between Boston and New York in 1884, and Washington, D.C., and Chicago in 1893. The latter represented the maximum effective long-distance telephone communication prior to 1900.

Long-distance service was vital to Bell Telephone's fortunes after expiration of Bell's patents, when it faced competition from numerous independent phone companies. By 1900, 6,000 independent phone companies had placed 600,000 phones in operation, while the Bell company had situated only 800,000. Bell lowered prices and improved service to counter independents, but its long-distance capabilities proved the most attractive to consumers. Competitors generally lacked the capital, rights of way, and organization necessary to establish competing long-distance lines. As a result of Bell's virtual long-distance monopoly, businesses and households often bought service from both Bell and an independent; independents gained an impressive share of local traffic, but only Bell subscribers received long-distance service. Its rivals repeatedly asked Bell for access to its lines and lobbied state legislatures for that right, but with little success. Their prospects dimmed further in the early twentieth century as Bell introduced loading coils. Essentially induction coils placed along transmission lines at distances approximating the signal's average wavelength, loading coils helped reduce signal attentuation and expanded possible long-distance transmission to about fifteen hundred miles.

Bell capitalized on its long-distance monopoly in the new century's first decade to reduce the number of items it manufactured and to standardize those it continued to produce. As the sole competent long-distance company, Bell was indispensable. New product models would not secure it a larger market share, so Bell sought instead to generate economy by minimizing supplies and parts needed for repair. It aggressively used this advantage to buy out independents, maintaining that with respect to telephony the nation deserved "one policy, one system, universal service."[3] These sentiments led Bell to favor government telephone regulation after 1906, which would preserve its entrenched position and guarantee its investors a dependable, substantial return; Bell would become America's telephone company.

Telephone and Radio

In 1907, Lee De Forest formed a company to contest Bell Telephone's long-distance supremacy. Unlike earlier competitors, De Forest planned to send messages through the air, not wires; radio was intended to rival the telephone.

De Forest's initiative built on the preceding quarter-century's labors. Heinrich Hertz's 1888 demonstration that electromagnetic radiation produced waves of specific frequencies that resonated with similarly tuned apparatus was soon followed by Oliver Lodge's and Guglielmo Marconi's longer transmissions. In 1896, Marconi sent Morse code several hundred yards. In 1900, Reginald Fessenden had transmitted speech nearly a mile. Marconi broadcast across the Atlantic Ocean a year later. England's John Ambrose Fleming opened the electronic era in 1904. He produced a diode (a two-element vacuum tube), which received and rectified radio signals, an application of Edison's earlier observation that a weak current passed in one direction through a vacuum between a light-bulb filament and metal plate. Two years later, De Forest placed a zigzag grid between the two elements, set a small voltage on the grid, and found that it created a much greater plate voltage; his three-element vacuum tube, the triode, amplified radio signals. That discovery, and Fessenden's invention of a dependable transmitter (a high-frequency alternator) the same year, gave De Forest impetus to challenge Bell.

Bell Telephone initially discounted the radio threat and concentrated on developing coast-to-coast service. It devised several electromagnetic repeaters, but they proved too slow and distorted the voice. Its lack of a workable continental system, coupled with the continuing efforts of companies such as De Forest's, led Bell after 1909 to launch a massive search for a suitable repeater. It turned to triodes in 1912, but found De Forest's tubes inadequate for power loads necessary for telephone transmission, and too short-lived. Rather than reject triodes, Bell personnel sought to improve them. Fortifying the vacuum was perhaps their most important contribution. It yielded an effective repeater, and on January 25, 1915, Bell inaugurated its first transcontinental line.

Bell's telephonic interest in triodes also introduced the company to radio. Its 1913 acquisition of De Forest's triode patents seemed to give Bell a commanding position in the radio market, but both General Electric and British Marconi contested the patent. In addition, GE had helped develop Fessenden's transmitter, and had secured control of the related patents. World War I brought another combatant: the federal government removed patent restrictions on radio and drafted Westinghouse along with GE and Bell to establish wartime communications. British Marconi withdrew from the American market after the war. It sold to GE control of its subsidiary, which was renamed the Radio Corporation of America (RCA), and GE authorized RCA to sell and distribute GE radio products.

These disputed claims stymied the various parties. No single organization had clear title to the critical radio-transmitter and receiver patents. Instead of waiting for the courts to adjudicate the matter, the rivals entered into a patent cross-licensing agreement in 1920 and 1921. Peace did not reign long; the creation of radio stations, with their potentially enormous market for radio receivers, produced discord. Although Westinghouse established the first station in 1920, Bell's entrance into the fray in 1922 made the situation untenable. Bell conceived of a nationwide network of Bell-controlled stations dominating American radio, and it refused to allow competitors to use long-distance lines for their feeds. RCA,

GE, and Westinghouse objected and took the question to arbitration. The arbitrator's decision became moot in 1925 when the U.S. Court of Appeals ruled that GE was entitled to patent protection for high-vacuum electron-tube use; Bell lost its claim to any fundamental radio patent.

Bell left the radio business after its legal defeat. It stopped manufacturing receivers, sold RCA its seventeen radio stations (Westinghouse had purchased a minority interest in RCA earlier), and agreed to refrain from commercial radio broadcasting. Bell granted its former competitors telephone-line access for network radio transmission in return for monopolies on ship-to-shore and transoceanic radio telephony.

SYSTEMATIZING THE TECHNOLOGISTS

America's reliance on technically proficient persons is quite old, as earlier chapters of this book indicate. Individuals demonstrating special skills or technical expertise have always abounded in America, and have performed vital functions. The nation's military academies, especially West Point, long had trained men in the building arts, and masters had transmitted their special knowledge to apprentices. By the mid–nineteenth century, broad technical experience had been recognized as apt preparation for business administration, and men engaged in technical occupations had occasionally formed organizations for camaraderie and to discuss objects of mutual concern.

Engineering Organizations

Not until about 1870, however, was the nation's first exclusive engineering organization, the American Society of Civil Engineers (ASCE), established to organize America's engineers. Because of the wide variety in American engineering, however, the ASCE soon proved unwieldy. It was followed by the creation of more specialized organizations, such as the American Institute of Mining Engineers (1871), American Society of Mechanical Engineers (1880), American Institute of Electrical Engineers (1884), American Institute of Chemical Engineers (1908), and Institute of Radio Engineers (1912). Each society catered to a particular constituency within the larger engineering profession.

Formation of the ASCE and other engineering organizations signalled the emergence of self-consciousness in America's engineers.

This emergence stood as part of the more general late–nineteenth-century division of American society in which expertise replaced the earlier concept of character as the *sine qua non* for evaluating competence. Engineers began to identify themselves as a group fundamentally different from other people. They considered themselves technically advanced planners, designers, facilitators, and administrators, not merely producers or artisans. They argued that special knowledge and skill—expertise—characterized each engineering division and made it distinctive, and mastery of that expertise entitled those engineers to special authority. Each of these societies would systematically press its case to demarcate a superior social role for its constituents.

But while self-identification was a critical step for engineers, gaining public sanction for their authority was paramount; without it they could never have jurisdiction in areas where they claimed entitlement. These societies sought public sanction from their inception in the late nineteenth century, but they did not gain it until the twentieth. Political participation was one avenue they explored, but the societies preferred to attack the matter internally; they gradually added rules and mechanisms to enforce them to guarantee that their membership would have the expertise congruent with the social responsibility they proclaimed, and to guarantee that future members possessed such expertise as well. The societies had various membership categories, each with its own requirements. They held regular meetings and published periodicals subject to peer review. They established standards of practice and codes of ethics to govern professional and private behavior. All these systematically worked to standardize each division's membership. The societies also sponsored or recommended curricula for the engineering schools, thereby systematizing entrance into the profession. These varied devices helped convince an often skeptical public that engineers met their self-announced criteria, and therefore deserved the authority that they sought.

The ASCE   But pathways to professional legitimacy were unclear in the 1870s and early 1880s, because engineers had few precedents to follow. Self-definition was the motivation behind building professional institutions, but that placed a premium on harmony, often at the expense of substance. Molding individuals into an effective national entity took time.

As the nation's first engineering society, the ASCE pioneered in working through these issues. Founded in New York in 1867, the group drew its initial constituency almost exclusively from the New York metropolitan area; of its 106 members in 1870, seventy-three resided within fifty miles of Manhattan. That equation had changed by 1880; the society's membership had grown sixfold, and only 114 members lived near the nation's greatest metropolis. Part of the society's success in attracting distant engineers stemmed from its activities; it formed numerous committees to investigate national issues about standards of practice and standarized methods. But the nascent national society's efforts rarely resulted in nationwide changes in engineering practice. In almost all instances, its committees either failed to articulate policy, or the society as a whole refused its endorsement.

The society lacked clout, not conviction. Public vituperative displays would have hurt the society's public image as well as its professional credibility. Dropping potentially divisive questions stopped the society from alienating the public and segments of its still-fragile constituency; concern for organizational homogeneity—and survival—was greater than concern for erecting standards. The society carried this sentiment to the extreme in 1870 by declaring that in every case "the society is not responsible for the opinions expressed by its members." Examination of technical and professional problems, not their adjudication, was its own reward as the ASCE demonstrated its awareness of the latest engineering concerns and provided a forum for discussion. That committee formation to investigate professional problems and concerns marked the end rather than the beginning of ASCE deliberations was less important than the fact that American civil engineers had a single national entity in which to voice their propositions.

The ASCE's committee on college engineering curricula exemplified this thrust. Established in 1874 to chart a standard set of college engineering courses, the committee issued its report later that year and was then excused; it offered no recommendations. It only congratulated colleges for having "done more to elevate the standard of the engineering profession than any other single agency." At the same time, the committee noted that an engineer's real worth was judged by his working-world performance. College engineering courses provided "a solid basis upon which to build the professional structure," but only "long and varied experience" in conjunction "with a body of practical engineers"—the ASCE—made a master.[4]

The ASCE's deliberations on engineering disasters reflected a similar penchant for peace rather than potential controversy. Its committee on disasters was to investigate catastrophes "not only to determine who were at fault, but also to assert the existence of a society as a body [whose purposes are] to exhibit the true causes of the failure of public works, and the engineering ability of those engaged in their construction."[5] But the committee and the society repeatedly neglected to apportion blame. The society's response to the 1873 Mill River Dam collapse proved typical; its committee met but did not deliver a report. Committee members produced their own conflicting documents and the society refused to discuss or act on them.

The disaster and college committees dealt with emotionally charged issues that threatened to tear the society apart. Gauging river and stream flow (1875), establishing a single masonry nomenclature (1876), deciding the best rail manufacturing and testing methods (1876), creating standardized railroad signals (1875), and urging federal adoption of the metric system (1874 and 1877) seemed less menacing, yet the ASCE did not advocate a clear position on any of these questions. After considering each subject, the committees either offered no report or articulated a statement that the society summarily dismissed. Any hint of controversy during the 1870s resulted in the society's tabling a matter.

The ASCE proved more willing to embrace conflict in the 1880s. A desire for harmony, crucial during the previous decade, dissipated as the society gained members. The ASCE attempted to mark its turf and risked offending some; it had arrived. Two mid-1880s incidents signaled the organization's new pugnacity.

The ASCE officers' 1884 decision to require members to wear emblems—a bubble level—at meetings and in public produced an unprecedented ruckus. Roughly a hundred members complained that the symbol did not accurately reflect the engineering profession. Engineers designed and managed projects. Assistants surveyed and used levels; the badge therefore denigrated engineers. The officers refused to budge despite the fracas, however, and the emblem remained the organization's official symbol until 1894.

The ASCE's effort to absorb local engineering associations within its organizational structure also met with contentiousness. ASCE members heatedly discussed the question from 1884, and a special committee drafted a new bylaw in 1889. Local engineering societies would retain the right to own property and make their own laws, but the ASCE would require members of locals to be ASCE members, and the constituent organizations to furnish the ASCE copies of presented papers. An extended, often nasty debate accompanied the proposal's introduction, and the ASCE membership refused its assent.

The ASCE flourished despite conflict. It had 878 members in 1885 and more than a thousand in 1890. The organization confidently had moved beyond cementing a constituency to exerting its influence to shape the profession. It continued that tack during the next several decades. The society adopted a code of ethics (1914), created technical divisions to establish engineering standards (1912), started student ASCE chapters at engineering colleges (1919), and advocated the licensing of engineers by boards of examination (1911). In 1916, the ASCE had nearly eight thousand members.

Systematizing Technical Education

As ASCE and other engineering societies gained firmer footing, they increasingly turned their attention to the next generation of engineering professionals. Shaping the collective future—defining and standardizing the nature of future engineers, and delineating their rights and responsibilities—was of great moment because it would enhance and cement engineering's public standing. Through their efforts, a formal collegiate engineering education would become *de rigueur* by 1910, and nearly all engineers accepting their first position after that date would hold engineering degrees. The nation's 126 engineering-degree-granting institutions would award more than 17,000 engineering degrees between 1911 and 1915.

Though engineering societies concentrated on the training of engineering college students from the 1880s, several engineers had examined engineering education earlier. What these purveyors of expertise uncovered no doubt shocked and disappointed them. Only about one in twenty practicing engineers in 1871 had received a collegiate engineering degree. Most had been trained on the job. In fact, only about eight hundred engineering degrees had been granted prior to 1870 at American military academies, land-grant colleges, polytechnic institutes, and scientific schools. Nor had these early engineers been prepared for collegiate training. Few high schools existed in 1870 and no established curriculum readied college students. Age was usually the most critical college admission criterion, with sixteen years of age often the minimum. These young men generally were so woefully unprepared that several institutions awarding engineering degrees felt compelled to create special preparatory departments to provide prospective engineers the rudiments necessary for collegiate studies.

The essence of creating a college-educated engineering profession was determining what the engineering colleges would teach. Questions of curricula were closely intertwined with the emergence of engineering self-consciousness, were matters of self-definition, and engendered substantial debate after 1870. Formation of the Mechanical Engineering Teachers Association (1892), Society for the Promotion of Engineering Education (1893), and the Joint Committee of Engineering Education of the National Engineering Societies (1907) manifested that concern. The last group invited the Carnegie Foundation for the Advancement of Teaching and the General Education Board to participate in its deliberations. Individual colleges also experimented and produced new initiatives. On one facet all agreed: the reliance on mathematical and physical science methods and determinations should form the backbone of engineering school curricula. These sciences made engineers distinctive and discriminated them from mere workers; professions required scientific bases. But although engineering schools offered students heavy doses of mathematical and physical sciences, and then moved into engineering theory and theoretical design, concerned engineers reached no similar consensus about the curriculum's other aspects. Practice, and its relationship to theory and engineering education, usually proved particularly troublesome.

Theory versus Practice   Before the mid-1870s, engineering educators held especially disparate views about the value of practice. One contingent maintained that practice had little or no place in higher education. Knowledge of science and theory marked professionals; practice was outside engineers' province or was to be learned on the job. For this group, laboratory apparatus to reinforce scientific or theoretical principles served as the extent of collegiate hands-on experience. Another camp stressed practice, and considered to be critical the creation of masterpieces, often with nods to efficiency and to its role in demonstrating production principles. Masterpieces were engineering benchmarks; their skillful manufacture enabled engineers to rationalize industrial processes.

Both sides modified their positions somewhat in the mid-1870s and after. Mechanical engineering education reflected those shifts. For example, in 1876 Massachusetts Institute of Technology's John D. Runkle adapted Russian-technical-school methods to American engineering education. He argued that machine manufacture could be analyzed into a number of typical operations and that those operations could be organized into groups, each of which depended on a distinctive machine tool. He required students to become expert in use of a tool and the group of operations employing that tool before permitting them to advance to the next tool. Runkle rejected the manufacture of finished products as vocational in nature, but affirmed that engineers needed competence in both theoretical design and practical manipulation to achieve their proper station.

Cornell University's Robert Thurston chose not to follow Runkle's lead. An early participant in ASCE and a founder of the American Society of Mechanical Engineers (ASME), he deemphasized practice at Cornell in the late 1880s and substituted specialized courses such as steam engineering design. Thurston accentuated movement into new areas, not mastery of established techniques, but he also recognized an engineer–industry nexus and fostered it by inviting industrialists to address students and faculty about manufacturing questions. These noted men provided the engineering curriculum's practical aspects. So too did frequent tours of Northeastern manufacturing facilities.

In 1906, University of Cincinnati's Herman Schneider went further. He divorced theory and practice completely by restricting college teaching to the former. But his cooperative system made practice a full engineering education partner; students attended classes half the school year and worked in industry during the remainder. Schneider claimed that his approach constituted "real practical work" because students' industrial training progressed from the least-skillful tasks to purely engineering activities; they gained a feel for the full spectrum of duties.

Schneider's cooperative education was the last major attempt before 1910 to harmonize theory and practice. Other issues also plagued engineering educators. They expressed dissatisfaction over the number of engineering-school dropouts (nearly 60 percent), inflexible curricula, and lack of standardized engineering curricula, both across the nation and within individual schools' engineering divisions. This ferment led the Joint Committee of Engineering Education of the National Engineering Societies to ask the Carnegie Foundation to investigate American engineering education; it selected Charles R. Mann, a University of Chicago physicist, to compile a report. Issued in 1918, the Mann report attributed engineering-school deficiencies to a "lack of coordination," especially an overemphasis on specialization. Mann urged colleges to limit enrollment by drafting a standardized admission test to determine engineering aptitude, and suggested that schools require entering students to take standardized general-engineering orientation courses. Under the present system, he complained, students usually took only science and drafting classes during their first two years. Their initial taste of engineering came in third-year specialized courses, a situation that undermined professional unity. Nor was Mann satisfied with the nature of specialized courses. He argued for the case-study method, sometimes known as the laboratory method of teaching. He wanted engineering faculty to encourage students to analyze machines and structures for their fundamental operating and structural principles; he insisted that students deduce from applications the principles themselves.

The Case of Chemical Engineers   Mann's report found substantial support in most engineering colleges, which implemented many of his recommendations. Few chemical-engineering educators considered the matter resolved, however. Chemical engineering faced a challenge unlike the other major engineering divisions. It had difficulty demonstrating its distinctiveness because both chemistry and mechanical engineering claimed it as a subdivision. Chemical engineers had not formed a national society (the American Institute of Chemical Engineers, or AICE) until 1908, and in that year only a dozen colleges had chemical-engineering programs, generally within chemistry or mechanical-engineering departments. Nor did the situation improve rapidly. As late as 1920, the AICE had fewer than 350 members. But the relative youth of chemical engineering, its problems gaining credibility, and its society's miniscule membership also worked to the AICE's advantage; members did not have to reform an extant professional structure. They needed either to define themselves and then work to put that definition in place at engineering schools, or to survey college chemical-engineering programs and abstract their self-definition from that analysis.

Chemical engineers engaged in both approaches to self-definition. Chemical-engineering education figured prominently from the first AICE meeting, but early deliberations failed to produce formal plans. The association reversed its course in 1914 and agreed to undertake a comprehensive study of American chemical-engineering education, but the combination of a lack of resources, World War I, and the anticipated Mann report postponed action. During the war, Arthur D. Little articulated the concept of "unit operations" to describe chemical engineering's uniqueness, a definition that became the professional standard. To Little, unit operations constituted the building blocks of industrial chemical processes and included "pulverizing, dyeing, roasting, crystallizing, filtering, evaporation, electrolyzing, and so on."[6] Only a few were involved in any industrial process. Each operation could be studied independently. Industrial production required the selection of appropriate unit operations, proper sequencing, and coordination; chemical engineers designed rational facilities that made provisions for the industrial process's particular conditions—temperature, pressure, and the like—and for construction constraints imposed by the reactants' physical and chemical properties.

The AICE quickly recognized the unit-operations idea's importance to chemical engineering, and it rewarded Little with its education committee chairmanship. Unhappiness with the Mann report, which ignored unit operations and mentioned chemical engineering only in passing, led Little's committee to begin the long-delayed chemical-engineering education survey. The committee completed its work in 1922, found great differences among American chemical-engineering college programs, and recommended that these programs adopt a standard nationwide curriculum organized around unit operations. This curriculum would be hierarchically arranged, omit other engineering division courses, and accentuate basic sciences as well as unit operations. The AICE accepted the recommendations, established a committee to persuade colleges to reorient their programs in accordance with the report, and promised to publish a list in 1925 of colleges meeting the organization's guidelines. Fourteen programs gained institute sanction following on-site visits. Also in 1925, the AICE voted to give this accreditation committee permanent status.

Creation of a chemical-engineering accreditation body pressured other engineering divisions to move on the question. The formation of the Engineers' Council for Professional Development in 1932 was the result. Composed of representatives of the national civil, mining, electrical, and mechanical engineering societies, the Council was empowered to enforce standard nationwide curricula for each division. Operating in conjunction with the AICE committee, it reigned as America's engineering-college accrediting agency. Colleges would be compelled to produce "a high grade but also . . . the same grade of men." They would function "in reality [as] a factory turning out engineers."[7]

Systematizing Technical Research

As it did with the formation of college curricula and professional organizations, the engineering self-consciousness of the last third of the nineteenth century spawned questions about the relationship between engineers and research, and between engineering research and industry. To be sure, American technologists always had tried to improve established processes or uncover new ones, and had a long, profitable marriage with American industry. (For instance, local, national, and international exhibitions had awarded prizes and premiums for novel designs and elegant devices, while Philadelphia's Franklin Institute had examined the causes of steamboat boiler explosions in the 1830s.) But these efforts were primarily ad hoc problem-solving inquiries, which differed in emphasis and usually in method from late–nineteenth- and early–twentieth-century research. After about 1870, research was inextricably linked to a kind of expertise, one characterized by knowledge not only of a particular subject but, especially, of the techniques of systematic, rational inquiry. Investigations rationally conceived and vigorously and logically pursued seemed to produce unambiguous, indisputable, nonpartisan results; they yielded new knowledge.

These investigations were based on precision and verification, which encouraged design of physically discrete, adequately equipped, and hierarchically governed areas (laboratories) to eliminate variability and variables. Investigators in these rigorously controlled laboratory environments measured, weighed, and counted to test and formulate their hypotheses. A few laboratories appeared in mid-nineteenth-century America, but early facilities were primarily pedagogical devices that generated little research. They served to demonstrate established principles to students or to teach technical competency—not to manufacture new knowledge. Not until the late 1870s did a substantial number of American colleges and universities adopt the research-laboratory idea, and provide permanent institutional homes for laboratories. Johns Hopkins University was perhaps the most influential.

The heavily endowed Johns Hopkins University was established in the mid-1870s as an experiment in higher education. Freedom from past practice permitted its president, Daniel C. Gilman, to emphasize original research. The university paid its faculty salaries higher than comparable institutions to attract the most promising people, and reduced teaching responsibilities to allow research time. Its laboratories were "adapted for exact measurement of physical quantities rather than for . . . qualitative illustration." Laboratory apparatus was "for investigation," not "for amusing children."[8] Those on fellowships worked in the professors' laboratories, as did graduate students preparing for research careers. The university supported several research journals and created a press to publish research monographs. Johns Hopkins University's prominence stimulated others to implement the research-laboratory idea. By the 1880s, research laboratories could be found in universities as diverse as Wisconsin, Cornell, Stanford, Michigan, Harvard, and California.

Definition of research as a particular kind of expertise placed the engineering profession in a perplexing situation during the several decades after 1870. Clearly, the vast majority of practicing American engineers in the late nineteenth century had no research training (or even a collegiate education), nor would terms of their employment ever permit them to engage in research activities. Yet laboratory research was increasingly being seen as the wellspring of technological innovation; it seemed that technological change often was the consequence of laboratory-generated knowledge and that systematically pursued laboratory investigation, though not immediately directed at practical applications, would nonetheless frequently yield major practical applications. This dilemma rarely found explicit extensive articulation within engineering societies because its formal consideration would surely have created an unwelcome, potentially divisive conflagration. Identification of research as an integral part of the engineering profession would grant engineers undertaking research greater status—and economic advantages—than those shunning it, while if the profession disassociated itself from research it would suffer a considerable loss of public esteem.

Engineering leaders attempted to smooth over the anticipated schism by scarcely mentioning engineering research but by treating it *en passant* as a legitimate professional activity, distinct from engineering practice proper, and justifying regular apportionings of funding, time, and respect. Engineering-society publications and honors reflected that approach. So, too, did engineering colleges, which began after 1900 to define research as a vital professional function, to select faculty with demonstrated proficiencies in research techniques manifested by research degrees—Ph.D.s—and to prepare the most promising students for academic life. Division between engineering researchers and practitioners also held predictable consequences. Equating research with a passionate search for truth engendered concern that those not thus engaged were serfs of industrial interests, which fostered cleavage between engineers, and between the profession and industry. Identification of research as the cradle of innovation, coupled with the profession's failure to go repeatedly on record embracing research as an aspect of its own activity, reduced practical engineering's overall public status. It suggested that the professional groups that had claimed systematic laboratory investigation as their fundamental activity—the scientists—stood preeminent, and that practicing engineers simply applied what scientists had uncovered; engineers seemed mere technicians. Scientists preached, and engineers then practiced what was preached.

Industrial Research Laboratories   A dropoff in public acclaim and simmering tensions between business and the profession did not prevent industries from employing engineers, of course. The industry–technical nexus remained as it always had been, immediately directed at productivity. Examination and refinement of production stood as the technicians' mandate. Not until around the turn of the century, when engineering colleges began to treat research as an important academic facet, did industry establish a new use for corporate-employed engineers. They were to work in industrial research laboratories, a new type of institution but clearly derivative of academia. Industrial research laboratories were not responsible to production facilities. Companies generally located laboratories away from manufacturing sites to insulate them from businesses' most pressing demands; laboratory scientists and engineers were encouraged to investigate natural phenomena, not merely to improve production.

Profitability, not altruism, led industrialists to form corporate research laboratories. They were convinced that systematic investigation was a critical determinant of technological innovation, and sought to systematize and privatize research. Original research should inevitably translate into new products, techniques, or processes. Corporate ownership of research institutes would speed and shape development and restrict access to discoveries until they were patented and implemented. Only the economics of industrial research laboratories gave pause, because they demanded steep initial investment in equipment and personnel and promised slow returns.

The situation became much clearer after 1900. Many corporations, especially chemical and electrical concerns, had nearly three decades of experience with laboratory-based technology and had a firmer sense of its economic prospects. Indeed, as early as the 1870s, American industries had looked to the nation's colleges and universities to provide private research and researchers. Companies had consulted with professors who used university facilities, and had employed academics and their students. Edison had been especially active, while college men such as Thurston had nurtured the industry–academia connection. During the twentieth century's first decades, several large firms with sufficient capital to undertake the venture and withstand the temporary lack of earnings decided that corporate-owned laboratories made sound economic sense. Industrial research laboratories would ensure the retention of market shares or help secure larger ones. As Willis Whitney, director of GE's pioneering research laboratory put it, "Our research laboratory [is] a development of the idea that large industrial organizations have both an opportunity and a responsibility for their own life insurance. New discoveries can provide it."[9] GE's laboratory was quickly followed by Du Pont (1902), Goodyear (1908), AT&T (1911), General Motors (1911), Eastman Kodak (1912), and American Cyanamid (1912). By 1931, 1,600 companies reportedly established industrial research laboratories, employing nearly 33,000 people.

Industrial research laboratories were attempts by corporations to systematically manage the production of knowledge, but they stood as only one example of a much larger cultural phenomenon. Late–nineteenth- and early–twentieth-century Americans in all walks of life acted as if their world was composed of systems, and worked to design or arrange these systems' diverse components so as to achieve peak efficiency. Such was the case in technical education and organization, but in the case of corporations, peak efficiency meant larger profits; communications technologies and electric lighting and power are clear examples of this. These spheres of activity, however, by no means exhausted the possibilities. Although Americans frequently disputed the elements or order of systems, the notion of system itself pervaded virtually every aspect of American life in the half-century after 1870. In sum, the country's inhabitants reinvented and remanufactured their nation.

NOTES

[1] Quoted in James D. McCabe, *The Illustrated History of the Centennial Exhibition* (1876; Philadelphia: National Publishing, 1975), pp. 277–78.

[2] Quoted in Herbert N. Casson, *The History of the Telephone* (Chicago: A. C. McClurg, 1910), p. 150.

[3] Quoted in Robert W. Garnet, *The Telephone Enterprise* (Baltimore: Johns Hopkins University Press, 1985), p. 130.

[4] *Proceedings of the American Society of Civil Engineers*, vol. 1 (May 6, 1874), pp. 96–97.

[5] *Proceedings of the ASCE*, vol. 1 (June 1874), p. 110.

[6] Quoted in Terry S. Reynolds, *75 Years of Progress: A History of the American Institute of Chemical Engineers* (New York: AICE, 1983), p. 12.

7 Quoted in David F. Noble, *America By Design* (New York: Knopf, 1977), p. 45.

8 Quoted in Hugh Hawkins, *Pioneer: A History of the Johns Hopkins University* (Ithaca: Cornell University Press, 1960), p. 46.

9 Quoted in Leonard S. Reich, *The Making of American Industrial Research* (Cambridge: Cambridge University Press, 1985), p. 37.

FOR FURTHER READING

Aitken, Hugh G. J. *Syntony and Spark: The Origins of Radio* (New York: John Wiley and Sons, 1976). An excellent technical history of early radio.

Beardsley, Edward H. *The Rise of the American Chemical Profession, 1850–1900* (Gainesville: University of Florida Press, 1964). A concise examination of the organization and development of a profession.

Bright, Arthur A., Jr. *The Electric-Lamp Industry* (New York: Macmillan, 1949). An economic survey of the subject.

Brooks, John. *Telephone: The First Hundred Years* (New York: Harper & Row, 1975). A popular history of the telephone.

Bruce, Robert V. *Alexander Graham Bell and the Conquest of Solitude* (Boston: Little, Brown, 1973). An affectionate biography of the inventor of the telephone.

Calvert, Monte A. *The Mechanical Engineer in America, 1830–1910* (Baltimore: Johns Hopkins University Press, 1967). A valuable discussion of changes in engineering education.

Cheney, Margaret. *Tesla: Man Out of Time* (Englewood Cliffs: Prentice-Hall, 1981). A popular biography of the father of the AC motor.

Clark, Ronald W. *Edison: The Man Who Made the Future* (New York: Putnam, 1977). A highly readable biography written for a popular audience.

Hammond, John Winthrop. *Men and Volts: The Story of General Electric* (Philadelphia: J. B. Lippincott, 1941). A popular company history with some useful information.

Hughes, Thomas P. *Networks of Power* (Baltimore: Johns Hopkins University Press, 1983). Considers electrical technology in America, Britain, and Germany.

Josephson, Matthew. *Edison: A Biography* (New York: McGraw-Hill, 1959). Perhaps the best biography of Edison.

Kevles, Daniel J. *The Physicists* (New York: Knopf, 1978). Discusses relations within a community of professionals.

Layton, Edwin T., Jr. *The Revolt of the Engineers* (Cleveland: University Press of Case Western Reserve University, 1971). An early history of American engineers.

Leupp, Francis E. *George Westinghouse: His Life and Achievements* (Boston: Little, Brown, 1919). An early, easily readable biography of Westinghouse.

McKay, John P. *Tramways and Trolleys* (Princeton: Princeton University

Press, 1976). Although primarily about Europe, includes much information about American streetcar development.

MacKenzie, Catherine. *Alexander Graham Bell* (Boston: Houghton Mifflin, 1928). Contains many selections from Bell's letters.

MacLaren, Malcolm. *The Rise of the Electrical Industry During the Nineteenth Century* (Princeton: Princeton University Press, 1943). An older transatlantic history, especially valuable for its breadth.

McMahon, A. Michal. *The Making of a Profession: A Century of Electrical Engineering* (New York: IEEE Press, 1984). The first book-length study of American electrical engineers.

Mann, Charles Riborg. *A Study of Engineering Education* (New York: Carnegie Foundation for the Advancement of Teaching, 1918). A thorough survey of the then current state of engineering education in America.

Passer, Harold C. *The Electrical Manufacturers, 1875–1900* (Cambridge: Harvard University Press, 1953). The best book on late–nineteenth-century American electrical technology.

Sandhurst, Phillip T. *The Great Centennial Exhibition* (Philadelphia: P. W. Ziegler, 1876). A contemporary recounting of the Philadelphia Exposition.

Sinclair, Bruce. *A Centennial History of the American Society of Mechanical Engineers 1880–1980* (Toronto: University of Toronto Press, 1980). The best institutional history of a professional engineering organization.

Smith, George David. *The Anatomy of a Business Strategy: Bell, Western Electric and the Origins of the American Telephone Industry* (Baltimore: Johns Hopkins University Press, 1985). Particularly valuable for the patent dispute and its aftermath.

Veysey, Laurence R. *The Emergence of the American University* (Chicago: University of Chicago Press, 1965). An examination of the origins of the modern university.

Wasserman, Neil H. *From Invention to Innovation: Long-Distance Telephone Transmission at the Turn of the Century* (Baltimore: Johns Hopkins University Press, 1985). Discusses the problems of developing long-distance telephony.

Wise, George. *Willis R. Whitney and the Origins of U.S. Industrial Research* (New York: Columbia University Press, 1985). An excellent biography of a pioneer of industrial research laboratories.

5

Systematizing the Fabric of American Life: the 1870s to the 1920s

I n the late nineteenth and early twentieth centuries, Americans systematically transformed the face of their land. Whereas telegraph lines had merely paralleled existing railroads, and ran only from station to station, electric power and telephone wires and cables proved massive (yet indispensable) intrusions into the landscape as they wound their way connecting generating plants, central stations, homes, and factories. The nation's inhabitants cut down forests, extracted ores and coal, and built roads at unprecedented rates and in unprecedented numbers. Steam- and then gasoline-powered farm machinery gradually changed the nature of American agriculture; farm size increased markedly and farmers moved forcefully into the Great Plains. Cities and city structures achieved great size and prominence, and spawned persistent suburbanization.

But the systematic transformation was not limited to the physical environs or the tapping of the country's natural resources. It extended to the social sphere as well. Late–nineteenth-century men and women identified society as a cauldron of diverse peoples, then acted as if that assessment was a fact. Concerned citizens, including educators, other professionals, and business-men, studied the groups they had already defined; reaffirmed that each group had particularistic capacities, capabilities, merits, and deficiencies; and then labored to see that their reaffirmations served as foundations of social and economic policy. The nation seemed complex and diverse, but these observers considered it a single entity, a notion that underscored the importance of uncovering the principles that systematized relations by ordering or arranging the many different parts. Those sentiments produced diverse social agendas. Perhaps the most obvious were the struggles for political power. Americans agreed that the nation was like a hierarchically arranged organism but disagreed about what segments of the population were its vital organs; most demanded to serve as the organism's brains or heart.

Not all Americans were consumed by disputes over who should hold power. Many were drawn to the less explicitly political question of leisure, itself a product of the rigid categorization and compartmentalization of the late nineteenth and early twentieth centuries, and therefore a fit object of social concern. Those concerned often combined physical and social elements to create leisure institutions of two major types; some were established to cater to a particular segment of society and were tacit acknowledgements of society's divisions, while others sought to appeal across group boundaries and to help organize the nation's fractious social groups. Americans generally found new technologies or technical processes instrumental in achieving either aim.

SYSTEMATIZING THE PHYSICAL ENVIRONMENT
The New City

The urban landscape was redesigned in the late nineteenth and early twentieth centuries. Horse-drawn street railways and then electric traction facilitated urban expansion and population growth, and frequent annexations kept much of this population within city boundaries. This rapid growth seemed to require the sorting out of peoples and activities; as great numbers of immigrants and rural Americans descended on cities, the correlation of a social type with a specific part of the city became an important urban feature. Residential neighborhoods became characterized by their inhabitants' class, religion, nationality, race, and occupation. The remaining city area also was differentiated, but by activity or function. Mercantile transactions stood as the central business district's focus, while heavy industry was relegated to other city sections. Like residential neighborhoods, both commercial and industrial districts seemed congregations of subdistricts, each of which exhibited more precise specialization. Service, pleasure, and entertainment districts also abounded as few cities failed to develop thriving tenderloin and tavern areas, or parks and parkways.

New Building Materials   Both the nature and increasing size of the new city created a persistent call for massive new construction, which provided a potent demand for building materials. Brick, cut stone, plain concrete, cast and wrought iron, and wood were standard building materials through the 1870s. Only in that decade did Americans begin to use iron and steel supports; the introduction of fireproof tiles lessened fear that fire would buckle and collapse ferrous members. Builders placed iron and steel in tile envelopes, which permitted entrapped air to dissipate heat, but even then generally employed ferrous materials only to brace floors. Solid masonry, brick, or stone exterior walls bore building weight. These load-bearing walls rested on stone or concrete pyramid footings and absorbed downward and lateral thrusts. Their thicknesses were directly proportional to building height, a factor that limited buildings to about ten stories, diminished interior light, and reduced interior space.

Concrete played a relatively insignificant role in these early buildings for two reasons. First, American concretes had traditionally depended on natural quicklime cements to bind together their constituent elements—sand, water, and aggregate. These cements, which were formed by heating broken limestone to drive off carbon dioxide, had worked adequately for concrete bridge footings from the 1810s, but they lacked the durability and binding power necessary for larger structures common to the new city. Americans looked to Europe for assistance and found that English builders long had used portland cement in public works. Identified in 1824, portland cement was manufactured by combining precise measures of chalk and clay, heating the mass to a temperature much hotter than other cements to fuse it, and grinding the klinker—the fused material—into fine powder. Some was imported to America in the 1860s, but large-scale American adoption of concrete using portland cement occurred after America began to manufacture it. Lehigh County, Pennsylvania, was the site of the nation's first portland cement works (1871); others were erected near South Bend, San

Antonio, and Kalamazoo by the mid-1880s. Portland cement's superiority quickly and continually won it adherents, and manufacturers repeatedly scrambled to satisfy demand. In 1888, the nation produced about 100 million pounds of portland cement, but still imported nearly eight times as much. The introduction of rotary kilns to heat and grind the cement in the late 1890s permitted domestic manufacturers to produce about 1,500 million pounds by 1898, and to multiply that output nearly sevenfold five years later. By 1913, the nation made 36,800 million pounds of portland cement and imported only a negligible amount.

Rapid adoption of portland cement in the 1870s and early 1880s did not, however, signal a sudden rise in concrete-based urban buildings. Portland cement was commonly poured or cast in blocks and used for piers, seawalls, and dams in the late 1870s and 1880s, but its lack of tensile strength rendered it unsuitable for structural elements subjected to forces other than compression. The persistence of this second constraint on concrete utilization was somewhat surprising because American builders had recognized that iron and steel withstood shearing and tensile forces, and had occasionally fortified masonry arches with iron bars to prevent cracking. Even more to the point, French and German architects had begun to use concrete reinforced with steel bars in the 1860s and 1870s. Not until the 1880s did American builders begin to experiment with reinforced concrete, and another decade passed before the combination became established as a primary building material.

Ernest L. Ransome assisted in the transition. A San Francisco concrete-block manufacturer, Ransome had difficulty selling his product in the Bay Area because concrete seemed susceptible to earthquakes; shocks stressed and often destroyed predominantly concrete structures. Relying on European precedents, Ransome reinforced his blocks with steel bars, built structures to demonstrate reinforced concrete's tenacity (varying the steel elements' shapes to conform to the anticipated stresses), and relentlessly promoted the material. Ransome's work showed a suspicious public that reinforced concrete added steel's tensile and shearing strength to concrete's ability to withstand compression, and that concrete provided steel, fire, and corrosion protection.

The Skyscraper   A new municipal building form, the skyscraper, accompanied the introduction of reinforced concrete in the mid-1880s, but was not dependent solely on that building material. Symbols of corporate wealth and success, and de facto expressions of urban diversity, skyscrapers quickly transformed the city sky-

line. These concentrated administrative centers dominated central business districts and testified to the large corporations' urban prominence.

The first two structures that broke with the load-bearing wall tradition and might be considered nascent skyscrapers, George B. Post's Produce Exchange in New York City and William Le Baron Jenney's Home Insurance Building in Chicago, were iron-framed. Interior iron skeletons or cages supported almost all the building's weight, exterior walls acted primarily as weatherproofing, and reinforced concrete rafts comprised the foundation.

These techniques were refined and extended over the next several decades to yield what historians have called "true skyscrapers." Steel T-beams and I-beams replaced iron in these new structures, and rivets supplanted bolts and were in turn supplanted by electric arc welding in the 1920s. Masonry gave way to reinforced concrete as walls that once bore the weight of a building evolved into mere curtains. Shelves extending from the steel skeletons bore curtain-wall weight, while reinforced concrete subflooring became standard.

These modifications produced sturdier, lighter, and taller buildings. Steel possessed greater tension and compression strength than iron or masonry, and greater resistance to fatigue, while weighing less than half as much as masonry or stone. The thin steel interior skeletons occupied less interior space, and curtain walls enabled architects to incorporate large glass windows.

Political Control of the Cities

Steel, and reinforced concrete made with portland cement, provided the materials to build skyscrapers and redesign cities, but who should control development of and govern these places was a political question. Newly identified groups fought to demarcate significant roles for themselves in cities in the 1880s and 1890s; each aimed to serve as the urban heart or brains. Bosses generally relied on the seemingly endless flow of immigrants for political support, often fortified by bribes and payoffs from influence-seeking electric-traction and power magnates. Ministers, small businessmen, and professionals frequently were the most contentious. Engineers staked their claim to power on their technical abilities

Skyscrapers with curtain walls dominated the urban skyline.

and experience with public works, while doctors maintained that public health was the most important municipal issue, construing it to involve virtually all facets of urban social and economic life, and asserting that medical expertise entitled them to urban leadership. Clergymen cited their moral authority, businessmen based their argument for ascendency on their fiscal management skills, and lawyers relied on their legal acumen.

City Planning   Others concentrated directly on the cityscape. Some architects and landscape gardeners called for humane, uncongested cities in which civic architecture was interspersed with parks and other nature settings. Daniel Burnham's White City at Chicago's 1893 Columbian Exposition epitomized that thrust. Others agitated for a functional or efficient city, arranged to encourage the flow of business and human services. These proposals hinged on similar propositions: rational development of the new city's physical plant required expert wisdom to plan it, and no single group possessed the expertise necessary to tackle the situation unaided. "City planning," wrote Cleveland's Frederick C. Howe in 1913, "treats the city as a unit. It anticipates the future . . . so as to secure orderly, harmonious and symmetrical development." It "means a city built by experts in architecture, landscape gardening, in engineering and in housing; by students of health, transportation, sanitation, water, gas, and electricity."[1]

Urban planning required concerted action. Each expert group needed to bring its particular expertise to bear on a designated part of the urban environment; each had to share authority and responsibility. Dividing the city into a series of discrete functions, the planning for each dependent on different experts, helped resolve tensions among competing interest groups. But it also produced demand for two new kinds of experts, both of whom specialized in administration and in coordinating the various experts' efforts: commission governments, and then professional city managers, to manage the present; city planning commissions, and ultimately a city-planning profession, to direct preparations for the future.

Formal city planning blossomed after 1900. In 1907, the Russell Sage Foundation sponsored the first citywide survey in America, the Pittsburgh Survey. That initiative was followed in 1909 by three landmark events: Burnham's Chicago Plan, the nation's first comprehensive city plan; the first National Conference on City Planning, held in Washington, D.C.; and Harvard's establishment of a city-planning course to train prospective planners, the first such course offered at an American university. Bureaus of municipal re-

*The late–nineteenth-century lament about the quality of
life in cities is given a certain poignancy by this photograph
of Pittsburgh.*

search—consortiums of experts to investigate municipal questions
and formulate policy—zoning commissions to control development, and comprehensive city plans became common during the
next decade as Americans began systematically to plan the urban
future.

The New Agriculture

Late–nineteenth- and early–twentieth-century American agriculture offers a compelling parallel to the rise of the new city. Around
1870, as city newspapers began to fret about municipal chaos and
inefficiency, agricultural newspapers and societies issued a recurring stream of complaints about farmers' isolation, the drudgery of
farm work, and farmers' lack of financial rewards and social standing. These indictments, like those in cities, engendered a wide variety of political, social, and economic campaigns, including
formation of farmers' clubs, the Grange, the Populist Party, and the

Good Roads movement. Each was an attempt to resolve the "farm problem," and presumed that American agriculture constituted a system: farmers must act collectively to remedy agriculture's plight because they depended on each other; each farmer's activities affected his neighbor. And each of these movements sought to overcome obstacles or to add what was missing or defective in agricultural life—to remove, add, or repair systemic elements—so that farmers could assume their rightful places in the modern American social system.

The Railroads and Agriculture   Railroads were among the earliest and most sustained objects of farmers' concerns, and certainly produced the bitterest feelings. Farmers, especially those in the South and Great Plains, accused railroads of an unabiding self-interest detrimental to farm life. Their argument was predominantly economic. Agriculturists deplored railroad pooling arrangements, rebate policies, and short-haul discrimination; maintained that these practices eroded farm profits and constituted gouging; and organized politically for redress. Several states passed rate-setting laws in the 1870s; more decisive was the federal government's enactment of the Railroad Valuation Act (1913), which empowered the Interstate Commerce Commission to review each carrier's financial history to establish "just and reasonable rates."

The farmers' lament was somewhat ironic because railroads had helped foster agriculture's westward march by selling land obtained from land grants to prospective farmers and speculators. Agricultural output west of the Mississippi was negligible as late as 1860, but by 1900 North Dakota and Minnesota had become the nation's largest wheat producers; Iowa, Kansas, and Nebraska (and Illinois, which is east of the Mississippi) its largest corn growers; and Texas its largest cotton producer. These states (Illinois excepted) had more than 42,000 railroad miles in 1900, a dramatic forty-five-fold increase in forty years.

Despite the vehement criticism, a railroad–agriculture nexus persisted, often to agriculture's advantage. Although railroad combination and consolidation, particularly in the two decades after 1890, did permit seven groups to control more than two-thirds of the nation's nearly 250,000 railroad miles, railroads reduced rates for agricultural products and enhanced service during that period. Technical factors and industry-wide administrative decisions enabled railroads to furnish cheaper, faster, and safer travel. Swifter, heavier, more powerful locomotives with greater traction moved longer trains and heavier cars, while adoption of the Janney automatic coupler, widespread air-brake employment, and use of iron and steel as universal materials in railroad-bridge construction speeded operations and reduced accidents. Mechanical refrigerator cars and special animal cars with feed bins and watering troughs ensured that farm products reached their market in excellent shape. An industry-wide standard railroad time (1883) facilitated transport, and a similarly derived standard railroad gauge (1890s) did away with superfluous reloading costs; any carrier's engines and cars could now travel on virtually any American track. Spurs were connected to feeders, and feeders to trunk lines, in America's truly integrated nationwide railroad system—factors often ignored by agriculturists as they denounced railroads.

Systematizing the Farm   Not all manifestations of the farm problem dealt as explicitly with economic concerns. A reproach to contemporary farm practice occurred as criticism of railroads was beginning, which in substance foreshadowed complaints about city administration. These critics of farm practice, many of whom were themselves farmers, contended that "guesswork, random efforts [and] 'cut and try' methods" constituted then prevailing agricultural practice. They maintained that these procedures were outdated, and argued that their perpetuation resulted in "waste of time and strength and substance"; these practices caused the farm problem and marked farmers as old-fashioned. Those offering this critique called on agriculturists to renounce antiquated practices, to let brains guide hands and machines, and to farm with "system." They defined system as "a collection of rules and principles" in which each individual activity had "a proper classification and a proper assignment." It was nothing more than "a place for everything and everything in its proper place," where everything was "done at the proper time and in the proper manner."[2]

These systematic agriculturists, men such as D. S. Curtiss, associate editor of the *Maryland Farmer*, and William R. Taylor, president of the Wisconsin Agricultural Convention, assumed that farming was composed of numerous discrete operations and that success was the consequence of rationally conceived and pursued methods. No activity seemed unimportant, and each required scrutiny and rationalization. Organization of farm space was a critical concomitant. The shape and location of farm buildings, utilization of tools and machinery, location of fields, and position of fences were predicated on their ability to enhance farm practice. For example, octagonal and circular barns increasingly replaced traditional rectangular buildings after 1870. While more complicated and expensive to construct, the new barn designs used space more efficiently and promised to speed farm work. Their self-supporting domes or rounded roofs required no space-consuming posts or support beams, and hay forks could operate unobstructed in their larger lofts. The two designs further reduced wasted space and facilitated farm labor by minimizing or doing away with corners; aisles were circular, not linear, and animal stalls were arranged around aisles, which lessened the number of steps farmers needed to take. Although these innovations seemed minor, their advocates hailed them as powerful examples of systematic farming's virtues and as demonstrative of the potential of efficient organization.

In a sense, these octagonal and circular barns possessed a symbolic function not unlike urban skyscrapers. Their unusual shape dominated the rural skyline, testified to their owners' prominence, and served as the central farm focus. Steam-traction engines played a similar symbolic role on the Great Plains, as well as a more practical one. Mounted on steel frames, these monstrous twenty-five-ton vehicles were powered by heavy steel high-pressure engines, which turned tracks or steel wheels with protruding lugs or rivet heads for better traction. Visible from miles away, their popularity stemmed from the plains' peculiar topography and climate: its firm, level, dry soil supported their great weight, while the lack of rainfall and need for irrigation made dry farming the only conceivably profitable agricultural form.

Based on the retention of capillary moisture, dry farming required deep plowing immediately after every rainfall—finely divided soil holds water nearer the surface where roots can tap it—as well as harrowing to create a dust cap to slow evaporation. Speed was essential because untrapped water would percolate through and compact soils or evaporate quickly. Russian wheat was one of the few crops that grew satisfactorily under dry-farming conditions. Its suitability converted the plains into the nation's breadbasket.

The requirements for dry farming wheat made steam-traction engines feasible. Pulling drawbar-attached steel disk plows and disk harrows, these engines provided the power necessary for frequent deep plowing, and the speed crucial for utilization of cropland moisture. The largest machine towed thirty plows and tilled one hundred acres a day. Their belt pulleys, which could be operated effectively only when engines were stationary, ran threshers. No other type of agriculture offered this potent combination of uses or this scale of activity.

Nor did the introduction of internal-combustion traction engines—true tractors—spark a nationwide tractor boom. First man-

*Early internal-combustion-engine tractors, such as this
1908 Caterpillar model, made inroads on the Great Plains
but remained a rarity on farms elsewhere.*

ufactured in the mid-1890s, tractors were lighter, less dangerous, and more powerful than steam traction, but no more useful. Although these factors enabled tractors eventually to supplant steam-traction engines—frameless, unit-driven tricycle tractors weighed less than five tons and developed more than twice the horsepower of large steam-traction engines—they were practical only on the Great Plains.

But establishing rational agricultural plans depended on more than fancy machinery and buildings. It required knowing the facts, the costs of farming. Systematic farmers stressed that a complete system of farm accounts—detailed double-entry bookkeeping—would enable agriculturists to learn each act's cost/benefit ratio. Informal time/motion studies—comparisons of time taken to complete various tasks—also served as a technology to systematize farming. Agricultural societies and newspapers were conduits for announcements of investigative results, and they satisfied the farmers' obligation to report findings to their peers for review, verification, and criticism. The bestowal of collegial sanction often led others to adopt the practices, which tended to standardize farm operations. Application of systematic farming techniques only sometimes increased farm profits and reduced drudgery, but its partisans always identified themselves as progressive.

The Challenge of Agricultural Colleges   Systematic farmers' contentions did not go unchallenged. The nation's agricultural college professors sympathized with the farmers' plight but generally opposed their methods to resolve it. They favored systematizing farm activities, but maintained that agriculturists were incapable of developing rational farming plans unaided. To these academics, the systematic-farmer approach was doomed because its methods were superficial and arbitrary; profit-and-loss statements and time/motion studies were inexact reflections of plant and animal growth principles. They argued that agricultural success depended on the application of these (incompletely understood) principles. Their agenda emphasized uncovering these rules, a task that they asserted required rigorous systematic investigation and that lay beyond farmers. Only experts schooled in investigative methods commanded the ability to decipher agricultural principles; agricultural college professors usually identified themselves as possessors of that expertise. Successful farm systematization demanded expert-derived knowledge expertly applied; farmers were to practice what agricultural scientists preached.

These claims were suggestive of those that also would be articulated by urban professionals, and had similar consequences. Systematic farmers reacted hostilely and attacked agricultural colleges as bastions of inadequacy, but agricultural scientists persisted. Farmers' institutes paraded the agricultural scientists' dictums from the 1870s; winter short courses carried their message a decade later. Agricultural scientists also agitated for creation of publicly funded agricultural experiment stations—places for systematic agricultural science investigation—and capitalized on opportunities to transform the U. S. Department of Agriculture (USDA) into the nation's leading agricultural science agency. These campaigns produced national legislation: the Hatch Act (1887) legitimated agricultural science research at agricultural colleges, the Adams Act (1906) extended it, and the Smith-Lever Act (1914) established a county agent system to deliver systematically the scientists' pronouncements. By the early twentieth century, agricultural scien-

*These early road-building machines were used to fashion
good dirt roads in rural and agricultural regions.*

tists had enlisted the great majority of farmers to their cause; farmers
had become dependent on agricultural scientists.

Rural road quality was another aspect of the farm problem that
systematized relations between farmers and agricultural college
personnel. Some farmers issued complaints in the 1870s, but it remained until the late 1880s for this crusade to gain systematic
expression, and another fifteen years passed before there was governmental action. Rural roads were closed by winter snows, and
turned by spring and autumn rains into impassable mud rivers,
which kept farmers from their markets and increased isolation.
Likening a complete country road system to "the arteries and veins
in the human system, through which the blood circulates,"[3] farmers established state Good Roads conventions in the late 1880s and
a national Good Roads convention in 1893. Chaired by General
Roy Stone, a prominent Civil War veteran, the national convention
was funded by A. A. Pope, a leading bicycle manufacturer, and joined
rural dissatisfaction over roads with bicycle enthusiasm.

These conventions often offered rural road construction plans.
But it was the land-grant college engineers, especially in the Midwest, who capitalized on good-roads sentiment to assert their pri-
macy in road construction and design. They maintained that public
works required experts to study, plan, and execute them, and that
the quality of rural roads depended on their talents. Iowa State Agricultural College's Anson Marston took the lead and was rewarded
in 1903 when the state established a state highway advisory com-

mission. Staffed by Marston's associates and students, the commission studied the road question, tested various construction materials and drainage techniques, and built experimental ways. It also issued recommendations (e.g., country roads required superior drainage and firm beds but not concrete paving) and construction guidelines, which were disseminated by college extension-agents. The state bestowed additional sanction in 1912 by empowering the commission to oversee all state road construction and to specify materials and techniques. The Iowa commission quickly became the training school for other states' engineers, which tended to standardize rural road construction nationwide. Establishment of the U. S. Bureau of Public Roads in the USDA in 1919, and appointment of Thomas MacDonald as the bureau's first chief, completed the standardization drive; MacDonald had been Marston's prize student, had worked on the state's initial highway commission, and had served since 1912 as Iowa State Highway Engineer.

Systematizing Nature

The agricultural and urban environments of the late nineteenth and early twentieth centuries struck emotional chords, but most citizens reserved their true passion for nature. Americans expressed an abiding love of and reverence for nature, even though they increasingly inhabited cities. The new ten-cent magazines repeatedly sung its praises, municipalities established park systems to preserve it, and urbanites moved to suburbs to celebrate it. Johns Hopkins's Herbert Baxter Adams located the genesis of American democratic institutions among the barbarian tribes in German forests, and Frederick Jackson Turner identified the frontier as American democracy's progenitor. A nature-study movement flourished, summer camps became "the customary thing,"[4] and Boy and Girl Scouts as well as Campfire Girls enrolled thousands as the nation's citizens hurried to imbue their children with nature's virtues. Adults watched birds, mourned the passenger pigeon's demise, and photographed natural settings.

These acts commemorated nature, but not pure wilderness. Few urged without qualification the preservation of land in its pristine wild state, and fewer asked contemporaries to forsake the present to return to the past. Most sought instead to combine nature with the industrial urban present to civilize the future. The wilderness was not inviolable, but nature was; nature was a cornerstone of the future.

Nature Photography   Broad agreement about nature's importance to American life inspired a wide spectrum of technological and technologically related activities, some of which are not immediately obvious. For instance, the desire to capture and preserve nature helped fuel an era of recreational photography. The well-to-do began in increasing numbers to photograph natural settings in the late 1870s, and by the mid-1880s most major American cities had one or more photographic clubs in which recreational photographers discussed their art. Nature photography secured a much broader base in following decades. New photographic processes made casual, part-time photography possible; mid-century photography had been a full-time occupation, requiring special facilities, know-how, and time.

Only with the introduction of factory-sensitized, gelatin-coated dry glass plates in the late 1870s, and production of gelatinized paper for positive prints a few years later, did photography become recreational, and even then it remained a somewhat restricted activity demanding skill, money, and care. Glass plates were awkward and easily broken, while developing negatives and printing positives necessitated special equipment and facilities, factors that kept photography outside the common person's means. George Eastman removed some of these obstacles in the half-decade after 1883. Conceiving of film on continuous rolls, not plates, Eastman and associates devised machines to produce flexible, gelatin-coated roll film, holders to advance and keep film in focus, and processes to diminish the effects of temperature and humidity changes. He also manufactured inexpensive, simple-to-operate cameras. But Eastman's solution to the developing and printing problem was most novel; his professional staff, not individual camera operators, would accomplish those tasks. Cameras with exposed film were sent to Eastman's Rochester, New York, plant. Experts there would develop negatives, print positives (Eastman also had built machines that coated positive paper), enlarge them if desired, reload cameras, and ship prints and cameras back to photographers, who therefore needed only artistic sense to photograph nature, not special technical skills.

Eastman further refined procedures during the next decade. Removable film cartridges soon eliminated camera shipment, while adoption of celluloid as the film base improved photographic quality. Eastman poured celluloid, a nitrocellulose compound soluble in camphor, amyl acetate, and other organic liquors, and allowed the solvents to evaporate; a transparent, tough flexible film, impervious to photochemicals, remained. A coating of silver salt in gelatin emulsion provided the dry photographic surface. The result was an inexpensive, easy-to-use photographic system accessible to virtually anyone.

Mining   Other technologies were more firmly grounded in nature. As the source of raw materials necessary for modern American life, nature was cherished for the bounty it furnished. The commercial mining and logging industries were among the most prominent natural-resource-based operations; faced with unprecedented demands for their products, they repeatedly increased productivity and profits by systematizing procedures and operating on economies of scale. Refiners dominated the mining industry and sought to control prices by gobbling up ore- and coal-bearing properties so as to eliminate middlemen and restrict competition. The construction of refinery-owned railroads and purchase of ships by refineries joined together mines, refineries, and markets, further lessening dependence on outside parties. Mining sites themselves exemplified centralization of authority and control of expenditures. Located away from major population centers, mining towns resembled colonial iron-making sites. Company employees provided support services, while miners lived in company-owned dwellings and bought articles in company stores.

An examination of late–nineteenth- and early–twentieth-century copper mining demonstrates how efficiently and systematically this industry tapped nature. It strove to extract as much ore as economically feasible, and to squeeze every ounce of metal from ores. From the 1840s until the 1920s (when Southwestern mines assumed importance), the nation's major copper mining occurred on Michigan's upper peninsula near Houghton. By 1870, the process was fairly well articulated. Miners sank narrow shafts that followed often-crooked veins, and bolstered them with timbers. They worked by candlelight, dug with picks and shovels, and placed ore in pushcarts running on underground iron rails. Iron ropes and steam-powered hoists lifted material from the shallow shafts (less than eight hundred feet deep), and steam-powered stamps crushed it. The ore was separated by hand-washing, and smelting was usually done at mine sites to reduce shipping costs. The crushed, washed ore was placed in small reverberatory furnaces and melted. The furnace top was retracted and slag raked off. The metal was heated again and splashed into the air to oxidize impurities. Hardwood charcoal combined with the excess oxygen to form carbon dioxide, and the nearly pure copper was poured in molds to form ingots.

Smelting remained virtually the same during the next half-century, while the other elements changed radically. Nitroglycerine and dynamite enabled miners to reach depths of more than 3,000 feet in the 1880s, and nearly 6,000 feet by 1900. Air pumps provided ventilation. Shafts became wider and straighter as miners dug tunnels into meandering veins. Steam and later electric pumps raised standing water from mine pits. First arc lights and then incandescent bulbs did away with candles. Miners in the 1880s used large, heavy, steam-driven compressed-air drills in place of picks and shovels. These cumbersome tools gave way by 1910 to one-man diamond-headed air drills.

Nor was that all; mining engineers explored every option and endeavored to rationalize each operation. They further straightened and widened shafts to facilitate conversion to compressed-air (later electric) hoists. Small compressed-air and electric locomotives carried ore on tramways. Larger efficient steam-driven stamps more thoroughly pulverized ores, while concentrators and regrinders replaced hand-washing. After 1900, even the steam from stamps was conserved. Mineworkers used it to run a low-pressure steam turbine, which powered an improved regrinder to agitate incompletely separated material in a revolving metal drum.

Systematizing Iron and Steel   Iron mining underwent a similar transformation. The geographical center of American iron mining had moved from the East to the Midwest in the mid-nineteenth century, and late-century miners worked the major deposits of upper Michigan, northern Wisconsin, and northeastern Minnesota. They opened the westernmost deposit, Minnesota's Mesabi, last, although it proved the largest, richest, and most easily extracted. Its emergence in the late 1880s and 1890s more than compensated for the declining Wisconsin and Michigan fields. These earlier iron fields required deep shafts to tap high-grade ore, but the Mesabi's deposits lay on and near the surface. Steam shovels easily lifted the ore directly and dumped it into railroad cars.

But unlike copper smelting, iron smelting rarely occurred at the mine site. Iron smelting required three times as much coal as ore by weight, and producers reduced transportation costs by locating blast furnaces adjacent to coal fields rather than mines.

Integration of the steel industry began at the iron mine. Although iron-ore extraction techniques were similar to those employed for copper—except at the Mesabi—iron mine operators faced the additional task of transporting the material to far-off blast furnaces; Pittsburgh, the Lehigh Valley, Chicago, and Birmingham,

Alabama, reigned as iron-smelting centers. Special thirty-ton hopper-bottomed railroad cars conveyed ore to the Great Lakes, where it was then lifted by electric hoist and dumped into ship holds. Steam-powered bumper beams distributed the load evenly. These 500-foot-long ships carried as much as 8,000 tons of ore to northern Ohio cities. Company employees unloaded ships with steam and electric cranes and placed ore in railroad cars. At the refineries the cars were hoisted and their contents dumped directly into furnaces.

Fuel for iron smelting was also subjected to scrutiny. The American iron industry rejected anthracite coal as it responded to heightened demand by systematically boosting production. Anthracite burned too slowly for maximum efficiency, and the industry increasingly turned to coked bituminous coal as a way to improve smelting speed and generate additional production economy. During the coking process, the coal was baked for two days in open-topped beehive ovens, but iron manufacturers in the late 1880s began to substitute closed "by-product" ovens to trap vaporized coal tar and coal gas, which they sold as industrial chemicals.

Coked coal was joined with iron ore for smelting in huge, highly mechanized blast furnaces. These furnaces were charged by conveyor-carried hoppers, and used recycled combustion-generated gases to heat blasts. Mud guns efficiently closed furnace tap holes, and conveyors facilitated pouring iron into molds and transporting the cooled metal.

Relatively little pig iron was made into wrought iron because the difficulties in mechanizing that process, coupled with steel's superior hardness, rendered wrought iron economically uncompetitive for most purposes. Steel became the preferred end product. Americans manufactured at least half their steel by the Bessemer process until 1908, but from the 1860s found it unacceptable when using phosphorous-contaminated pig iron; phosphorous exploded when exposed to the air blast and yielded brittle steel. French and German engineers had encountered that problem nearly a decade earlier—much continental iron was loaded with phosphorous—and had designed a regenerative or open-hearth furnace to combat it. Air was channelled over a network of superheated bricks (reverberatory furnaces were much cooler) before it came to bear on the metal. Experiments demonstrated that this open-hearth technique worked especially well on charges containing pig iron and scrap steel. Although it proved slower than the Bessemer process, steelworkers could extract samples periodically from the hearth, test them, and, if necessary, adjust the charge's composition. Ability to control and standardize steel contributed to the open hearth's popularity. In addition, these furnaces had the capacity to neutralize acidic phosphorous iron when lined with a basic substance such as dolomite. This technique consistently secured a larger market share from its introduction in America in 1868, and dominated American steel in the 1910s and after. Well before that time, manufacturers had further mechanized steel making. Electric cranes and hoists lifted containers of molten steel from regenerative furnaces and poured it directly into molds mounted on railroad cars for shipment to highly mechanized rolling mills.

Logging Systematic mechanization of iron and steel production, and adoption of new technologies, permitted industrialists to increase output and lower unit costs. Late–nineteenth- and early–twentieth-century loggers also modernized to exploit nature more fully. Water transport and power requirements had limited the pre-1870 American lumber industry. Lumberjacks cut trees only near waterways and floated logs to small mills where waterwheels moved saws. Unfinished lumber was rafted down rivers to wholesalers in St. Louis and Chicago for finishing and commercial distribution. Mill owners' attempts to consolidate forest-to-market activities and to benefit from economies of scale guided the industry's post-1870 growth. A massive shift to steam power and railroad expansion enabled them to effect these changes. Railroads themselves consumed lumber for road beds, bridges, piles, and depots. New railroad spurs into dense woodlands far from watercourses permitted lumberjacks, often under contract to mill owners, to open new timberlands. Mill owners situated mills close to new logging sites—cut boards were cheaper to ship than corresponding tree trunks—and began to send millwork directly to retail yards to eliminate wholesalers. This procedure resulted in the development of a nationwide lumber-grading system and made mills responsible for finishing lumber. Steam-edgers, shingle, lath, and slab saws, and planing, flooring, matching, and molding machines became standard mill equipment. So, too, did artificial drying kilns, in which fans passed air over steam-filled pipes in enclosed chambers to cure green lumber. Artificial drying proved much swifter and more complete than traditional open-air methods, decreased warping, and killed insects. Moisture removal also lessened lumber weight and railroad freight costs.

In addition to steam power (especially self-regulating Corliss engines), innovation marked the new late-century lumber mills. Logs transported to mills by horsedrawn sledge, or occasionally by railroad, were deposited in man-made hot ponds. Heated with steam

Steam engines powered the lumber industry.

exhausted by the mills' engines, these ponds kept timbers from freezing to permit year-round sawing. Electric lights made night milling possible. Each log was attached in turn to a steamjacker, which hauled it from the water and carried it up a slide to the second floor of the mill proper. A mechanized log-stop disconnected the jacker and rolled the log down a slight incline, from which a steam-powered device lifted it and positioned it on the carriage. Two iron levers held the log in place. The steam-powered, cylinder-run carriage brought the log to head saws, which removed bark, and to a band saw, which cut it into boards. The band saw had replaced the circular saws popular at mid-century because it made a much thinner, less wasteful cut. By steam-propelled live rolls, the boards went first to an edger, who squared the boards' edges and made them standard width; then to the trimmer, who trimmed the ends and cut boards into standard lengths; and finally to graders, who sorted lumber into grades. Automatic loaders lifted the lumber and placed it in carts to be taken to kilns. Steam cranes raised the kiln-dried boards and loaded them onto railroad cars.

Even mill wastes did not escape scrutiny as lumbermen tried to squeeze out all conceivable profit. Farmers purchased some sawdust for mulch and fertilizer, but mills also used mechanized carriers to collect sawdust and transport it immediately to steam-engine fireboxes. Imperfect boards also became fuel, and a few mills converted defective logs into charcoal and other substances. Nevertheless, the waste problem seemed inimical to the idea of efficient "modern" industry, and the U.S. Congress established a Forest Products Laboratory in Madison, Wisconsin, to resolve it. The laboratory recommended for mill wastes a process called destructive distillation and maintained that one cord of hardwood would yield sixty bushels of charcoal, ten gallons of methanol, and forty-five pounds of calcium acetate. Mill owners would reap increased revenues from the sale of destructively distilled wastes, and other American industries would gain crucial raw materials.

Political Control of Nature   Congressmen were hardly the first federal officials to raise formally the question of the systematic use of nature. Men employed in the West as federal administrators from the 1870s argued that nature was public property requiring federal regulation. In the 1880s and 1890s, proponents of that agenda identified themselves as a group defined by their special knowledge, and maintained explicitly that their conclusions were nonpartisan and that their methods were scientific—the products of careful study, rational thought, and verifiable techniques. They also asserted that the public interest dictated that they set forest policy. It remained until about 1900, however, for this new group to establish schools to produce forestry-management professionals. Their contention rested on the notion that America needed rational planning to promote systematic, orderly development that would yield raw materials for contemporary life yet protect nature for posterity, that only they could effectively plan that program, and that implementation required national legislation. And they frequently portrayed mining and lumber concerns as systematically ravaging nature and as operating out of selfishness without regard to the future; they cited corporate irresponsibility as demonstration of the imminent need for federal law.

Gifford Pinchot parlayed his experience managing timber at a private North Carolina estate with his political connections to emerge as the most prominent spokesperson for Federal regulation of natural resources. In the two decades after 1890, Pinchot and his followers convinced Congress to establish them as the nation's forest-management bureaucracy and to designate vast woodland tracts as national forests. Using their bureaucratic skills to shape national policy, these men stressed sustained-yield forest management, which meant the systematic utilization *and* replenishment of national forests. A trained forestry force tackled disease problems, fought fires, and supervised the cutting and sale of timber and the planting of seedlings. Mining was also permitted and regulated in national forestland. Both mining and lumber interests generally favored national-forest management policy. It guaranteed them future raw materials and profits, while expenses incurred by compliance could be passed on to consumers.

SYSTEMATIZING LEISURE

In the late nineteenth century, Americans made as rigid a distinction between work and leisure as they did between city and farm. The ideas that citizens ought to have leisure and pursue leisure activities were new to the period, as was the notion that work and leisure constituted radically different but essential elements of mankind's developmental system. Leisure activities were very seldom pursued to enhance working performance or tied directly to it, which is not meant to suggest that all leisure pursuits lacked purpose or were simply for relaxation. Zoos, parks, and the more general attempts to involve one's self with nature, including recreational photography, were considered "rational amusements," activities in which systematic observation actually improved the well-being of observers. Sports were often linked to physical improvement, a necessary concomitant of mental development. Nor did each form of leisure appeal, nor was it established to appeal, to every American social group. Some were clearly elitist, while others catered to the working class or a specific ethnic group. Nevertheless, a few of these activities cut across social divisions. And on one aspect of leisure Americans agreed: systematization and rationalization of leisure endeavors would yield the most efficient, and therefore most satisfactory, leisure. In that sense, the precepts surrounding leisure were virtually indistinguishable from those of work.

Sports and the Growth of the Middle Class

America's upper crust interested itself in horse racing quite early, but the middle class was most responsible for the boom in the late nineteenth century in participant sports. Taking advantage of the new inexpensive summer resorts established precisely to provide this group with leisure activities, the middle class lustily pursued archery, lawn tennis, track and field, croquet, and golf. This enthusiasm was soon channelled into the creation of national associations to codify rules and regulations, prescribe appropriate garb and equipment, sponsor national competitions, and adjudicate disputes. The United States Lawn Tennis Association (formed 1881), Amateur Athletic Union (1888), and the United States Golf Asso-

*Baseball became a popular form of late–nineteenth-century
leisure. Crowds as large as that shown above often
attended games at New York's Polo Grounds.*

ciation (1894) functioned in these capacities. Other participant sports
took place indoors, generally in the winter months. The college,
YMCA, and private-gymnasium building rush of the 1880s and 1890s
provided facilities for gymnastics, handball, basketball, and
volleyball.

The rise of spectator sports paralleled that of participant sports.
Construction of special-purpose stadia or fields generally required
supporters to commit for leisure significant funds and urban space.
Implicit in the commitment and enthusiastic public response
was that performances offered were worth watching; these ath-
letes were expert practitioners. Organized baseball, which acquired
the sobriquet the "national pastime," and intercollegiate football, commonly known as "king football," reigned as the most

popular spectator sports in the late nineteenth and early twentieth centuries.

Organized baseball was quick to capitalize on the willingness of the middle class to pay for leisure, and was from the National League's creation in 1876 a for-profit enterprise. Sound late–nineteenth-century business principles marked the incorporation. The league instituted among its members uniform ticket prices, predetermined schedules, and a single set of rules. It stipulated minimum city size for a franchise, demanded financially stable ownership, and penalized violators of league precepts. The league expelled teams for failure to complete schedules, or players for gambling. It adopted a reserve clause in 1882 making salaried players the exclusive property of the club they played for; players were unable to change teams unless sold, traded, or released. A league-wide salary ceiling kept costs down and profits substantial. Using negotiations, agreements, mergers, threats, and more aggressive tactics, the league beat back challengers and oversaw the organization of subsidiary, or minor, leagues. The National League's monopoly was overturned in 1900 by the Western League (later renamed the American League), whose well-heeled owners raided National League rosters and offered players salaries far in excess of the senior circuit's limit. Both leagues bore the brunt of unrestrained competition poorly, and in 1903 agreed to affiliate. The accord, which also gave us the World Series, enabled the owners to control salaries and to prevent others from forming competing leagues.

"King football" also owed its popularity to the middle class. As expertise replaced character and a college education became essential for entry into the professions and business, the children of those Americans wealthy enough to afford college attended in unprecedented and constantly increasing numbers. Graduates and their parents developed an identification with these schools, which spread to the gridiron. That these sporting events had so little competition from other spectator sports certainly contributed to their success. Princeton and Rutgers played the first intercollegiate football game in 1869, and this student-controlled effort almost immediately led to college-administered contests and salaried coaches. By 1881 football had gained its first national organizational entity, the Intercollegiate Foot Ball Association, which established a system of football rules for games among member institutions. Football conferences quickly formed within the association, rivalries intensified, All-American teams were selected, and crowds in excess of fifty thousand were not uncommon. Nor were the consequences unpredictable; colleges sometimes paid their coaches more than their presidents, while emphasis on victory replaced sportsmanship and led to increasingly violent incidents. Commentators noted unethical, dangerous play from the early 1890s, but the Association refused to make major changes in a popular game. Several colleges discontinued football around 1900. President Theodore Roosevelt, an enthusiastic sports advocate, was among those appalled by the carnage. He called representatives of Harvard, Princeton, and Yale (then football powerhouses) to the White House in 1905 and threatened to ban football unless colleges took dramatic action. Eighteen college players died on the gridiron within two months of the White House meeting. Finally, in December 1905, representatives of sixty-two colleges and universities gathered to form the Intercollegiate Athletic Association of the United States (IAAUS, renamed the National Collegiate Athletic Association in 1910) to clean up the game. Dominated by college administrators, not coaches, the IAAUS radically changed football. It created a neutral zone between teams, legalized the forward pass to spread out players, and raised first-down yardage from five to ten yards to encourage passing. The organization also mandated chalk lines on the field to assist referees, and later forbid locking arms and the flying-wedge formation.

A Market for Sporting Goods   The rise of spectator and participant sports created a corresponding demand for sports equipment. Albert G. Spalding parlayed careers as a star baseball player in Boston and Chicago, as promoter of the Chicago White Stockings, and as a founder of the National League to achieve a position in the sports-equipment industry comparable to that of Eastman in photography. His firm, A. G. Spalding and Brothers, was created in the same year as organized baseball. Among its first assignments were publication of the official league rule book, and exclusive manufacture of the official league ball. Spalding doggedly publicized himself and his business, claimed falsely that its products were the result of interchangeable manufacture, and maintained that standardized sports equipment was essential "to insure uniformity and guard against fraud." His company retained its exclusive agreement with major-league baseball into the twentieth century. It also produced the various minor leagues' official balls.

A. G. Spalding and Brothers' National League ties earned it enough money by 1880 to develop a three-acre industrial site, and to erect separate factories for the production of baseball bats, croquet implements, ice skates, hunting and fishing gear, roller skates, and sundry other sports apparatus. In 1887, it turned out more than a million baseball bats. Spalding also made baseball uniforms; exclu-

Spalding's advertising prowess enabled the company to capture the lion's share of the sporting-goods market.

sive supplier of apparel for the National League, the company initially attempted to designate field position by jersey color. It abandoned that effort after a year, but continued as the league's sole uniform supplier, and gradually added the minors. Spalding further expanded operations in the 1880s and early 1890s. By 1895, it was the nation's leading manufacturer of golf clubs, tennis racquets, bathing suits, footballs and football apparel, basketballs, catchers' masks and chest protectors, uniform bags, baseball gloves, and ball-and-strike indicators, which umpires used. Its Chicopee, Massachusetts, plant produced bicycles. Spalding was awarded an exclusive contract to provide all the equipment and uniforms for the 1900 and 1904 Olympic games.

Spalding's connections and genius for publicity, coupled with the firm's deserved reputation for quality, enabled it to dominate the sports market. Its owners pumped profits back into the organization to secure new, better machines, modernize facilities, and explore new production lines. Spalding also relentlessly purchased patents and competitors. The company deflected antimonopolistic sentiments in several instances by selling sporting goods under labels of secretly acquired firms. This equipment was often produced at the same plants as regular Spalding apparatus and was in every way identical except for the manufacturer's name. Spalding had the appearance of competition without its liabilities.

Bicycles

Spalding pioneered in sports equipment, but he became a bicycle manufacturer only after dozens had preceded him. The Centennial Exhibition first had brought bicycles to America's attention. These expensive English imports with large front and small rear wheels were popular only with the upper class; A. A. Pope manufactured an American version for the same market in the late 1870s. To protect their interests in bicycle racing as sport, upper-class bicycle enthusiasts formed an organization, the League of American Wheelmen. Created in 1880, the League sanctioned races and generally opposed efforts to restrict cycling. This upper-class monopoly was broken by the introduction of cheaper, chain-driven safety bicycles about a decade later. It touched off a middle-class boom, and adoption of rubber pneumatic tires a few years later continued it. Only then did Spalding manufacture bicycles. A slew of cycling academies rose to teach riding, more than eighty-five periodicals regularly followed the craze, and hundreds of cycling clubs opened their doors. Bicycle manufacturers sponsored annual New York City trade shows, while entrepreneurs operated rental establishments. An estimated four million Americans owned cycles in 1895, and several million more rented them.

As cycling became a mass activity in the late 1880s, the League began to reflect its new middle-class constituency. Its redirected thrust was agitation for responsible cycling. It developed road signs and traffic signals and favored legislation requiring bicycle lights and bells. Two League-supported comprehensive city ordinances, in New York and Chicago respectively (both passed in 1897), served as models for other municipalities, and marked the zenith of its influence. The New York City code mandated a maximum cycle speed of eight miles an hour, required cyclists to keep feet on pedals and hands on handlebars at all times, and granted pedestrians right of way. It stipulated that cyclists ride on the right side of roadways, use lights from a half hour before sunset to a half hour after sunrise, and sound bells before passing. In Chicago, the League backed cycle registration. The law specified that bicyclists pay registration fees, record descriptions of their machines at police headquarters, and affix numbered identification tags to their vehicles.

The introduction of the chain-driven safety bicycle after about 1890 sparked a middle-class cycling boom.

Ironically, the bicycle boom burst in the same year that these landmark ordinances were enacted. The League within three years had abandoned bicycles, concentrated on improving roads, and adopted a new name, American Road Makers. The nation had discovered automobiles. (For comments on the relationship between bicycle manufacturing and automobiles, see Chapter 6.)

Automobiles

The automobile was a European invention, but Americans expressed a sudden interest in them in the late nineteenth century. Yet they initially had difficulty deciding among machines powered by steam, electric, or internal-combustion engines. Each was propelled by improved varieties of long-familiar engines.

Automotive internal-combustion engines generally were four-stroke cycle engines in which flywheels carried pistons through the three non-working strokes. Patented in 1866 by Nikolaus Otto, a German inventor, to utilize extensive urban coal-gas distribution systems for domestic and industrial power, Otto engines were adapted to automobiles in the mid-1880s by Gottlieb Daimler, who had served as Otto's production manager. Daimler and associates substituted gasoline for coal gas and introduced float-feed carburetors. Their competitor, Karl Benz, devised electric-induction coil-sparkplug ignition systems to facilitate safer, steadier combustion. The French engineering firm of Panhard and Levassor acquired rights to manufacture Daimler's motors in 1890, positioned them in front of chassis, ran crankshafts longitudinally, and employed combination clutch, gearbox, and transmission systems.

Automobiles quickly became the darlings of French high society, which sponsored several competitions in the mid-1890s among the three major engine types. Each had advocates. Steam vehicles accelerated smoothly but started slowly and required copious amounts of water and fuel. Electric cars operated simply and quietly, but their huge batteries limited range and utility. Internal-combustion automobiles started immediately and carried large power reserves, but proved noisy and difficult to repair. These contests testified to the superiority of internal-combustion-engine vehicles—their speed and dependability usually translated into victories—and publicized automobiles internationally. An attentive American public followed these races. By 1895, the U.S. patent office had received more than five hundred automotive applications, almost all of which had been filed within the previous two years.

New England bicycle men, Central Michigan carriage-makers, and Lake Erie marine-engine machinists led the American drive to produce automobiles. Their previous involvement with transportation, familiarity with large-scale production, and ownership of machine tools eased the transition. Wealthy Americans constituted their market. Early manufacturers sought to build fast, large, luxurious, and dependable cars, and merchandized their products accordingly. They entered endurance, dirt-track, and cross-country races, attempted to set land speed records, and secured testimonials from famous political and cultural figures.

America's first automotive organizations served only the very wealthy, those individuals who could afford to purchase automobiles. More than twenty-five city-based clubs were formed prior to 1901 to provide members social amenities compatible with their station. But hundreds of Americans produced steam, electric, and gasoline automobiles by the early twentieth century, and some began to dabble in the middle-class market. Ransom E. Olds was probably most successful in tapping the middle class. The nation's largest automobile producer, Olds manufactured several thousand vehicles a year as early as 1902. Henry Ford's yearly sales did not surpass Olds's until Ford permanently left the luxury-car field some years later.

The middle class's entrance into motoring produced a change in automotive institutions. Those established in 1902 generally retained their local orientation but welcomed a more heterogeneous clientele and aggressively boosted motoring; they dispensed with social activities and instead sponsored state and city legislation and pushed for good roads. One extant organization, the Automobile Club of America, took a different route. Formed in 1899, the New York City club had national aspirations, and although it initially admitted only the well-to-do, it accepted a broader constituency after 1901. In 1902, it unveiled its program to create a system of national motoring organizations. It expected city-based automobile clubs to become its local affiliates, but these bodies balked at ceding autonomy to the New York group. Negotiations continued and a new organization, the American Automobile Association (AAA), was born to sit at the apex of a pyramid of state and local automobile societies. Open only to clubs, not individuals, the AAA championed laws protecting and regulating motoring as well as good roads. A competitor, the American Motor League, tried to recruit unaffiliated individuals, but the AAA's organizational base enabled it to withstand the challenge. The League, which was run by leaders of the now defunct League of American Wheelmen, lost all influence after 1904 as the AAA expanded its membership to include individuals. Only then did the AAA attain the position suggested by its name.

The AAA benefited from earlier state and local agitation as it attempted to encourage responsible motor-vehicle ownership and operation nationwide. No measure proved as controversial as vehicle registration. New York State enacted a registration law in 1901; by 1905, twenty-five other states had followed suit. Many cities and some rural counties previously had required motorists to register vehicles and display identification tags, or license plates. These laws enabled governments to identify vehicle owners and ensure that property taxes and damage claims were promptly paid. But problems resulted when motorists journeyed from one jurisdictional unit to another. Vehicles had to be registered in each place, a factor that slowed interstate travel and cost motorists money. As early as 1902, automobile clubs lobbied the U.S. Congress to establish a single standardized national registration, and in 1905 an AAA-inspired bill was introduced on the floor. Fearing national intervention and a loss of revenue, state legislatures opposed the move and took action. Starting in 1905, they reined in localities, passed single statewide vehicle-registration statutes, and recognized other states' registration policies. By 1912, nationwide reciprocity had obviated the need for national law.

Operator licensing proceeded more slowly. Chicago instituted an eighteen-question test in 1900, the first such move. Administered orally by an examination board, the annual test covered automobile servicing—owners repaired their own cars—motorists' responsibilities, and operators' past driving records. Other localities soon followed, and a situation similar to that surrounding registration ensued. Auto clubs disliked the confusion, but supported registration to keep unprepared drivers off streets. Their involvement led to two pioneering state-licensing laws, both of which superseded local arrangements. The 1906 New Jersey statute required operators to sign affidavits testifying to their driving experience, physical condition, and knowledge of state motor-vehicle laws. Massachusetts enacted more comprehensive legislation a year later. Motorists needed to pass both written exams and driving tests. By 1909, ten states had adopted licensure laws combining affidavits and testing and had entered into reciprocal agreements.

Laws systematized nationwide the pursuit of the leisure activity of motoring. A similar sorting out of automobile and automotive-parts manufacturers occurred at roughly the same time, which reduced consumer choice and confusion by rationalizing the automotive field. But this rationalization was neither inspired by automobile owners or operators, nor was it accomplished through legislation. The automobile industry's relatively small sales and its many poorly capitalized manufacturers encouraged speculators to bank on its vast potential late in the new century's first decade, and to attempt to seize control through combination. The result was that by the 1910s a relatively few large firms dominated the industry.

The escapades of William Crapo Durant and Benjamin Briscoe are especially enlightening and demonstrate how comparatively simple it was to form a large automotive corporation in the early twentieth century. In 1908 Durant established by New Jersey charter a holding company and began to trade the firm's stock for automotive-company titles. Durant's holding company had only $2,000 capital. Yet desperate automotive manufacturers were willing to exchange corporate control for a chance to share in the anticipated domination of the marketplace. Buick, which Durant already owned, and Olds, which had fallen on hard times, quickly capitulated.

Durant gave out $60 million in unsecured stock within a year and acquired more than one hundred small firms. He also gained patent rights to numerous devices to forestall future competition. Many were designed for steam or electric cars and went undeveloped. Others were inferior to those already in use. Two potential acquisitions demanded cash: Durant satisfied Cadillac's terms, but Ford offered stiffer resistance. Henry Ford requested $8 million immediately. Durant countered with a proposal of $2 million right away, a similar amount a year later, and the balance the following year. Ford turned down Durant's firm, General Motors Company.

Briscoe, who had participated with Durant on two earlier ventures, was much less successful on his own. His initiative nonetheless helped order the industry. He set up a holding company, the United States Motor Company, in 1910 and started to trade paper stock for corporate titles. One hundred thirty companies were "purchased" for $42.5 million in stock within a year. Briscoe soon found that with the exception of his own car company, Maxwell-Briscoe, none of the U.S. Motor Company's acquisitions was profitable. He subsequently ran short of operating money and sought bank loans, which were granted with the stipulation that he hire Walter Flanders (see Chapter 6) to streamline the holding company. Flanders closed virtually all 130 acquisitions, keeping alive only those firms directly related to the production of U.S. Motor's one moneymaker, the Maxwell-Briscoe.

Briscoe's failure, like Durant's then-modest success, nevertheless had the effect of eliminating well over one hundred automobile companies. Although the industry never was unified under a single corporate banner, its streamlining placed a comparatively few large corporations active nationwide in control of production of a leisure commodity. By the mid-1910s, a generous handful of firms made the overwhelming majority of the nation's for-pleasure motor vehicles.

Motion Pictures

The motion-picture industry also was rationalized in the early twentieth century; several large combines came to dominate picture production and distribution. This predominantly urban leisure form was accessible to all social classes, but its origins rested in nature photography and among the upper-class horse-racing set. To settle a wager in 1877, Eadweard Muybridge, an English émigré and photographer of nature settings, brought photographs to motion; twenty-four wet-plate cameras lined up side to side captured

Eadweard Muybridge prepares his cameras to determine if all four feet of a horse leave the ground when the horse is galloping.

Muybridge's zoopraxiscope projected his horse photographs and was a progenitor of the modern motion-picture projector.

a galloping horse. Muybridge used bright light, a white background, and fast shutter speeds to record the horse's movements. He printed these photos on a glass wheel which, when rotated, passed them sequentially in front of a strong light and projected them on a distant surface. Earlier European work alerted Muybridge that continuous projection did not yield the illusion of motion, so he placed a shutter on his projector's lens to stimulate a strobe effect. This produced arrested motion, which human eyes translated into "true" movement. Muybridge's "zoopraxiscope" was within a few years projecting 200-photograph sequences.

Nature study, not motion pictures, fascinated Muybridge, and he did not commercialize his device. But his demonstrations, lectures, trips to Europe, and subsequent examinations of aviary and human movement publicized his techniques. Others investigated cinematography's possibilities, adopting gelatin emulsion film in the early 1880s, roll film and roll holders by 1887, and celluloid in the early 1890s. Thomas Edison was the most noted and commercially successful pioneer. His familiarity with continuous-roll processes—telegraphic printers and sound-cylinder phonographs—prepared him to appreciate cinematography's mechanical elements. Edison initially hoped to join phonographic and photographic cylinders for use by individual viewers, but he failed to develop adequate coordination mechanisms. Edison abandoned the modified glass-wheel principle but remained committed to individual viewing even as his group patented in 1889 the essentials of the modern motion-picture projector: reel-to-reel perforated film, sprocket wheels, revolving shutter, and constant light source. Nonprojecting kinetoscopes were Edison's first commercial cinemagraphic triumph. Announced in 1891, electric-motor-powered kinetoscopes pulled film loops forty to fifty feet long past electric lights at forty-six frames per second; shutters positioned between film and viewer produced arrested motion.

Edison licensed kinetoscopes to commercial establishments in 1894. Individuals apparently enjoyed peering into oculars, and kinetoscope parlors, like phonographic parlors a few years earlier, became common urban institutions nationwide. The kinetoscope's popularity stimulated interest in projection cinematography, and several projectors competed for the late 1890s market. The Biograph projector, designed by William K. L. Dickson, a former Edison associate, debuted in a Pittsburgh auditorium in September 1896 and quickly spread across America. Before the century's close, others had improved projection techniques by synchronizing shutter and film motion, and by positioning a slack loop in the film to reduce tearing. A new recreational institution, movie houses, followed in the projectors' wake. Americans supported five thousand in 1907 and twice as many three years later.

Movie houses required a constant supply of material to present, as Edison anticipated before he leased his first kinetoscope. In 1892, he established a motion-picture studio at his West Orange site. Other manufacturers followed Edison's example; about a dozen New York or Chicago producers of projection equipment were America's pri-

*New York and Chicago were the initial homes of the
motion-picture industry, but feature filmmakers soon
gravitated to sunny Southern California.*

mary early filmmakers and dominated in the years immediately
after 1900. Movie houses formed exchanges to distribute films among
member theaters.

Competition among producers of equipment and films lowered
profits because it permitted exchanges to play producers against
one another, while similarity of projection devices raised the specter of patent-infringement suits and other costly litigation. Major
producers organized themselves into the Motion Pictures Patents
Company (MPPC) in 1908 to ameliorate both problems. The MPPC
cross-licensed patents among member producers and gained an
agreement from Eastman, by far the largest celluloid supplier, not
to sell the nitrocellulose-based product to unaffiliated firms. Distribution exchanges and individual exhibitors had no choice but to
fall in line, for without access to films, they would go out of business. The MPPC formalized this hierarchical relationship in 1910
by organizing exchanges and exhibitors into the General Film
Company, which was restricted to handling MPPC products.

A few independent picture-producers survived the MPPC on-slaught. Buying celluloid from Europe and a handful of small American firms, they managed to generate motion pictures. Independents also attempted to ensure a market for their product by binding together unaffiliated exhibitors in 1910 to create the Motion Picture Distribution and Sales Company. Their success in lining up exhibitors led Eastman to reconsider his MPPC agreement, and he began to sell celluloid to independents in 1911. Easy access to film buoyed independents, and in 1912 they established a new consortium, the Universal Film Manufacturing Company, to challenge the MPPC.

Independents' tactics differed from the MPPC. They tried to shape demand rather than explicitly control outlets. Independents emphasized product and hoped to attract audiences through recognition of their performers' and directors' names and efforts. Their ultimate success in creating a star system produced a scramble for celebrities in the late 1910s and paved the way for additional competition. Fox, Loew, Paramount, and Warner Brothers each was formed by the early 1920s.

The early motion-picture industry profited through establishment of a technologically based leisure activity designed to attract diverse social groups. In that sense, it differed from automobiles, bicycles, and spectator sports, which achieved profitability by serving the desires for leisure of more circumscribed markets. But the manner in which these several leisure forms were shaped and shaped themselves—why they were successful late–nineteenth- and early–twentieth-century "inventions"—was fundamentally the same. Productions, technologies, institutions, and industries that seemed sensible did so because those who viewed them recognized them each as constituting an element in a system of discrete yet interdependent, hierarchically arranged parts, or as the system itself. This happened even as Americans disputed among themselves the parameters of any particular system. Nor was the nation's physical environment immune from this conceptual onslaught. The revamped urban and agricultural environs were tangible products of these ideas, as was the infatuation with nature. Work would prove no exception.

NOTES

1 Quoted in M. Christine Boyer, *Dreaming the Rational City* (Cambridge: M.I.T. Press, 1983), p. 78.

2 Quoted in Alan I Marcus, *Agricultural Science and the Quest for Legitimacy* (Ames: Iowa State University Press, 1985), pp. 13–14.

3 Quoted in Marcus, *Agricultural Science*, p. 9.

4 Quoted in Peter J. Schmitt, *Back to Nature* (New York: Oxford University Press, 1969), p. 201.

FURTHER READING

Ackerman, Carl W. *George Eastman* (London: Constable, 1930). Contains selections from this photographic pioneer's letters.

Bledstein, Burton J. *The Culture of Professionalism* (New York: Norton, 1976). A provocative interpretation of the meaning of professionalism.

Boorstin, Daniel J. *The Americans: The Democratic Experience* (New York: Random House, 1973). A comprehensive, challenging examination of American society from the Civil War to the early 1970s.

Condit, Carl W. *The Rise of the Skyscraper* (Chicago: University of Chicago Press, 1952). The best single volume on the subject.

Condit, Carl W. *American Building* (Chicago: University of Chicago Press, 1968). A popular survey, joining design and technique.

Dulles, Foster Rhea. *A History of Recreation* (New York: Appleton-Century-Crofts, 1965). One of the earliest books on the subject.

Flink, James J. *America Adopts the Automobile, 1895–1910* (Cambridge: M.I.T. Press, 1970). A highly readable scholarly analysis.

Friedel, Robert. *Pioneer Plastic* (Madison: University of Wisconsin Press, 1983). A short, informative history of celluloid.

Fries, Robert F. *Empire in Pine* (Madison: State Historical Society of Wisconsin, 1951). Concerned with the socioeconomic and political aspects of the lumber industry.

Gates, William B., Jr. *Michigan Copper and Boston Dollars* (Cambridge: Harvard University Press, 1951). Joins economic and technological history effectively.

Glaab, Charles N. and Lawrence H. Larsen. *Factories in the Valley* (Madison: State Historical Society of Wisconsin, 1969). Discusses the rise of wood-based industries.

Gray, R. B. *The Agricultural Tractor: 1855–1950* (Beltsville, MD: USDA, 1954). A year-by-year survey of new tractors.

Hays, Samuel P. *Conservation and the Gospel of Efficiency* (Cambridge: Harvard University Press, 1959). An excellent examination of the politics of nature.

Hoogenboom, Ari and Olive. *A History of the ICC* (New York: W. W. Norton, 1976). The definitive history of the early years of the Interstate Commerce Commission.

Hotchkiss, George W. *History of the Lumber and Forest Industry of the Northwest* (Chicago: George W. Hotchkiss, 1898). A catalogue of lumbering ventures and sites.

Jenkins, Reese V. *Images and Enterprise* (Baltimore: Johns Hopkins University Press, 1975). An effective discussion of Eastman and the early motion-picture industry.

Kolko, Gabriel. *Railroads and Regulation, 1877–1916* (Princeton: Princeton University Press, 1965). Considers the influence of railroad owners on the establishment of railroad regulation.

Larson, Agnes M. *History of the White Pine Industry in Minnesota* (Minneapolis: University of Minnesota Press, 1949). Accentuates the technical aspects of the lumber industry.

Lesley, Robert W. *History of the Portland Cement Industry in the United States* (1924; New York: Arno Press, 1972). A useful book, written by a member of the industry.

Levine, Peter. *A. G. Spalding and the Rise of Baseball* (New York: Oxford University Press, 1985). An excellent biography.

Lewis, W. David. *Iron and Steel in America* (Greenville: Eleutherian Mills-Hagley Foundation, 1976).

MacGowan, Kenneth. *Behind the Screen* (New York: Dell Publishing Co., 1965). Contains some useful information about the early motion-picture industry.

Maxim, Hiram P. *Horseless Carriage Days* (New York: Harper and Brothers, 1937). Reminiscences of an early-American automobile builder.

Mussey, Henry R. *Combination In the Mining Industry* (New York: Columbia University Press, 1905). An economic treatise, written in a muckraking style.

Parsons, A. B., ed. *Seventy-Five Years of Progress in the Mineral Industry, 1871–1946* (New York: American Institute of Mining and Metallurgical Engineers, 1947). A compendium of developments in mining and refining.

Pound, Arthur. *The Turning Wheel* (Garden City: Doubleday, Doran, 1934). Paid for by General Motors, this popular history of the company is a treat.

Rader, Benjamin G. *American Sports* (Englewood Cliffs: Prentice-Hall, 1983). A brief survey of some aspects of the topic.

Rae, John B. *The American Automobile* (Chicago: University of Chicago Press, 1965). A solid, single-volume study.

Schlebecker, John T. *Whereby We Thrive* (Ames: Iowa State University Press, 1975). An all-too-brief history of American farming.

Scott, Mel. *American City Planning Since 1890* (Berkeley: University of California Press, 1969). A comprehensive catalogue of city-planning initiatives.

Scott, Roy V. *The Reluctant Farmer: The Rise of Agricultural Extension to 1914* (Urbana: University of Illinois Press, 1971). Excellent on the late–nineteenth-century debates about farming.

Segal, Howard P. *Technological Utopianism in American Culture* (Chicago: University of Chicago Press, 1985). Examines the plans of writers who envisioned a perfect, technologically produced society, and shows how these blueprints reflected then-contemporary American culture.

Smith, Duane A. *Mining America* (Lawrence: University Press of Kansas, 1987). A critical assessment of the nation's mining practices.

Smith, Robert A. *A Social History of the Bicycle* (New York: American Heritage Press, 1972). An excellent and enjoyable history of the bicycle in America.

Stover, John F. *American Railroads* (Chicago: University of Chicago Press, 1961). A quick overview of railroad economics. Weak on technical detail.

Webb, Walter Prescott. *The Great Plains* (1931; Boston: Ginn and Company, 1974). A masterful introduction to the history of that region.

6
Systematizing Workers and the Workplace

THE SYSTEM IN THE HOME

I n 1841, Catharine Beecher published her classic *Treatise on Domestic Economy*. The book outlined a set of manners, morals, and obligations that Beecher identified as characteristically American, as defining an American homemaker. In that sense, Beecher's volume provided an illuminating parallel to Emerson's American Scholar address. Both authors announced a new type of person, one typified by a behavior unlike that of citizens of other nations. Only the locus of concern varied. Emerson called for a distinctly American scholar, Beecher for a distinctly American woman.

Beecher's 1869 effort, *The American Woman's Home*, expressed a radically different perspective. Coauthored by her sister, Harriet Beecher Stowe, the book described in great detail an efficient home's organization and operation. No item seemed too small or insignificant for analysis. For example, the authors scrutinized the kitchen, enumerated its constituents, and explained their proper alignment in terms of the requirements of kitchen work. Food preparation, cleanup, and storage were designated subsets of kitchen work. Flour bins, sinks, dishes, strainers, and the rest each received the Beechers' attention as they attempted to reduce kitchen form and activity to a series of specific rules. Nor was their treatment of the kitchen and its elements unusual. To the Beechers, efficient housekeeping was the product of rationally conceived designs, plans, and procedures; it was a consequence of study, not a manifestation of character. Study would result in systematized households and household work.

This sentiment appeared repeatedly after 1870. The establishment of domestic science, domestic economy, or home-economics courses, usually at land-grant colleges, was perhaps its most obvious manifestation. Iowa Agricultural College offered its first course in 1875, as did Illinois and Kansas. College kitchens, dining rooms, and laundries supplemented lectures and served as laboratories in which students experimented in and learned housekeeping science. Division of home economics into subunits quickly followed. Foods, household administration, textiles and clothing, and the like emerged as distinct fields of inquiry within the broad rubric of home economics. The identification of domestic economy as an area of study—one composed of discrete and differentiable subdisciplines—and of its investigators/teachers as a specific group, led to annual conferences, beginning in 1899, at Lake Placid, New York, and to creation of the American Home Economics Association in 1909. It also produced a definite periodical literature. The *Journal of Home Economics* (1908) catered to researchers; it was preceded by general-circulation magazines such as *Good Housekeeping* (1885), *American Kitchen Magazine* (1894), and *Sanitary Home* (1899).

New Household Machinery

Women were certainly not overlooked in the late–nineteenth- and early–twentieth-century reconceptualization of American society. The household, especially the kitchen, became their factory, while they served as production experts and, if unable to afford servants, laborers. Indeed, the home required rationalization as much as the factory. Brains were to guide hands to systematize households and housework. Reduction of drudgery and speeding of operations were the goals; the amount of work, whether great or small, was less important than how that work was pursued. Working patterns derived from study were cherished, as were machines that enabled women to accomplish household tasks more quickly and more sci-

American women embraced electrical machinery in their zeal to reduce housekeeping to a science. This 1911 electric kitchen epitomized that effort.

entifically. In a sense, machines had an intrinsic virtue; they were the consequence of rational, systematic thought put into practice. That they were composed of physically discrete mechanical parts precisely arranged for a particular purpose clearly resonated with the notions of late–nineteenth- and early–twentieth-century Americans.

All machines were not regarded equally, of course. Machinery had to enhance or at least not detract markedly from human performance. But consonance between machines and the organizational ideas of the post-1870 period surely eased their introduction into all facets of American society, and, coupled with the nation's human and material resources, fostered their incredible proliferation. Machinery was a metaphor for society that held true no matter what the locus; it was as accurate for the household as it was for large-scale production.

In the case of household machinery, patents for various devices were issued regularly from the 1850s, but the American public gen-

Farm women often gathered to do the laundry. Note the crank-operated wringer.

erally accepted these household contraptions only after 1870. By the 1890s, the market for these machines was sufficiently great to support enough manufacturers to finance trade papers such as *House Furnishing Review*, while a decade later a manufacturer-sponsored Home Furnishing Goods Exhibition was held at New York's Madison Square Garden to join retailers with consumers and manufacturers.

The Washing Machine   The mechanical incursion into the household was led by washing machines and vacuum cleaners, which mechanized two of the most labor-intensive housework activities. The earliest washing aids mimicked the time-honored practice of rubbing fabrics over ribbed washboards; clothes were cranked by hand through beds of rollers turning in opposite directions. Not until 1851 did a washing-machine patent employ different principles. James T. King received protection for a device that exploited the natural circulation of steam and boiling water. Two concentric half-full cylinders constituted the machine's working parts. A fire was set beneath the outer cylinder, while soap and clothes were placed within the inner, mobile cylinder. Steam and boiling water caused the inner cylinder to move, and stirring the outer cylinder's water by hand-crank reinforced the movement. Agitator-type machines, devised about 1870, were similar but included a crank-activated plunger operating in the vertical plane.

Washing machines proved immensely popular from the 1870s, but few families bought their own. Commercial laundries served urban areas while cooperative laundries (several families purchasing a single machine together) made inroads in rural regions. Rinsing generally occurred in separate vessels at these locations. Clothes were then rung either by hand or between hand-powered rollers. The persistence of hand rinsing and wringing until the twentieth century stemmed not from a failure of vision but from a lack of a cheap mechanical prime-mover. A crank-operated wringer, which spun clothes at high speed to take advantage of centrifugal force, was patented as early as 1878, but it required great amounts of manual labor. Wringing by centrifugal force became feasible only with the advent of small variable-speed electric motors.

Competition was brisk as vacuum-cleaner manufacturers sought to capitalize on an untapped market.

The Vacuum Cleaner   The history of vacuum cleaners in America followed a somewhat similar course. People washed or beat rugs and carpets by hand throughout the nineteenth century. Although two mechanical-cleaner patents were awarded around 1860, neither proved commercially practicable. Both devices were essentially the same process; as people rolled the mechanisms over carpets, the revolving wheels turned fan blades, which created suction, lifted dust, and deposited it inside the apparatus. One machine also employed contact brushes to stir up carpet dirt and promote deeper cleaning. Both worked poorly; they lacked the suction necessary for the task. Electric cleaners promised to solve the suction problem, but the early electric motors' great size and weight restricted machine movement. Designers of electric vacuums either situated the motor in a central place, such as the basement, and moved tubes and hoses from place to place, or built a large truck on which to roll the cumbersome cleaning unit. Not until the 1910s did all-in-a-piece portable vacuum cleaners with small motors come on the market. By 1920 they had become standard household equipment among the wealthy.

SYSTEMATIZING FACTORY WORK

Americans accorded production machinery new prominence in the half-century after 1870, but the nature of the workplace, the factory, engendered the earliest concern. Correspondents to and essayists in the *American Machinist, Mechanical News, Engineering Magazine, Scientific American,* and other technical periodicals complained that these structures were poorly designed, disorganized, or not organized to best advantage. They discussed whether hollow square, L-, or H-shaped factories best facilitated work, contemplated the merits of one-story versus multifloored buildings, and debated proper locations of foundries, drafting rooms, and foremen's offices. These commentators, many of whom called themselves mechanical engineers, also maintained that factory administration was outmoded, chaotic, or wasteful, and therefore a detriment to productivity. They identified successful manufacturers as those with the most highly refined factory environments and administrative methods, and urged others to adopt these practices.

Suggestions, debates, and complaints intensified after 1900. By 1920, few factories resembled their 1870 predecessors. Factories and factory work had been reconceptualized and rearranged. Mass-production techniques and scientific management methods were products of that reconceptualization.

Building Systematic Factories

The Predecessor to the Systematic Factory   But that lay in the future. Factories in 1870 looked and functioned very much like their mid-century counterparts. Line shafts and countershafts traversed rectangular multistoried buildings and transmitted power from waterwheels or steam engines to pulleys and belts to drive machinery. Machines requiring the most power were placed closest to prime-movers. Work processes generally moved from the top floor down. Raw materials or disassembled pieces emerged as finished products on the ground floor; gravity helped reduce handling. Individuals who had demonstrated mastery in many factory tasks—foremen— doled out work (often contracting with employees for a specific number of items by a specified date) and supervised its completion. A building boom in the 1870s and 1880s encouraged by fire insurance companies led to a wave of new factory construction, but did little to change the traditional physical layout; the substitution of tile-protected iron for timber merely reduced fire hazards. Nor did the introduction of electricity result in changed factory organization. Electric factory lighting began to replace gas in the 1880s, and was common a decade later. And although electric companies championed electricity's other industrial applications, electric motor use was rare in the 1880s, employed only to operate stationary cranes or as replacements for steam engines or waterwheels to turn line shafts.

Systematizing Factory Administration

Attempts to coordinate factory activities, not modify factory hardware, had a more telling impact in the nineteenth century's three concluding decades. Cries for "systematic management" culminated in unit cost analyses for specific products. These detailed factory cost-accounting systems yielded production and inventory control plans, as well as wage plans, to stimulate production and lower costs. Rationalization of factory endeavors stemmed from the proposition that each factory operation contributed to the system's efficiency, as measured by profitability. It required experts to develop central factorywide plans, which were based on information generated by cost accounting. Job cards and time clocks helped produce the requisite data. Establishing central offices staffed by clerical forces was among the fruits of rationalization, as were centralized purchasing and standardized materials. Storing supplies in designated areas and permitting only foremen to requisition materials also checked waste and promoted control. Premiums granted to workers who exceeded production quotas and piecework rates were aimed to spur production and productivity. Foremen in these facilities were reduced to head-worker status—responsible for supervising their departments, gangs, crews, or shops, but not for routing materials or instituting procedures. Their job was to implement central policy, not conceive or tamper with it.

The Central Office   The central office stood at the apex of this new factory system. Few enterprises possessed anything structured or functioning like it prior to the 1870s. Until then manufacturing establishments generally had consigned paperwork to small "cages" on factory floors. Industrialists' offices had resembled mid-century parlors, not places of work; people were greeted and deals were negotiated there. Clerks had been more nearly apprentice businessmen (virtually all were males) than office help. They had copied contracts, delivered missives, written letters, and the like to learn about business. They purportedly saved their money, ultimately to begin their own businesses. The term "secretary" had referred to either rolltop desks with pigeonholes, or to managers of entire businesses. Letterpresses and simple ledgers had constituted record-keepings' mainstays.

Ad hoc, undifferentiated office practices gave way in the decades after 1870. Offices became businesses' brains, the sites at which material was gathered, examined, and analyzed to derive future business needs and strategies. Their significance was reflected in spatial arrangements. Post-1870 offices grew larger, more populous, and (in the case of manufacturing) physically distinct. Although every office operation dealt with record generation, collection, maintenance, or retrieval, each required its own specialist; filing, shipping, and billing clerks had different tasks, as did typists and stenographers. Each specialist, many of whom were women, was counted on to standardize the appropriate record forms and establish precise record-keeping systems. Routinization, argued one commentator, "is the machinery of progress." It is like "oil, which if properly applied, will make the machine run smoothly."[1]

The introduction of office equipment and specific office procedures emerged from that milieu. Desks, filing cabinets of standard dimensions, standard ledgers, standard-sized paper and envelopes, carbon paper, and typewriters (which spawned an entire class of office equipment including addressographs, mimeographs, adding machines, and check-makers) were the implements that rationalized the office.

Analyses of Efficiency   Among the tasks often undertaken in central offices were economic analyses of power generation and transmission. Until the 1880s analysts had presumed that lineshafting/belting systems were the sole viable means of power transmission. Their examinations were restricted further, moreover, by nationwide adoption in the 1860s of cold-rolled wrought-iron and steel shafting, and in the mid-1870s of standardized bearings and supporting hardware. Pulleys and belts were generally the foci of their concern, and most investigators zeroed in on the latter (bearing lubrication also attracted attention). Using various torque measuring devices—dynamometers—they sought to quantify the power lost in transmission and to develop the most efficient belt design, thickness, and width. Inquiries into the "belting problem" produced numerous calculations of belt variables—areas of contact, tension, adhesion, even effects of atmospheric pressure and temperature and the like—but little agreement about its resolution. Factory lineshafting/belting systems remained troublesome.

Design of Systematic Factories   Factories designed and erected after about 1900 were usually significantly larger than earlier manufactories. Most were several buildings, not single massive structures, and suburban locations often replaced more expensive city sites. Structural steel and reinforced concrete were predominant construction materials, which encouraged new factory forms, larger windows, and flexibility of design. Using information gathered by systematic-management crusaders, architects tailored facilities to the type of industrial manufacture to be performed there, and by the 1910s buildings were designed to enhance work flow between departments. So, too, were factory offices. Managers restructured them to make paperwork proceed efficiently and to enable office workers to focus exclusively on their jobs. Flow charts simulating paper routes became the basis for locating desks, which also were positioned to discourage conversation and peering out of windows.

New Use of Electric Power   Electric power did not play a significant role in turn-of-the-century factory design. To be sure, from the late 1880s, trade associations such as the National Electric Light Association (a consortium of electrical manufacturers) and the American Institute of Electrical Engineers pushed electric power for industry as cheaper, cleaner, quieter, and more adaptable. But electric power accounted for only about 5 percent of mechanical drive capacity in 1900, and only about 25 percent in 1909. Not until 1920 did electric motive power surpass steam and water. By

*Laying out factories to facilitate flow and reduce handling
were two goals of late–nineteenth-century managers.*

1929, 78 percent of industrial power was electric. The ways in which
factories utilized electric drive power are telling: until about 1905
most facilities employing electric motors used them merely to turn
factory lineshaft/belt systems; a large electric motor simply replaced a waterwheel or steam engine as the plant's prime-mover.

Electric group drive superseded single large-motor arrangements
and remained the predominant electrical form through World War
I. Introduced in the early 1890s at a few facilities, group drive offered industry somewhat more flexibility but did not free it of lineshafting. A group-drive factory housed several shorter, separate
lineshafts—generally from thirty to 150 feet long—each powered
by its own motor. These electric engines acted independently; each
turned its line at a different speed and enabled manufacturers to
match machine and line speed. In practice, similar machines
(grinding, stamping, milling, or the like) ran on the same line. Group
drive's shorter shafts led to innovative factory design, simplified
expansion, and operating economies. Only lines in use, not whole
factory lineshaft systems, required power. Malfunctions idled only
single lines, not complete enterprises.

Unit drive did not become popular until about 1920, although
several leading industrialists had introduced it nearly a decade earlier. The union of motor and machine in a single device permitted

factory owners to remove all shafting, belts, and pulleys, leaving overhead space for traveling cranes. Motor failure disabled only a single tool. Machines became truly portable, and factory design and expansion knew no restraints. Most manufacturers chose to minimalize handling and emphasize production flow; they organized factories by the natural sequence of manufacturing operations.

Systemization of Large-scale Production

The late–nineteenth-century passion for efficiency and standardization worked to reduce the human variable in production's equation. Men and women were inconsistent, wasteful, and fallible. Machines more closely approached perfection, and manufacturers revamped production systems to stress precision and exactness. The substitution of special-purpose machines for filers, and establishment of rational jig/fixture and model-based gauge systems, objectified fitting. High armory practice and interchangeable parts became tests of expert production and badges of technological sophistication.

Mechanical engineers justified these practices on other grounds. Machines speeded production and, because they increased volume and capacity, lowered unit and labor costs—factors that compensated for their steep capital investment and higher operating expenses. Demand existed for additional production units. Standardized production granted benefits of economies of scale. Utilization of special-purpose machine tools imposed order on manufacturing plants by emphasizing sequential flow.

Whether these claims of the benefits of systemization for industry had economic bases, or were merely rationales to encourage manufacturers to recognize the legitimacy of production specialists, may be debated. It can be argued, for example, that sequential flow was virtually impossible in late–nineteenth-century line-shafted or group-driven factories without extensive product handling and movement, which would raise costs astronomically. Similarly, several industries would have been better served by adding personnel to satisfy temporary demand, thereby keeping capital costs down. Also, economies of scale depended on markets: demand must be and remain high.

Adoption of Systematic Techniques   Nonetheless, many late–nineteenth-century American industries adopted these techniques. Northeastern machine-tool- and bicycle-makers proved among the most successful. Several factors worked to machine-tool and bicycle manufacturers' advantage. Both capitalized on heightened demand. The equation of precision with efficiency sparked a special-purpose machine-tool boom, and the bicycle craze of the late 1880s and 1890s opened a strong market for a new consumer good. Both industries required comparatively little new capital investment because they owned many of the requisite machines. From mid-century, machine-tool manufacturers had had a slow, steady market for their product. Northeastern bicycle-makers came from the ranks of firearm, sewing-machine, or agricultural-implement producers, or contracted for parts from these firms; they merely adapted their special-purpose machine tools to a new product. Their success also stemmed from refining established techniques and instituting strict quality-control systems. Bicycle-makers, most notably A. A. Pope, modified drop-forging to reduce machining, and created inspection and testing departments. Gauged several times before sale, Pope bicycles gained a reputation among professionals for expert manufacture.

High armory practice—a system of jigs and fixtures and special-purpose machine tools to yield manufacture of interchangeable parts—reigned preeminent among the Northeast's technologically advanced. But not all manufacture of interchangeable parts depended on special-purpose machine tools. Western carriage- and wagon-makers competed for the national bicycle market from about 1890, but did not adopt high armory practice. Their skilled English- and German-born mechanics replaced drop-forging with sheet-steel stamping or pressing, which produced interchangeable parts and minimalized machining; the more precise stamping and pressing employed special-purpose dies that eliminated the need for most special-purpose machine-tooling—gear-milling machines sometimes proved an exception. Machinists sat fixed at presses while runners carried material from station to station.

Stamping and high armory practice yielded interchangeable parts, but neither achieved the pinnacle of efficiency. Both were constrained by lineshafting or group drive's apparent physical limitations. Left unmodified, these power transmission systems would never permit optimal sequential flow. As important, however, neither practice addressed the problem of efficient assembly. Indeed, there was no reason to do so. Without the speed of production generated by sequential flow, no assembly bottleneck existed. Henry Ford created a facility that maximized flow. Only at that time did he confront the need for rapid assembly.

Henry Ford's Systematic Techniques   Ford began to rationalize factory operations and prepare for high-volume automobile manufacture as early as 1906. His advisers included several gifted machinists and mechanics steeped in European production techniques, plus Walter Flanders, a machine-tool salesman who, as a Vermont machinist apprentice, had become well-acquainted with high armory practice. Together these men emphasized interchangeable parts, special-purpose machine tools, and rational jigs, fixtures, and gauges, and impressed upon Ford the importance of sequential machine-tool arrangement. Lineshaft requirements hampered attempts to introduce sequencing, however, and Flanders left in 1908, just after Ford's announcement of the Model T, the car for the masses. Ford's crew outlined in great detail the Model T production process before situating machinery or beginning manufacture. Their central plan called for use of pressed steel rather than cast or drop-forged parts whenever possible to reduce machining, as well as the renewal of efforts to place machines in sequential order to facilitate flow.

Demand soon exceeded production capacity, and in 1910 Ford opened a new plant in Highland Park, Michigan. The extensively planned facility would make only a single standard product, the Model T. That decision encouraged Ford's staff to design and employ machine tools especially suited to Model T manufacture within the new factory setting rather than adapt tools from earlier sites; Ford's new special-purpose machine tools were also generally single-purpose machine tools. The freedom to create new machines that fitted precisely to both the task and environment proved critical. It enabled Ford to overcome the problem of power transmission, and by 1913 to organize machinery according to true sequential flow.

Highland Park machinery was group-driven. But Ford's mechanics capitalized on the rigidity implicit in single-purpose machine tools and in a single standard product to make the plant's lineshafting/belting systems simulate unit drive. Indeed, each production machine was envisioned as unique; it would fill one and only one place on one and only one line. Relationships to neighboring machines were static. Sequential positions were fixed. These assumptions encouraged Ford operatives to customize lineshaft-to-machine power transmission. A specialized system of pulleys and especially of gears for each machine delivered the proper power from lineshaft to machine, which ensured correct cutting-tool speed. Customized gearing and pulley arrangement mediated machine- and shaft-speed differences. This revolution in power transmission led to a corresponding revolution in the organization of production, which drew plaudits as early as 1913. One commentator marvelled that "so thoroughly is the sequence of operations followed that we not only find drilling machines sandwiched in between heavy millers and even punch presses, but also carbonizing furnaces and babbitting equipment in the midst of the machines. This reduces handling to a minimum."[2]

It would be wrong to suggest, of course, that Ford never varied machine sequence or developed improved techniques and devices. His staff seized every opportunity to replace men with machines and to eliminate skilled labor. It divided tasks into constituent parts, built new machines, and adjusted processes to speed output. Simplicity of use and accuracy of operation characterized Ford's production machines. Ford would later pay tribute to his mechanics' precision machinery by insisting that mass production had no need for fitters.

Ford's ability to arrange the new facility to better advantage speeded production, but it also created assembly bottlenecks. The company initially responded to these problems in two ways. It moved subassembly tables from their traditional position against walls to the middle of floors and placed part bins at table centers; parts were accessible to workers on both sides. Final assembly was prosecuted by a series of assembly gangs. Each gang was responsible for a particular assembly aspect and each member had a small, specified task. Automobile chassis were mounted on sawhorses in rows; the gangs moved sequentially down the rows, and runners distributed parts just prior to use.

These procedures quickened assembly but failed to match the pace of production. Assemblers worked at their own speeds and lost time traveling from station to station. The introduction of moving assembly lines, first used in 1913 for flywheel magneto subassembly, increased speed, eliminated human variability, and reduced waste. Pulled by endless chains at set rates, moving assembly lines required workers to remain stationary and to repeat a single straightforward operation. By July 1914, Ford used gravity-slide and conveyor-fed assembly lines for all sub- and final assembly. Production assistants persistently adjusted lines, seeking to determine points of best advantage. As with other facets of Ford's operation, the company publicized its assembly-line methods. Its aggressive publicity encouraged other industries to attempt to reproduce its techniques.

Moving assembly lines became the archetypical expression of Ford's methods.

SYSTEMATIZING THE HUMAN ELEMENT

Scientific Management

Scientific management fervor swept America during the new century's first two decades. Its appeal stemmed from its promise to promote industrial harmony and efficiency by impartially systematizing employer–employee relations and production. Its partisans cloaked themselves in a mantle of science by claiming that their method was scientific and its determinations therefore beyond reproach. Application of scientific management techniques required experts steeped in its precepts, and its proponents (generally American Society of Mechanical Engineers members, until about 1910) offered themselves to industries as consultants.

Scientific management summarized those themes fundamental to America in the half-century after 1870. The term originated with Frederick W. Taylor, its foremost proponent and publicist. Taylor and his apostles took the systematic management crusade of the previous decades as their starting point and sought to objectify it. Rather than base wage plans on an individual's or a group's past productivity, Taylorites wanted to use detailed job analyses, and time and motion studies of the work involved, to determine "scientifically" the standard time and output required. Those standards would be used to establish differential piece rates. Workers not meeting these standards received lower rates, while those exceeding them were paid more per piece. Taylorites boosted the establishment of central planning departments in factories to administer these practices, schedule work, and coordinate purchasing. Eight "functional foremen"—each with different specialized responsibilities—implemented and helped shape planning-department policy.

Despite excitement about scientific management's potential, and Taylor's success in publicizing his principles, few industrialists adopted the master's system exactly as he outlined it. Labor often chafed under scientific management methods, and managers guarded against usurpation of their authority. Owners complained about the lengthy stay of consultants and frequently found their suggestions impractical. As important, Taylor and his disciples sometimes disagreed. For example, Frank Gilbreath incurred Taylor's disfavor in 1911 by rejecting stopwatch-determined and recorded-time studies as imprecise. Gilbreath instead advocated "micromotion." According to Gilbreath, micromotion—filming workers and dividing each minute into 100 equal units—was the correct method for "standardizing performance of labor" because it eliminated "all error due to the human element or to differences in mental reaction times";[3] it was scientific, while stopwatches were subjective. Finally, Taylorism did not suit mass-production facilities. A shop-based management system was superfluous in plants in which work tasks were separated into constituent elements, and where moving assembly lines guaranteed regularized execution.

Taylorism proved short-lived, but a general desire to systematize factories and factory operations scientifically was more enduring. It took shape in a class of men identifying themselves as unique and expert in production-management techniques. These individuals left ASME in 1911 and established the Society for the Promotion of the Science of Management, which later merged with the Society of Industrial Engineers to form the Society for the Advancement of Management.

Industrial Social Welfare and Social Science

Large industrial concerns established libraries, choral groups, and clubhouses for laborers after about 1870. They hired counselors, instituted wage incentive plans, and improved work environments. These initiatives aimed to harmonize employer–employee relationships, to reduce turnover and soldiering (pretending to work while actually loafing), and to quash unionization and job actions. Manufacturers recognized that labor would continue as a vital aspect of the industrial system, and that efficient production depended on its efficient performance, even as they endeavored to reconstitute or replace it with machinery. The creation among laborers of positive, predictable behavior would forge them into an efficient cohort marked by reliability and productivity. Many industrialists understood that economics was only one facet of the labor question; proper job performance was coterminus with and a product of proper living. As one commentator noted, "family quarrels have an almost immediate effect on the output of lathes and drill presses."[4]

Ford's Approach to Industrial Sociology   Although Henry Ford came to that realization rather late, he quickly moved to the vanguard of those trying to systematize employee–employer relations. In 1914, he launched a landmark three-pronged program to increase labor efficiency: the Five Dollar Day, the Ford Sociological Department, and the Ford English School. These efforts attracted almost as much notice as the moving assembly line.

The Industrial Workers of the World's militant unionism, a daily absentee rate at the Highland Park plant in excess of 10 percent, and a 370-percent yearly turnover rate (requiring Ford to spend nearly $2 million to train new workers in 1913) were the immediate impetus for his novel decision. He paid $5 a day to every "qualified" Ford worker, "even the lowliest laborer and the man who merely sweeps the floors." This "profit-sharing plan" more than doubled the average Ford factory worker's pay. Rather than increasing expenditures, however, the $5 day constituted "efficiency engineering," Ford maintained, and would stand as "one of the finest cost-cutting moves we ever made."[5] Statistics backed up Ford's

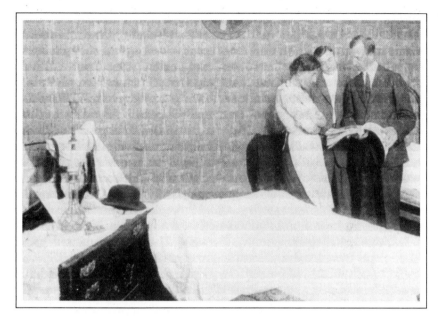

Ford Sociological Department operatives inspected the
homes of workers and offered families practical advice.

contention. Turnover fell to 16 percent in 1915, and absenteeism
dropped to 0.5 percent.

The Ford Sociological Department determined who was qualified
to receive this wage. Its operatives, some 250 by 1920, visited homes,
interviewed families, neighbors, and clergy, and sought to corrob-
orate each statement workers made. Using a code of conduct inspired by Henry Ford stressing family, community, thrift, and
character, they checked and rechecked whether employees met the
profit-sharing template. Unmarried men under age twenty-two,
married men living by themselves or involved in divorces, new employees with less than six months' seniority, and any worker "living unworthily as a profit sharer" were deemed unsuitable. Later
the list was expanded to include those taking in male boarders,
frittering away evenings unwisely, sending funds to Europe, spend-
ing money foolishly, using alcohol, and speaking a language other
than English. A full 40 percent of the Ford force failed to qualify.

The Sociological Department revised its determinations every six
months. Those not initially qualified were to reorient their lives in
the intervening period, but continued "unworthiness" led to dismissal. Yet the Sociological Department did more than investigate;

it also provided Ford employees and their families with information useful for daily existence. Department employees discussed such rudimentary activities as family budgets and savings accounts; gave elementary lessons in home management and hygiene; taught how to shop to best advantage; and showed how to distinguish among cuts of meat.

The Ford English School followed a similar educational premise. Teaching English was its announced purpose, but the school also instructed Ford's non-English-speaking employees in the virtues of efficiency, thrift, industry, and economy. It offered mandatory courses in domestic, community, and industrial relations. The school's mission, like that of the Sociological Department, was to weld systematically the diverse groups comprising Ford's labor force into a standardized, dependable cohort. The school's commencement exercises, held in the city's largest hall, reflected that thrust:

> On the stage was represented an immigrant ship. In front of it was a huge melting pot. Down the gang plank came the members of the class dressed in their national garbs and carrying luggage such as they carried when they landed in this country. Down they poured into the Ford melting pot and disappeared. Then the teachers began to stir the contents of the pot with long ladles. Presently the pot began to boil over and out came the men dressed in their best American clothes and waving American flags.[6]

Systematizing Hiring   Both Ford's program and scientific management attempted to systematize workers once employed, but others tried to systematize the employment process. From about 1900, a few corporations established departments to investigate job applicants. Rather than try to mold disparate individuals into a cohort, these industrial concerns instituted policies "of questioning, of research, of careful investigation . . . of seeking exact knowledge and then shaping action on discovered facts."[7] These deliberations presumably produced a more homogeneous labor force, one selected for its likely dependability and productivity. But systematic examination was necessarily subjective and, despite its pretenses, lacked science's rigor.

An emerging group of psychologists stepped into the void. In 1912, Harvard's Hugo Müsterberg maintained that psychology would revolutionize industry by detecting "those personalities which by their mental qualities are especially fit for a particular kind of economic work."[8] Several businesses attracted by the idea of a precise, nonpartisan, indisputable approach to worker selection, and the promise of lower turnover, freedom from labor unrest, and enhanced efficiency, used industrial psychologists' tests. By 1914, American companies had by test selected traveling salesmen, motormen, conductors, telegraphers, typists, stenographers, and telephone operators. Industrial psychologists invaded factories in 1915 when Northwestern University's Walter Dill Scott devised intelligence, dexterity, and general-ability tests for industrial workers. The federal government's support of psychological testing of soldiers and workers in selected industries during World War I further bolstered industrial psychology's credibility and prestige. Scott and others capitalized on this increased familiarity and formed industrial-psychology consulting firms. Scott's company alone had more than sixty business clients in 1923.

New Relations between Labor and Owners

By hiring industrial psychologists, scientific and systematic managers, and others to expertly and systematically select, train, and keep a labor force, industrialists acknowledged the utility and

These workers' homes near Pittsburgh steel mills would never have passed the muster of Ford's Sociological Department.

necessity of the human element in late–nineteenth- and early–twentieth-century production. These facts also indicated that manufacturers recognized the existence of a "labor problem." Labor required fitting; it had to be shaped in such a manner as to enable it to pursue dependably its specific task within the general production system. Certainly there was no longer any role for skilled artisans; mechanization and extreme division of labor had rendered them obsolete. But the problems between labor and their employers were not products of new or revamped means of production. Instead, they emerged from the same premise from which those new and revamped means of production blossomed: namely, that systems, including social systems, were composed of diverse fixed (and limited) elements hierarchically arranged.

This notion led to a radical disjunction among capital, managerial, and labor elements, a fact which labor recognized and institutionalized in the new trade unionism of the late nineteenth and early twentieth century. Prior to the 1870s, Americans had formed unions sporadically, usually for ad hoc local situations. Most had served as social clubs or self-insurance agencies. Those few organizations involved in more worldly pursuits had usually attacked the status quo and sought a return to an idyllic past. The majority of those activists had considered the nation as composed of two groups: workers and monopolists. Mid–nineteenth-century unionists defined employers, managers, and virtually everyone else as workers, and identified their interests as coterminous with those of the overwhelming majority of Americans; America was a social unit, not a social system, and should rid itself of those structures and individuals not acting uniformly, as barriers to individual opportunity. These unions attracted suffragettes, as well as prison, land, currency, and utopian reformers, because they had detected similar problems and advocated similar goals.

A different perspective emerged in the 1870s and predominated by the late 1880s. Labor as a class became self-conscious; it ceased to identify its interests as identical with those of its employers. Owners and management were its natural antagonists, yet it recognized that its relationship to these classes dictated its members' well-being. Labor proved willing to accept loss of status and social mobility in return for better working conditions and more money. This view led to repeated confrontations, generated state and national legislation, and produced new organizations.

The American Federation of Labor   Formation of the American Federation of Labor (AFL) in 1886 and its rise during subsequent decades resulted from labor's new self-consciousness. Its president, Samuel Gompers, argued that the AFL was a product of the realization by "wage-workers of this continent" that they constitute "a distinct and practically permanent class of modern society; and consequently have distinct and permanent interests."[9] Only "true" workers—those who labored with their hands—could join the organization, which emphasized labor's economic power to counterbalance organized capital. Deeming work stoppages of primary importance, the AFL engaged in politics only to defend labor's economic position, and backed, for example, immigration-restriction legislation in order to limit the labor supply and protect wages. Begun as a consortium of unions organized by trade, the AFL gradually extended its influence to become in the early twentieth century the head of a system of local, district, and state affiliates. It stood as labor's spokesperson and defender, a manifestation of labor's industrial and social importance, but also an acknowledgment that labor was neither the brains nor the heart of the socioindustrial system, only its hands and feet.

The Growth of Industrial Education   In its role as workingpeople's protector, organized labor often butted heads with industrial leaders. Such was the case with the question of vocational education after 1900. The vocational-education movement dated its formal start to 1880, when C. M. Woodward established St. Louis's Manual Training School, which prepared high-school-aged students to become machinists or woodworkers. Similar high schools were opened before 1886 in Chicago, Baltimore, Philadelphia, Toledo, Cincinnati, and St. Paul. Bostonians had erected a comparable institution, but for elementary school children, a decade earlier. New York quickly followed Boston's lead. By the late 1880s, virtually every American city had created special industrial schools or offered industrial courses of study within more traditional schools.

These early industrial schools had been formed under the auspices of church missions and private philanthropic agencies to solve the social problems of poverty and pauperism. Alarmed by the high dropout rate and the apparent inadequate supply of industrial workers, reformers considered traditional schooling irrelevant for poor children, and sought to train them systematically to become useful, productive citizens able to fill a niche and participate in the American socioeconomic and political system. City and state governments eventually adopted and funded these programs. At their heart were two assumptions: identification of a segment of the American population (designated by their heritage and their lack of

Vocational and technical high schools trained future generations of laborers to work machinery essential to American industry.

wealth) as inherently incapable of developing beyond a specific point; and the propriety of selecting and planning the direction of a class of people to a particular life's work, not necessarily with their consent. The poor could become neither managers nor employers, only laborers. Vocational guidance, coupled with specialized education, would ensure that they achieved their highest possible station. Democracy's meaning had changed from freedom of individual opportunity to freedom for groups to assume their appropriate place within American life and sample its fruits.

Labor unions initially supported industrial education but grew concerned when after 1900 industrialists became explicitly involved. Generally fearing that manufacturers backed vocational education to expand and cheapen the labor supply, unions opted for publicly supported and controlled institutions, usually night schools, for young men and women already employed; limited to those under eighteen years old, night schools would train laborers in the fine points of their daily work. Large industrialists instead often championed corporate apprentice schools. Located at production facilities and run by industrial firms, these shop schools taught young laborers how better to operate their machines. Items manufactured there were sold.

State laws mandating that children attend public school or its equivalent until age sixteen had an impact on the shape of industrial education, as did the various post-1905 state commissions

Training laborers to fit the American industrial machine became a principal goal of American public education.

seeking its most prudent form. The laws usually caused manufacturers to release young employees for a few hours daily to attend general-purpose trade schools, while the commissions frequently resulted in creation of more specialized industrial courses and additional public funding. Wisconsin enacted the most daring measure in 1911. It required places with populations in excess of five thousand to erect industrial schools. Local boards of industrial education, composed of employers and laborers in equal number, levied taxes and framed instruction. Passage of the Smith-Hughes Act in 1917 nationalized vocational education by providing matching federal money for precollegiate industrial education, but required states to gain project approval from a national vocational education board.

WORLD WAR I: A SYSTEM OF THOUGHT'S LAST BATTLE

The United States entered World War I at about the time of the Smith-Hughes Act's passage, and the war effort dwarfed other concerns as the nation was thrust into a situation it had never before encountered. It needed to raise, equip, and send an army across the ocean, and to provide support as well as furnish matériel for its beleaguered allies. That America remained undaunted by these prospects stemmed from a belief in the morality of its cause and from nearly a half-century of experience rationalizing and systematizing virtually every facet of life. Participation in World War I placed new demands on the country, but not fundamentally inconsistent or unfamiliar ones. Only the scale would be different; Americans treated the war as a characteristic late–nineteenth- and early–twentieth-century event, as the culmination of previous efforts, not the inauguration of a new agenda. The nation would manufacture what it needed; it would become like a factory to prepare and fight "a war to end all wars." Systematization of industrial organization, management, and production, as well as transportation and communication, and scientific and technological efforts, would "keep the world safe for democracy."

Marshaling Resources for War

The nation waited for a formal declaration of war to mobilize, but began preparations for mobilization after the *Lusitania*'s sinking in 1915. The Naval Consulting Board (NCB), created by the secretary of the navy and composed of Edison and other industrial leaders, solicited nominations from the national engineering societies for state industrial preparedness committees, which compiled preliminary resource inventories. The National Research Council (NRC) was formed a year later to coordinate scientific research. Its mission included inventories of scientific personnel, equipment, and research; integration of government, university, industrial, and foundation science and scientists; and communication among investigators. The Council of National Defense (CND), also established in 1916, had a broader mandate. Its six officials, all members of President Wilson's cabinet, coordinated the entire premobilization campaign.

The CND increased its authority after Congress declared war on April 6, 1917. The declaration empowered Wilson to commandeer essential industries and mines, seize and operate all transportation and communication systems, and control distribution and fix prices of war-related commodities. Wilson delegated these powers to several boards and placed them under the CND's supervision. The CND completed the NCB's inventory and assigned the naval board the task of screening inventors' suggestions and devices for their military potential. The NRC was designated the CND's research arm, and was directed to coordinate war-related scientific research and resources. Antisubmarine weapons, precision rangefinders, and dependable gas masks resulted, as did psychological tests used both to determine potential officers and to weed out unacceptable draftees.

The War Industries Board, headed by Bernard Baruch, controlled industrial production. It regulated all extant industries that manufactured war matériel; developed new industries, plants, and supply sources; established prices; and determined production and delivery priorities. Some 30,000 different articles fell under its purview. Baruch ruthlessly slashed product variety and set rigorous stipulations to free resources. Little escaped his axe. For example, he reduced the number of typewriter-ribbon colors from 150 to five, mandated a lower maximum length of shoe uppers, and regulated the quantity of steel in corset manufacture. The last initiative saved an estimated 8,000 tons of steel annually. Baruch's systematic frugality, coupled with reallocation of resources and production capability (he had, for instance, fifty munitions factories built in the war's first few months), enabled Americans during the war to produce 2.5 million rifles, 181,000 machine guns, 3,000 artillery pieces, 3.5 billion rounds of small-arms ammunition, 5 million gas masks, and 8,000 training planes. CND-sponsored fuel, food, and transportation boards received similar authority and worked comparable miracles.

New War Technology on Land

American manufacture of machine guns was a response to the changed nature of land warfare. These killing machines made traditional military advances foolhardy, and impossible without massive loss of life. Rifles, bayonets, and hand-to-hand combat no longer carried the day, and horse cavalry was relegated to history's scrapheap; warfare ceased to be considered glorious and became impersonal.

Machine guns came of age in World War I, but their invention dated to the late nineteenth century. Hiram Maxim, an expatriate American who had lost the incandescent-light race to Edison, devised the first fully automatic machine gun in 1884 and demonstrated it in 1887. The force of recoil from Maxim's gun operated the ejection, loading, and firing mechanisms. Maxim also developed a smokeless powder when he found that black powder residue clogged his new gun's barrel. Military men marveled at Maxim's display, but failed to grasp the machine gun's revolutionary killing potential. Most continued as late as 1914 to regard machine guns as nothing more than adjuncts to conventional weaponry, even after arms manufacturers such as Colt produced lighter, improved ver-

Howitzers became a trench-warfare staple during World War I.

sions. This misapprehension had vanished before America entered the war, and the nation's soldiers quickly familiarized themselves with the weapon.

Industry also contributed new armament designs. Machine-gun-supported, barbed-wire-fortified trench warfare predominated in the European theater, while thick metal fragmentary hand grenades further bolstered defenses and gas masks warded off poisonous phosgene. Assaults on these congealed lines proved difficult. Artillery was one manner of attack. Trench mortars lobbed bombs, and glycerin recoil artillery, sometimes mounted on mobile railroad cars, fired streamlined tracer, incendiary, and armor-piercing shells. Internal-combustion vehicles spearheaded the other primary land assault mode. Caterpillar tractors, and armored tanks mounted with machine guns and equipped with pneumatic tires, attempted to burst through lines and drive the enemy from trenches.

New War Technology at Sea

Torpedo-armed German submarines transformed traditional sea warfare practices as drastically as the machine gun changed land war. Rather than engage in conventional naval bombardments, submarines killed quietly, and battleships, even the awesome *Dreadnaught,* quaked at the thought of these underwater death merchants. Submarines also disrupted war matériel shipments and blockaded ports, both of which were impediments to conducting a transatlantic war.

Americans assumed the lead in developing antisubmarine measures. Its subchasers were small wooden craft with radio listening devices to detect a submarine's presence. They carried depth charges, which detonated upon reaching a certain depth to force submarines to surface, and guns to finish them off. Subchasers sank more than two hundred submarines. Henry Ford's mass-produced Eagle Boats made a late entry into the war and served as subchasers, but also came equipped with antiaircraft guns. American destroyers were sleeker and swifter than their European counterparts. Their five-inch guns and torpedoes proved potent antisub weapons. New electrically detonated deep-water mines prevented submarines from moving through mined areas.

Convoys were the most effective antisubmarine measures. Systematic assemblages of subchasers, destroyers, cruisers, and other ships, convoys tied up many ships and were expensive but protected transports and merchantmen. Submarines attempting to penetrate the convoy's shell were detected by subchasers and attacked by numerous vessels. Convoys negated submarines so completely that by war's end Germans refrained from assaulting them. Allied shipping losses had become minimal.

New War Technology in the Air

World War I was the first major conflict to see a new dimension, the air, added to warfare. The American military was especially unprepared for this type of fighting; it possessed a scant 109 airplanes, mostly trainers, and had only eighty-three pilots when Congress declared war. European combatants operated far larger air forces and had highly refined the art of aerial warfare. Scouts, machine-gun-fitted fighters, and multi-engine bombers filled the Western European sky as air duels, strafing, and bombing runs, as well as aerial reconnaissance and photography, were commonplace. That air power would assume such prominence was surprising; the first manned flight of a powered, heavier-than-air craft had occurred a mere thirteen and one-half years earlier. That Americans would lag so far behind Europeans was equally surprising; airplanes were an American invention.

The Advent of Powered Flight   Orville and Wilbur Wright made the first successful airplane flight on December 17, 1903, at Kitty Hawk, North Carolina. Others had flown earlier. The French had pioneered hot-air–balloon flight in the late eighteenth century, and gas and hot-air balloons had been popular nineteenth-century entertainment. Maxim's railbound, steam-engine-powered vehicle and Otto Lilienthal's gliders had attracted considerable attention in the 1890s. Samuel Pierpont Langley, the Smithsonian Institution's head, had helped galvanize American interest in flight. He parlayed his scientific reputation (and Theodore Roosevelt's support) to land, in 1898, a $50,000 governmental grant to build and fly a heavier-than-air-powered craft. Relying on design studies that he had begun in 1886, Langley constructed the gasoline-engine-driven *Aerodrome*. Several attempts to launch the plane proved fruitless, the last occurring nine days before the Wrights's flight.

A comparison of Langley's and the Wrights's notions of the problem of flight indicates that Langley lagged far behind. He sought to "urge a system of rigid planes through the air at great velocity."[10] To Langley, flight was similar to running a knife through butter. Propeller and wing shape were crucial, as were centers of gravity; proper angles of incidence would enable the system of rigid planes to slice through or "deform" air, which was to Langley a homogeneous elastic substance—almost identical, perfectly elastic cubes. Langley's design of his inclined-plane launch mechanism stemmed from that consideration. The Wrights held a more sophisticated view, perhaps because of their bicycle-racing and -making experience, which also familiarized them with strong lightweight materials and wood- and metal-working tools. They regarded propeller and wing shape as important, but conducted their own wind-tunnel tests and emphasized the air's fluid nature. Flight could never be reduced to a system of rigid planes; continual coordinated compensation to the various forces acting on the aircraft was essential. The Wrights created aerial stability through constant adjustability born of flexibility. Their conception of control was three-dimensional, and its most telling features were their plane's roll and yaw stabilizing devices. A mechanism that, through a series of lines and struts, provided immediate compensating helical twists across the entire wing reflected their commitment to fluid dynamics, as did their decision to link the rudder to this wing-warping apparatus. Coupled with a hand-operated forward elevator, their control system suggested a much more complex idea of the requirements of flight.

The Development of American Aviation   Neither the Wrights's Kitty Hawk success, nor their repeated flights near Dayton, Ohio, during the next few years, triggered an American aviation boom.

The first company specifically formed to manufacture airplanes did not produce one until 1908, and the U.S. military purchased its first aircraft a year later. It had only six in 1914. Stunt-flying appealed to the American public after 1910, and in the early teens a few entrepreneurs established commercial air service between nearby but hard-to-reach places. By that time, however, European airplane development was already in full swing.

America trailed Europe in airplane design, but its infatuation with automobiles and industrial rationalization helped it pull ahead in airplane engine construction. In 1916, Packard Motor Car's J. G. Vincent and Hall-Scott Motor Car's E. J. Hall created the Liberty Engine for airplanes, and developed techniques for its mass production. This powerful, dependable internal-combustion engine came in eight- and twelve-cylinder models and became the American staple. Many were shipped overseas and fitted in English and French aircraft during the war, but the premobilization engine blitz did not extend to planes themselves. In the year prior to Congress's declaration of war, the American aviation industry produced a mere 411 aircraft. The slim total did not dissuade the nation from planning to make more than 8,000 training planes and 12,400 fighters and transports by spring 1918. It managed to produce only half that number, almost all of which were trainers. Fighters were purchased in Europe. Despite this drawback, America amassed an air force on Old World soil of more than 5,500 planes by Armistice Day. Its pilots had shot down an estimated 850 enemy aircraft.

Production of Munitions During the War

The nation was almost as inadequately prepared to produce munitions as airplanes. The American chemical industry long had relied on Germany and, to a much lesser extent, other European powers for the bulk of its organic chemicals, generally coal-tar derivatives. Connections between the German government and the German chemical industry had undercut American domestic prices, discouraged production, and enabled Germany to establish a virtual international monopoly. With the war, however, Germany refused to ship coal-tar derivatives (such as benzene, phenol, toluene, xylene, and naphthalene), and its submarines prevented other nations from taking up the slack. Acetone and glycerine were also in short supply.

These several chemicals had many industrial uses. All were essential to production of explosives. Nitroglycerine and dynamite required glycerine. Some smokeless powder used acetone as a solvent. Benzene and phenol were picric acid intermediates. Toluene was TNT's basic component. Other modern explosives employed xylene and naphthalene. The German boycott did not sting only America; England and France were acutely affected. When America entered the war, its allies increasingly looked to the New World for munitions.

America was woefully unprepared to serve, but quickly mobilized its resources. The steel industry rapidly completed conversion from beehive to by-product ovens, and chemists distilled the resulting coal tar into fractions.* Others cracked† petroleum for toluene and other products, or attempted to synthesize these materials. In 1915, America produced only 2.5 million gallons of benzene, 623,000 gallons of toluene, and 196,000 pounds of naphthalene. In 1918, it manufactured eighteen times more benzene, nearly eleven times more toluene, and more than thirty-two times more naphthalene. Soap and candle industrial refuse was recovered and purified into commercial glycerine. Destructive distillation of seaweed and timber yielded acetone. Government money built the massive new ordnance plants, Nitro and Old Hickory. Dynamite production was four times as great in 1917 as it had been in 1913 (peacetime uses had included mining, and canal, tunnel, and road building), and black and smokeless powder manufacture skyrocketed during the same period—from 866,569 pounds to more than 438 million pounds. American picric acid and TNT production was valued at $651,000 in 1913 but reached $394 million four years later.

The End of a System of Thought

America's ability to mobilize its material and human resources contributed to the allied victory. Armistice Day received an enthusiastic welcome, but the nation did not return to the way of life it had previously known; there was no return to "normalcy." Several social, demographic, and economic changes had accompanied the war effort. Unions had gained affiliates and strength. Rural southern blacks who had made the pilgrimage North to work in war plants frequently stayed in northern cities. Women had logged long hours replacing men in factories. The nation's essential industries

---

* *fractional distillation:* a technique for separating liquids based on differences in boiling point.
† *cracking:* the use of pressure to break long-chain hydrocarbons into more valuable short-chain hydrocarbons.

*Airmail took off after World War I. Here two Americans
deliver mail from Canada.*

had become familiarized with mass-production techniques, the war
had awakened American entrepreneurs to the potential of new industries, and the public anticipated new goods and services. The
aviation industry began its takeoff; air mail and passenger service,
crop dusting, and express delivery of parts and perishable commodities were postwar products. The American organic-chemical industry also came of age; governmental confiscation and sale of
German patents and property protected wartime investment, while
stiff tariffs shielded the nascent industry from postwar foreign
competition. American sources of supply, tapped during the war
for munitions, continued to produce essential chemicals, which were
transformed into plastics, solvents, gasoline additives, dyes, and
medicines. These new sectors of industry became increasingly important in the postwar world.

But America also seemed changed in a different way. Its pursuit
of the war had been conducted according to an organizational idea

that had begun to be implemented some five decades earlier. Land, air, and sea warfare were separate and separable types of combat. Mobilization, production of ordnance, and manufacture of aircraft were distinct activities, united only in their contribution to the war effort, and given priorities according to their importance. This notion of systems as composed of discrete circumscribed elements hierarchically ordered connoted more than assignment of position or place, of course. It defined processes and procedures, as well as relationships among components. As important, it assumed that these elements were virtually inviolable, and that each possessed an almost intrinsic property, quality, or ability; identification of each element was without reference to others. In that sense, World War I marked the culmination of that mode of thought. The portrayal of components as rigid and encapsulated gave way in the 1920s, and with it went the post-1870 view of the nature of systems. The technologies of the 1920s and after would be built upon or explained according to different premises.

NOTES

1 John S. Fotherington, "Systematizing the Routine Of An Office," *System*, vol. 7 (1905), p. 426.

2 Quoted in David Hounshell, *From the American System to Mass Production, 1800–1932* (Baltimore: Johns Hopkins University Press, 1984), p. 229.

3 Quoted in Daniel Nelson, *Workers and Managers* (Madison: University of Wisconsin Press, 1975), p. 66.

4 Quoted in Stephen Meyer III, *The Five Dollar Day* (Albany: State University of New York Press, 1981), p. 124.

5 Quoted in Keith Sward, *The Legend of Henry Ford* (New York: Rinehart, 1948), pp. 52, 56.

6 Quoted in Olivier Zunz, *The Changing Face of Inequality* (Chicago: University of Chicago Press, 1982), p. 312.

7 Quoted in Nelson, *Workers and Managers*, p. 149.

8 Quoted in Loren Baritz, *The Servants of Power* (Middletown: Wesleyan University Press, 1960), p. 37.

9 Quoted in Gerald N. Grob, *Workers and Utopia* (Evanston: Northwestern University Press, 1960), p. 140.

10 S. P. Langley, *Experiments in Aerodynamics* (Washington: Smithsonian Institution, 1891), p. 5.

FOR FURTHER READING

Allen, Edith. *Mechanical Devices in the Home* (Peoria: Manual Arts Press, 1922). Written for vocational education classes.

Armstrong, David A. *Bullets and Bureaucrats* (Westport: Greenwood Press, 1982). Contains some interesting material on the machine gun.

Baldwin, Keturah E. *The American Home Economics Association Saga* (Washington: American Home Economics Association, 1949). An affectionate history sponsored by the organization.

Bennett, Charles A. *History of Manual and Industrial Education, 1870 to 1917* (Peoria: Manual Arts Press, 1937). The standard history of the topic.

B'evier, Isabel and Susannah Usher. *The Home Economics Movement* (Boston: Whitcomb and Barrows, 1918). Written by persons in the movement.

Bilstein, Roger E. *Flight in America* (Baltimore: Johns Hopkins University Press, 1984). An easily readable, thorough treatment.

Bliven, Bruce, Jr. *The Wonderful Writing Machine* (New York: Random House, 1954). A popular biography of the typewriter.

Boorstin, Daniel J. *The Americans: The Democratic Experience* (New York: Random House, 1973). Useful on office equipment.

Chandler, Alfred D., Jr. *The Visible Hand* (Cambridge: Harvard University Press, 1977). Discusses the organization of management.

Cowan, Ruth Schwartz. *More Work for Mother* (New York: Basic Books, 1983). A provocative interpretation of the role of technology in the household.

Current, Richard N. *The Typewriter and the Men Who Made It* (Urbana: University of Illinois Press, 1954).

Davies, Margery W. *Woman's Place Is at the Typewriter* (Philadelphia: Temple University Press, 1982). A social history of the secretary as female.

Ellis, John. *The Social History of the Machine Gun* (New York: Pantheon, 1975). A disjointed but interesting narrative.

Eppright, Ercel Sherman and Elizabeth Storm Ferguson. *A Century of Home Economics at Iowa State University* (Ames: College of Home Economics, 1971). A history of one institution. Useful for the early years.

Frederick, Christine. *The New Housekeeping* (New York: Doubleday, Page, 1912). A later volume extolling the virtues of systematic housekeeping.

Freudenthal, Elsbeth E. *Flight into History* (Norman: University of Oklahoma Press, 1949). Emphasizes the Wright Brothers' contribution.

Haber, Samuel. *Efficiency and Uplift* (Chicago: University of Chicago Press, 1964). Discusses ideas of efficiency in late–nineteenth- and early–twentieth-century America.

Hallion, Richard P. *Rise of the Fighter Aircraft, 1914–1918* (Annapolis: Nautical and Aviation, 1984). Excellent study of aircraft in World War I.

Haynes, Williams. *The American Chemical Industry*, vol. 3 (New York: D. Van Nostrand, 1945). Comprehensive treatment of the munitions industry.

Holley, I. B., Jr. *Ideas and Weapons* (New Haven: Yale University Press, 1953). A provocative interpretation of military armaments.

Lifshey, Earl. *The Housewares Story* (Chicago: National Housewares Manufacturers Association, 1973). Discusses the rise of trade associations.

Macksey, Kenneth and John H. Batchelor. *Tank* (New York: Charles Scribner's Sons, 1970). A popular illustrated history of armed fighting vehicles.

Mayr, Otto and Robert C. Post, eds. *Yankee Enterprise* (Washington: Smithsonian Institution Press, 1981). Useful on the Eagle boat.

Nevins, Allan. *Ford, the Times, the Man, the Company* (New York: Charles Scribner's Sons, 1954). The first of a three-volume study.

Rae, John B. *The American Automobile* (Chicago: University of Chicago Press, 1965).

Rodgers, Daniel T. *The Work Ethic in Industrial America, 1850–1920* (Chicago: University of Chicago Press, 1978). An analysis of how middle-class reformers thought of work and the workplace.

Taylor, Frederick Winslow. *Scientific Management* (New York: Harper & Brothers, 1947). A compilation of several of Taylor's earlier works.

Turck, J. A. V. *Origin of Modern Calculating Machines* (Chicago: Western Society of Engineers, 1921). A technical history of their development.

Ware, Norman J. *The Labor Movement in the United States, 1860–1890* (New York: D. Appleton, 1929). An older but effective discussion of the origins of late–nineteenth-century unionism.

P_{art}
T_{hree}

From Industrial America to Postindustrial America

THE 1920S TO THE PRESENT

The notion of system, such a crucial aspect of late–nineteenth- and early–twentieth-century thought, underwent drastic revision in the decade after about 1920. Whereas systems in the earlier period were static entities, composed of diverse, fixed, and limited parts hierarchically arranged—the system equalled the sum of its different parts—the new systems of the 1920s and after were dynamic, predicated on a much more complex relationship among the parts. Each part seemed to acquire a share of its definition from its interrelationships with the other parts in the system; the parts seemed to reflect on one another. Flexibility, adjustment, and compensation characterized these modern integrated systems, which were greater than or different from the sum of their parts.

This idea of systems as complicated dynamic processes opened new technological possibilities. It produced a reconsideration and reorientation of technological processes and the acknowledgement that previously unforeseen factors might affect technological decisions. But it also made it conceivable to seek technological solutions to virtually every problem, including social problems. The introduction of new technologies into a system would vary the dynamics or increase integration and shift equilibrium to a more favorable spot. From this perspective came efforts as diverse as introducing electricity into rural regions to conquer poverty and isolation, automobiles into cities to reduce urban crowding, and agricultural and industrial technologies into Third World countries to combat communism.

This unabashedly optimistic assessment of technology's influence began to wane in the 1950s and continued to dwindle during the next several decades. The previous era's notion of system remained in place, but the individuation of American society, coupled with a perception that the nation's (and the world's) resources were limited, contributed to a disillusionment with traditional decision-making apparatus. A critique of professionalism and expertise generally emerged, and technology as a product of experts became a subject of immense concern. Most Americans conceded that technology was a powerful force, and that its introduction usually produced both positive and negative consequences. Technology after mid-century was increasingly seen as both a cause of and cure for social problems; it was a social question. But in an era shorn of the aura of expertise, the means to decide the relative merits of a particular technology were not clear. Each technological initiative, even those that enhanced individuation or furthered social integration, was subjected to individual scrutiny.

7

Technology as a Social Solution: the 1920s to the 1950s

EMERGENCE OF A NEW NOTION OF SYSTEMS

The Hawthorne Experiments

I n 1924, the National Research Council selected Western Electric's Hawthorne Works, a facility of nearly 30,000 workers located on Chicago's west side, as the site for an investigation of the relationship between labor productivity and factory illumination. Carried out by Massachusetts Institute of Technology (MIT) electrical engineers, this rather modest old-fashioned study merely identified two small groups of workers, then subjected one to various lighting conditions and left the other constant. Much to the engineers' surprise, productivity of both groups increased no matter what their lighting conditions. This discovery led the researchers to conclude that lighting was at best only a minor factor in worker performance, and that numerous other variables needed to be controlled before an estimate could be made of its precise importance.

Western Electric, understandably curious as to why worker output increased in both groups, sponsored additional tests in

April 1927. Rather than seek a simple correlation between lighting or any other single feature and productivity, these and later tests were aimed at understanding the dynamics of the work environment. The tests continued for several years, cost more than $1 million and tens of thousands of worker hours, and ultimately produced new methods to bolster worker productivity. They would become the basis of a new managerial technology of the workplace.

MIT researchers representing a number of specialties, not just electrical engineering, conducted these new experiments. The intention of this interdisciplinary research team was to isolate six female relay assemblers in a small room located at the corner of the regular relay assembly room to investigate relationships among working conditions, monotony, and fatigue. The team began the two-year study by observing the women under normal conditions to achieve a base line, then situated them in the test room to acclimate them to their new surroundings. They next exposed the women to a variety of working situations, recorded their conversations, and noted modifications in productivity. Observers found that productivity increased in the new environment. New wage systems were instituted and output again moved ahead. Productivity increases also followed the introduction of rest pauses, variations in the length of the working day and week, and provision of free nutritionally balanced lunches. Only when all the perks (except the new wage system) were withdrawn did productivity level off. It moved upward again as each perk was reinstituted. These findings puzzled researchers. It seemed that each factor in the working environment "was so dependent on its relations to other factors that it was impossible to consider it a thing in itself having an independent effect on the individual."[1] They also recognized that the women had formed a social network among themselves, and had developed a sense of participation in critical determinations.

Such social networks provided the focus for the next set of tests. Management wanted to know the implications of social networks and the pressures and sanctions networks imposed on members. The Harvard Business School's Elton Mayo supervised the research team's investigations, which began in November 1931 and lasted six and a half months. Hawthorne's fourteen male bank-wiring operators were observed during normal activities. From that examination, Mayo's staff reported that worker networks were very intricate social organizations that erected their own sets of rules. Identification with a network prevented deviant behavior—behavior outside the network's norm—and protected the network from any outside interference or control. Worker social networks tended to adopt perspectives different from management's—they disputed, for example, management's conception of a fair day's labor, and repeatedly produced at lower network norms. Researchers then concluded that network standards determined individual productivity. The traditional nexus between ability and performance (measured in the working world by industrial psychology firms like that of Walter Dill Scott) was of only peripheral significance.

Western Electric's management pondered methods to stop soldiering, while Mayo compared results of the assembly and bank-wiring room tests. He argued that the women's productivity had accelerated in each instance because they had an *active* role in the experiment, and therefore wanted it to "succeed." The women considered themselves taken into management's confidence, as partners in the enterprise, not antagonists. Another battery of tests was then occurring that confirmed Mayo's assessment; Western Electric had begun to interview its employees about management programs but found it impossible to keep them on the topic because workers wanted to discuss what interested them. Researchers altered the interview process, made it nondirective, and permitted laborers to talk about whatever they wished. During follow-up contacts some time later, workers often mentioned improved factory working conditions, and a few even applauded the new wage system. But their impressions were falacious; the company had changed neither working conditions nor wages. Employees seemed to have gained from their interviews feelings of belonging and recognition. Management seemed to value their opinions; that attitude appeared to hold important implications for worker well-being and, ultimately, productivity. Put more boldly, worker grievances seemed often to have no integrity of their own, but were instead manifestations of a sense of alienation from management, a posture that could readily be modified.

The company at last had an answer it could apply. It established "personnel counseling" in 1936, and staffed these posts with long-term, well-liked Western Electric employees. It instructed them to deal "with attitudes towards problems, not the problems themselves";[2] counselors were to listen. Worker productivity seemed to blossom under this arrangement as workers felt accepted by management and placed the company's interests within their social networks. So, too, did the number of counselors blossom. Western Electric's five counselors in 1936 soon expanded to ten in 1938, twenty-nine in 1941, and sixty-four in 1954.

Personnel counseling quickly spread outside Western Electric. A generation of human relations experts, specializing in small group and industrial social dynamics, invaded American factories and succeeded the scientific managers of an earlier day. Questions of fatigue, working conditions, lighting, monotony, and the like receded as attitudinal adjustment of workers emerged. Worker productivity became a state of mind, not a result of physical situations, but it still could be engineered.

The New Notion of Systems

The results of the Hawthorne studies were at odds with late–nineteenth- and early–twentieth-century American industrial experiments. Earlier investigators had sought to increase worker productivity by adding performance incentives or penalties, selecting or training a work force appropriate to the tasks at hand, refashioning the physical environment, or reshaping the assigned job. These programs had been based on a particular understanding of the nature of work, workers, and the workplace that stemmed from the notion of system as a static entity composed of discrete, fixed, and limited parts that were hierarchically ordered. The Hawthorne experiment proceeded from very different assumptions, which marked the mid-twentieth century as fundamentally dissimilar from the late nineteenth and early twentieth centuries. "Interdependence of relationship" seemed to identify the more modern system. Its parts appeared "so closely integrated as a whole that no change can occur in any of its phases without affecting other phases in some measure."[3]

Much more complex than the earlier representation, the post-1920 notion of system accentuated relationships among things previously considered only peripherally related, and suggested interconnections where none had been established formerly. Parts within these new systems seemed to have little integrity of their own, and to lose whatever intrinsic character they had possessed when examined outside the system. Each part received a portion of its definition from its relationships to other parts; its definition was relative or indeterminate because parts interacted with each other. Introduction of new stimuli upset the precise balance and resulted in adjustment and compensation throughout the entire system, a factor that rendered the idea of a permanent or natural hierarchical order meaningless.

A new vocabulary accompanied this transition, as did new definitions of old terms. *Ecology, multipurpose, interdisciplinary, populations, decentralized, culture, networks, dynamics, pluralism, federal, interdependence, communities, region,* and *dynamic equilibrium* were words frequently employed to describe and explain the nature of the new post-1920 system and its elements. Before 1920 very few Americans had conceived of systems in roughly the same manner. Arthur Little's unit operations and the Wright brothers' conceptions about the nature of flight are perhaps the most prominent examples, but these scattered individuals never extended their analyses' terms. As articulated by post-1920 Americans, these ideas blurred distinctions between technology and other endeavors and made it conceivable to seek technological solutions to social questions. They also produced reexamination and reconceptualization of established technologies. Scrutiny was applied to devices, processes, and activities that had served for nearly half a century. Technical design likewise underwent revision: machines designed in the art deco style were conscious attempts to merge aesthetic and technological interests. Technology had implicitly become an art form, among its many newly perceived interactions.

THE GOVERNMENT AND SOCIAL ENGINEERING

Government involvement in technological activities rose exponentially as Americans increasingly looked to technologies as a means to solve social questions or to ameliorate social situations. Its participation in this social engineering movement was characteristically managerial and accomplished in either of two ways. In some cases, particularly in the 1920s and early 1930s, government served as coordinator, gathering together a diverse but appropriate team of experts to consider a problem that was by definition multifaceted and to design a technological solution that took into account numerous social, cultural, economic, and political factors. Governmental authority in these instances was informal; it held legal power only to form and convene teams, and relied on its power of persuasion to implement recommendations.

Hoover and National Planning

This governmental role was favored by Herbert Hoover, whose 1928 election to the presidency epitomized America's burgeoning love affair with social engineering. Trained as a mining engineer, Hoover had directed the World War I food relief effort and later served as secretary of commerce. His experiences convinced him of modern life's complexities, and he viewed their resolution as an engineering task. Systematic planning comprised Hoover's agenda, but unlike the engineers of earlier decades he regularly recognized that technical solutions had profound social and cultural implications. To Hoover, planning—and he was most interested in national planning—required close cooperation among leaders of business, labor, agriculture, the professions, and other segments of society. These men and women would confer, analyze demographic trends, and formulate consensual, multidimensional, nonpartisan strategies to guide pluralistic America's future.

These views guided Hoover's attempts to regulate key emerging industries such as aviation, radio, electric power, and highway construction. Cooperation between public and private sectors of the economy, not direct governmental intervention, was his touchstone. Hoover wished to take at least the executive branch out of politics (an arena despised by most engineers as corrupt and imprecise), and limit its role to evaluating and implementing policies and programs formulated at conferences by consortiums of leaders from the private sector. Efficient private regulation, in effect, was his ideal, with heavy reliance on technical expertise and cooperative associations.

Roosevelt and the New Deal

The Great Depression discredited Hoover and undermined his approach to government involvement in social engineering. But while Franklin D. Roosevelt and his "brain trust" dismissed Hoover's apolitical, noninterventionist stance, they remained committed to social engineering's prospects. Acting during a time of crisis and backed by congressional support, Roosevelt's administration employed federal funds and authority in an effort to relieve short-term economic distress and to ensure future prosperity. Roosevelt's concept of a "New Deal" demanded that his forces engineer a planned, well-managed economy.

Roosevelt's New Deal exemplified the second type of connection between government and social engineering. Like Hoover's national planning, it, too, was managerial, relied on the wisdom of specialists (many of whom held governmental posts) operating in tandem, and depended on government-inspired coordination. But most crucial, government was accorded the responsibility of active participation, of functioning as a counterbalancing agent to compensate for various disruptive forces, one of which seemed often to be the greed of the private sector. Government power loomed large in this formulation. Only government possessed the all-encompassing perspective necessary to produce effective policy, and its legal authority could compel disinclined segments of the population to comply with its programs.

Critics of the New Deal—The Technocrats   Critics attacked Roosevelt's New Deal vision, but not the idea of using government as a social-engineering agent. His assessment of the amount of governmental authority necessary to restore equilibrium between private and public interests was usually the focus of concern. Hooverians found the New Deal too radical, while radicals (such as Upton Sinclair, Father Coughlin, and Huey Long) argued that it failed to go far enough.

The Technocrats, a group of engineers who wished to eliminate politics and politicians from governmental decisions, fell into the latter category. At its peak in the mid-1930s, the Technocratic movement had enrolled only a few thousand members, but many more sympathized with its aims. A national government that made decisions on purely technical grounds constituted its program.

Technocracy's leaders were apparently inspired by their association between 1919 and 1921 with the iconoclastic social critic Thorstein Veblen. Veblen offered a plan for ridding America of waste and extravagance that, with modifications, the Technocrats later adopted as their own. It included the voluntary abdication of all absentee and greedy owners of big business; their replacement by diligent, reform-minded technicians and workers; creation of a national directorate to supervise the reallocation of all goods and services; and elimination of the artificial price system based on the equally artificial monetary system. These changes, Veblen contended, would increase America's industrial output by 300 to 1,200 percent.

Veblen, however, became disillusioned with the very engineers and technicians he had hoped would capture American industry and government, for they instead readily accommodated themselves to corporate capitalism. "By settled habit," he lamented in 1933, "the technicians, the engineers and industrial experts, are a harmless and docile sort, well fed on the whole, and somewhat placidly content with the 'full dinner-pail' which the lieutenants of the Vested Interests habitually allow them."[4]

By contrast, the Technocrats, led by Howard Scott, an enigmatic New Yorker with a mysterious past, were more optimistic and believed that less drastic efficiency measures than Veblen had prescribed would produce a veritable utopia in the near future. Technology would provide abundance for all in a system in which relative values of commodities would be determined by amounts of energy necessary to create them, and in which all citizens would receive "energy certificates." But Technocracy's appeal was limited by its technical jargon and complex charts, which perplexed more than impressed potential supporters. Scott and his associates grew increasingly embittered and threatening as they failed to gain widespread approval. Their militaristic demeanor and structure, special insignia and salute, and grey uniforms and fleet of grey automobiles scared away many technical professionals, who otherwise were attracted to the pitch. It became, then, another fringe group of the Great Depression, and the term "technocracy" has become one of derision.

The Administration of New Deal Technologies

Criticism such as that offered by the Technocrats did not measurably alter the emphasis of Roosevelt's New Deal. Removing people from relief rolls was its immediate objective and economic revitalization its long-range goal; social engineering was its means to accomplish both. In New Deal thought, American pluralism was a strength rather than a detriment; New Dealers attempted to incorporate different social segments and specializations whenever they initiated projects. Its unprecedented support for the social sciences, arts, and humanities stemmed from a belief that modern life was constructed of more than new structures—buildings, dams, and the like. America's diverse peoples would interact and change one another, the result of which would be an America greater and healthier than the sum of its constituent social elements.

Carried to its logical conclusion, this argument proved contradictory. Successful efforts to merge the same dissimilar elements in every project according to New Deal premises would produce an America of almost inconceivable homogeneity, destroying the country's pluralistic vigor. This irony was particularly clear in New Deal programs that sought to modify the nation's physical landscape. Defining cities as blighted and rural areas as backward, New Dealers hoped to create new environments "fit for life and the living." These environments would merge the best of the urban and the rural; ideally, they would yield a pristine, homogenized, nationwide, quasi-suburban existence (except that these places would not be suburbs because they would not be tied to central cities) in which America and Americans would achieve free and full expression. Two relatively new technologies were to be instrumental in this transformation: electric power and automobiles. The former would offer benefits available in cities, while the latter would pro-

The Army Corps of Engineers participated in many multipurpose projects in the 1930s. The Bonneville Dam was among the largest.

vide opportunities for decentralization of population comparable to those in suburbs and on farms.

The Army Corps of Engineers   This prescription for America's ills sparked public-works engineering projects of unprecedented scale and complexity. Numerous federal bureaus such as the Forest Service, Geological Survey, National Weather Service, Wildlife Service, and Bureau of Reclamation lent their expertise to these endeavors, as did state and local entities. Among the most frequently involved was the Army Corps of Engineers. Until the New Deal, the Corps had been restricted to improving only one aspect of river utilization—navigation. It built dams, levees, piers, and other facilities, and removed obstacles from rivers and harbors, but it was authorized to plan other river uses. By the late 1920s, the Corps was empowered to devise flood-control plans as a regular adjunct to projects it had undertaken, to establish a hydraulics laboratory for study of flood prevention, to survey America's 200 major river

basins, and to consider the hydroelectric power-generation potential of its navigation projects. Under Hoover, this information was
turned over to private enterprise for action. Only after Roosevelt's
election did Congress grant the Corps power to engage in river development other than navigation. Congress did so in a peculiar way;
it mandated that each new Corps project was to have multiple purposes. Flood control measures, for example, were also to aid navigation and to generate hydroelectric capacity. This sentiment led
to the Corps receiving primary responsibility for flood control along
all American rivers and the ability to construct major hydroelectric
plants. The huge Bonneville Dam on the Columbia River near
Portland, Oregon, which was begun in 1933 and completed five
years later, was its first significant effort. Within a decade, the Corps'
dams were producing billions of kilowatt-hours of electricity.

***Rural Development—The Tennessee Valley Authority and Rural
Electrification***   Some New Deal efforts were even more ambitious. The Tennessee Valley Authority (TVA), established in 1933,
was a conscious attempt to use government power and funds to
make over a poor, predominantly rural seven-state region traversed
by the Tennessee River. Technology—in this case the introduction
of cheap, plentiful electricity—was to generate new government-
regulated industries and thereby to improve the lives of the region's inhabitants. It was most definitely "an experiment in social
reconstruction." According to its leading proponent, Arthur E.
Morgan, a prominent civil engineer and president of Antioch College, the TVA's purpose was to create "an integrated social and
economic order" in the valley; "the improvement of that total well
being, in physical, social and economic condition, is the total aim."[5]
Morgan envisioned a comprehensive, detailed regional plan that
overlooked no aspect of life in the region. It would even include
inducements to encourage residents to stop smoking and drinking.

The TVA failed to fulfill all of Morgan's social objectives. Dis-
sension among supporters, as well as opposition from other sectors,
prohibited the TVA from implementing its broad political, social,
and cultural reforms. Although it did not result in cooperative
communitarianism, the TVA did produce numerous technical and
economic achievements: dam construction, flood control, navigation, land reclamation, crop diversification, and above all, cheap
electric power generation. Establishment of employee hiring and
training programs, homes, schools, libraries, small local cooperative industries, recreational areas, and the model town of Norris,
Tennessee, enabled the TVA to effect some social change.

The TVA model was popular among New Dealers, who sought to export it to other regions. To be sure, electricity was in some demand in rural environs. In 1930, only 13.4 percent of farms were electrified; the Midwest and South trailed far behind New England and the West. Electrification seemed to urbane New Dealers a potent cure for the "disease" of ruralness. Electricity appeared likely to foster new segments of economic growth; reduce rural health problems, loneliness, and cultural deprivations; and enable rural inhabitants to employ the modern electrical conveniences long available to their urban brethren. But efforts to create elsewhere public regional agencies similar to TVA—especially in the Missouri Valley and the Pacific Northwest—failed. So, too, did ambitious plans for Pennsylvania and the Northeast. The reasons behind these failures were rarely technical. Private utility companies objected to government intervention and formed holding companies to pool profits and power. Even those power engineers backing government involvement often disagreed on the precise mechanism for cheap electricity generation. Some favored outright public ownership, others supported public regulatory commissions, and still others preferred limited public ownership with "yardsticks" for measuring private utility performance. These engineers sometimes advocated comprehensive plans for efficient and nonexploitative natural-resource use, and considered themselves scientific managers of the environment comparable to scientific managers of industry.

Inability to create new TVAs did not stymie New Dealers. Power engineers gained crucial federal assistance in implementing their various proposals in 1935. Roosevelt's Rural Electrification Administration (REA—headed by the mechanical engineer Morris Cooke, a former disciple of Hoover and Frederick W. Taylor), renewal of the Federal Power Commission, and passage by Congress of the Public Utility Holding Company Act to regulate and reduce electric holding-company power spearheaded the national government's campaign. REA assistance helped many farmers form cooperative power pools and build electric lines themselves. A quarter of the cooperatives purchased TVA power, but the overwhelming majority bought power wholesale from private suppliers at higher rates. Federal agencies and regulations, as well as court challenges, ultimately forced private concerns to lower electric rates and to contribute actively to rural electrification.

By World War II this hodgepodge network was virtually complete, and nearly all rural dwellings received electricity. With electric power came good lighting, running water, indoor plumbing, refrigeration, electric irons, washing machines, and radios, as well as electrified farm tools and machines from grinding wheels to wood saws.

New Urban Landscapes   Electricity and regionally planned projects such as the TVA remade much of rural America. New Dealers were less successful in the redesign of urban America. Their failure was not due to lack of effort, however, for they created numerous model communities across the nation to demonstrate what city life in America could become. Their "garden cities" program was among the most clearly articulated and sustained. New Dealers hoped to dot America with comprehensively planned communities of approximately thirty thousand people, with balanced industrial and agricultural economies, that would be circumscribed by fields and forests, or greenbelts. Residents from older decaying industrial cities would be willing to relocate to the garden cities. Modern technologies—automobiles and electricity—would eliminate urban overcrowding and poverty. The economically, religiously, and racially diverse populations of these new cities were to be linked by rapid-transit systems and superhighways to other nearby garden cities, and joined in turn to regional centers with populations of some sixty thousand apiece.

The Roosevelt administration managed to construct only three garden cities: Greenbelt, Maryland (1937); Greenhills, Ohio (1938); and Greendale, Wisconsin (1938). None proved popular. Each lacked an industrial base to complement their greenbelts and pastoral settings. All eventually lost government sponsorship and became purely commercial and largely residential private enterprises.

A wave of suburbanization, which began about 1920, contributed to the failure of garden cities. Ironically, modern technologies, especially automobiles and buses—two transportation forms treasured by New Dealers because of their flexibility—had provided a potent alternative to inner-city congestion for urbanites of even modest means. Widespread adoption of automobiles permitted development of new sectors of suburban growth. Formerly, suburbanization had been constrained by the need for mass transit and had congregated along streetcar and rail lines, but automobiles enabled urbanites to flee inner cities and settle wherever there were roads. Suburban desirability translated into an unprecedented commitment for building roadways for automobiles and for new public expenditures to make cities suitable for cars: from new traffic arrangements and equipment such as one-way streets, automatic traffic signals, and garages, to suburban spinoffs such as freeways and parkways. These freeways offered fast, safe driving conditions, thanks to controlled-access roads, wide median strips, strong divider fences, and cloverleaf intersections.

Urban mass transit became a casualty as its emphasis on standardized travel (mass-transit vehicles generated economies of scale by carrying numerous passengers on regular routes at scheduled times) conflicted with the rise of suburban sentiment. In 1929, Detroit's voters rejected a $280-million proposal for a new subway system plus additional streetcars, while four years earlier residents of Los Angeles were not even given an opportunity to vote on a proposed subway and elevated track systems because the proposal was considered doomed from the outset and was withdrawn from consideration. Local transit companies themselves sometimes opposed major new mass-transit initiatives, for decreased ridership in the wake of automobiles made them fearful that increased fares to pay for bonding measures would further reduce the number of passengers.

General Motors (GM) also apparently gave the decline of urban mass transit an assist. For, as alleged in a 1974 Senate antitrust hearing, between 1932 and 1956 that automobile company had helped destroy 100 street-railway systems in forty-five American cities. GM, Standard Oil of California, and Firestone Tire and Rubber, three industrial giants who would directly benefit from an increased number of motor vehicles, had formed a holding company, National City Lines, which in turn bought street-railway companies in sixteen states. These companies were then converted to operating small, more flexible GM buses, which played to the suburban trend. The companies were finally resold to operators who promised to purchase only GM equipment.

It was the frustrated garden-city vision, rather than the suburban-based accentuation of automobiles, that gave what became the interstate highway system much of its early inspiration. Commentators had boosted the economic value of a nationwide system of interconnected paved roads from the 1920s, but regional and national planners both inside and outside the Roosevelt administration openly pressed in the 1930s for construction of a government-financed national network of highways as much for its human as economic benefits. To New Dealers, these superbly designed roads would run through sparsely populated areas and replicate a phenomenon produced by other forms of transportation a century earlier: communities would spring up along these roadways, and urbanites appalled by big-city living would flock to these places. The result would be urban deconcentration and the rise of presumably balanced, integrated small cities, which joined diverse peoples harmoniously with nature.

REVITALIZING RURAL AMERICA

Balancing Agriculture and Industry

Not all attempts to revitalize American life came from the government, or depended on extensive governmental involvement. Privately undertaken and financed efforts to change America both pre- and postdated New Deal programs, and originated in a number of sectors, including industry. Most of these would-be reformers concentrated on rural revitalization.

The situation in rural regions seemed to demand immediate attention, while the lack of a rigid rural bureaucracy compared to cities made social-engineering prospects more appropriate and promising. As important, the reformers recognized that rural and urban America were inextricably intertwined, and that a program aimed at one would necessarily modify the other. New industrial technologies or processes would always serve as the agents of change, and manipulators generally established small pilot projects to show the utility of their vision of the new American rural life. They presumed in each instance that the force of their demonstration of the virtue of a new balance between agriculture and industry would inevitably attract a significant popular following and even government support.

Ford's Village Industries   Henry Ford concocted perhaps the most novel rural revitalization scheme. Using his great wealth, Ford inaugurated about 1920 a plan to improve rural America through industrialization, but sought to keep that industrialization modest and in balance with agriculture. During the next quarter-century, he opened nineteen small plants at picturesque rural settings to manufacture components for Ford products. He hoped to establish many others throughout rural America, but never was able to do so. One rustic factory employed only nineteen persons. Located in tiny river communities within sixty miles of Dearborn, Michigan, these village industries were each initially outfitted to match the area's hydroelectric capacity. Rather than build new facilities, Ford converted existing structures into small factories whenever practical (gristmills proved particularly attractive), stocking them with the latest tools and machines. Conveyor belts and assembly lines reigned supreme within these small industrial sites. Raw materials

*Henry Ford's nineteen village industries produced small
parts for Ford vehicles and enabled farmers and other rural
Americans to supplement their incomes while maintaining
their traditional occupations, residences, and values.*

were trucked to these plants each morning. Every evening finished
products were collected and shipped to Ford's huge Highland Park
and River Rouge complexes.

These rural communities were explicitly not company towns. The
plants were the only Ford presence. Ford always selected his labor
force from within the community and often tried to reflect the local population demographics; he even reserved comparable per-
centages for the elderly, handicapped, and other disadvantaged
persons. His notorious Sociological Department had no village industry jurisdiction. Ford's sole stipulation was that these employees of his rural facilities also farm. They were asked to purchase
plots of land, and to grow crops during their spare time. That generally meant before and after work and on weekends, but Ford initially permitted his rural laborers to absent themselves from factories

for months for harvesting and other sundry agricultural chores. Those employees unable to acquire farmland were strongly encouraged to take up gardening.

As with almost everything done by Ford, his village industries attracted attention. Roosevelt's Farm Security Administration even attempted to establish a variant of the program. It established the village of Jersey Homesteads in 1933 in the midst of 1,200 acres of woodlands, hired unemployed New York City garment workers to run a garment factory there nine months a year, and had them work on a government-owned farm during three spring and fall months. Jersey Homesteads quickly failed, apparently because the garment workers strenuously objected to farming. So, too, did Ford's village industries prove an economic catastrophe; estimates suggest that Ford spent millions of his own money sustaining the project. When Ford died in 1947, the company swiftly moved to sell off most village-industry plants.

The Chemurgy Movement   Village industries were not Ford's sole attempt to engineer a new rural America. He also was involved with the chemurgic movement, which probably came closest to securing the place its founders thought it deserved. William J. Hale (chairman of the National Research Council's Division of Chemistry and Chemical Technology, and Dow Chemical founder Herbert Dow's son-in-law) coined the term *chemurgy* in 1934. It was derived from two Greek words: *chemi*, the art of transforming materials, and *ergon*, work. Its proponents sought to blur traditional distinctions between agriculture and industry and to establish instead an industrial–agricultural continuum; agriculture would become the predominant source of industrial chemicals.

Although the word chemurgy entered the American lexicon in 1934, and a national organization was formed the following year, its practitioners had begun chemurgic labors more than a decade earlier. Orland R. Sweeney, head of Iowa State College's chemical-engineering department, was among chemurgy's first investigators. During the period of 1920 to 1933, Sweeney, his colleagues, and students engaged in chemurgy—the transformation of agricultural products into industrial materials—on an unprecedented scale. They derived chemicals and made products from soybeans, barley, flax, wheat, rice, oats, bananas, beets, tobacco, pecans, peanuts, cotton, pineapples, artichokes, peat, sorghum, and even chicken feathers. But corn was their favorite raw material. Sweeney's associates destructively distilled, pulverized, chemically digested, and fermented corn cobs to produce furfural; charcoal; acetic, formic, butyric, and oxalic acids; methanol; calcium acetate; xylose; pentosans; and ethanol. They treated cornstalks with pressurized steam, pulverized them, or had them chemically digested to yield lignin, glue, nitrocellulose, rayon, paper, and boards of varying densities. Sweeney's wide range of boards could be used for everything from a cork substitute to a replacement for steel in automobile bodies or Pullman cars. From the cornstalk processing the Ames group recovered liquid natural gas, flotation oils, polyhydroxyl alcohols, and adhesives.

Word of Sweeney's work brought to Iowa State representatives of the governments of Australia, Argentina, Britain, Hungary, Germany, and the Soviet Union. The United States government also took notice. In 1928 the National Bureau of Standards established a facility and research team in Ames, while the U.S. Department of Agriculture (USDA) located an agricultural by-products laboratory there five years later. Henry A. Wallace, later Roosevelt's secretary of agriculture and his vice-president during Roosevelt's third term, invested in a Sweeney-designed cornstalk board-production plant in 1929. That factory produced the insulating boards used at the 1933–1934 Chicago World's Fair.

Wallace's backing of Sweeney's initiatives probably stemmed from shared sympathy for the farmers' plight. The Roosevelt administration as a whole certainly endorsed his chemurgic efforts. But Sweeney embraced a vision of America's future that made him an anathema to most industrialists and the Republican party. Taking a kind of economic ecological approach, Sweeney surveyed ancient civilizations, concluded that their demise resulted from "the city system," and worried that America was following the same route: urbanization upset the natural economic balance. Urban populations required dramatically higher support costs than comparable rural peoples. He criticized mass production as fostering urbanization, and maintained that if industrialists ever determined mass production's *total* costs—the urban support costs—they would agree that handcrafted goods were cheaper. Massive but voluntary relocation of urban people to new rural communities was America's only chance for salvation. There the new rural inhabitants could erect small hydroelectric-powered plants to convert agricultural commodities into chemicals, and to produce industrial goods. America would become a continuum of family farms and small rural industrial communities.

Hale's ideas differed radically from those offered by Sweeney. Hale wanted industrial stability and looked to agriculture to provide it. Defining "an agriculturist simply as an organic chemical manufacturer . . . nothing else," he sought to incorporate agriculture as an integral part of his new industrial order; agriculture would become the cheap, dependable source of chemicals necessary to fuel the American industrial machine. These precepts led Hale to propose a revolutionary new agricultural system. Corporate-owned "agricultural supply centers," not privately owned farms, constituted its essence. Hale envisioned each "agricenter" as encompassing several thousand square miles and each operating under "a real industrial leader." Working in conjunction with lawyers, bankers, scientists, and other industrialists, each agricenter's leader would negotiate contracts with chemical manufacturers to set next year's production goals.[6]

Hale's agricenters would be scientifically managed, and efficiency experts would review and modify farm practices. Superintendents would implement policy and ensure that laborers, who would be paid hourly wages, followed it. Agricenter corporations would supply seeds, fertilizers, machines, and other necessary items as agricenter scientists and engineers worked to improve agricultural machinery and yields. Processing plants would be established nearby to transform agricultural products into industrial chemicals.

It was a vision closer to Hale's than Sweeney's that resulted in formation of the National Farm Chemurgic Council. The organization got its start in 1935 when Hale and Carl B. Fritsche, a Detroit industrial engineer, prevailed on Henry Ford to sponsor a "Conference of Agriculture, Industry, and Science" to discuss chemurgy's prospects. Held in Ford's company headquarters at Dearborn, Michigan, the meeting attracted 300 participants. Most represented large industrial concerns, or were academics with close industrial ties. Physicist Robert Millikan of California Institute of Technology and Karl T. Compton, president of Massachusetts Institute of Technology, were the organization's most noted academic partisans.

Ford provided the council invaluable national publicity and legitimacy. He had long advocated closer agriculture–industry ties and was no stranger to chemurgy. Ford's automobiles employed soybean-derived plastics for upholstery, body finishes, horn buttons, and distributor covers. The Ford Motor Company had made extensive use of products based on corn, cotton, flax, sugar cane, and wood in automotive manufacture. Ford's chemurgic interests also had led him to display an "Industrialized American Barn" at the Chicago World's Fair. This exhibit described how entrepreneurs could utilize empty or deteriorating barns to process crops for industrial uses. Sitting in the midst of a soybean patch, Ford's barn was filled, not with hay, but with various soybean-processing tools and machines.

Neither the Council nor chemurgy generally gained the massive public support that its proponents had anticipated. Farmers certainly disliked Hale's vision and were concerned about more immediate forms of relief. Most processes to transform crops into industrial raw materials proved more costly than alternate means. Political factors also complicated the issue; the movement's predominantly corporate, Republican leadership continually criticized New Deal programs, especially those taking land out of production or catering to small farmers. "In the midst of the chemical revolution we are chemically disorganized; in the depths of depression we are chemurgically incapacitated," complained Hale.[7] Unable to effect the changes deemed necessary to implement the new agricultural–industrial order, the Council contented itself with reporting on chemurgic processes, objecting to government policies, and publicizing the chemurgic idea.

Failure of explicit social-engineering attempts, such as chemurgy or the TVA, to produce new settlement types, or to redress the balance between industry and agriculture, or between the urban and the rural, in a fashion consistent with their proponents' objectives did not discredit those objectives or planning. Paradoxically, lack of success reaffirmed to both social engineers and the American public generally the need to design comprehensively planned, integrated environments "fit for life and the living." Failure was attributed to the same phenomenon that made planning so desirable—modern life's complexity. Modern life was such a complicated system of interrelationships that social engineers had failed to grasp all the various relevant factors, or to understand completely how the system's different elements influenced and changed each other. That stance, coupled with the assumption that those difficulties could be ironed out soon, made planning even more compelling, and intensified efforts.

CHANGES IN PRODUCTION TECHNIQUES

The repeated inability of social engineers to foster the precise modifications that they sought did not mean that traditional American modes of production persisted unchanged. Just the opposite was the case. The post-1920 notion of system was not restricted to social engineering, but permeated virtually every facet of American life. As Americans considered and pursued their endeavors, they incorporated the idea of system as an interlocking network of mutually interacting parts, although not necessarily such an idea's potential social and cultural ramifications. In some cases, that incorporation simply resulted in a reconceptualization of established practice without physical change, while in others processes or activities were altered. It was from this intellectual context that mid-century technologies—and economic and business principles—were conceived and applied. The redefinition of salient points led to identification of new relationships, which created new understandings, desires, and needs. These in turn translated into investigation into new areas, development of new human productions, and formulation of new marketing strategies. In essence, virtually no enterprise remained unchanged.

Changes in Agriculture after 1920

Manifestations of this reconceptualization were quite evident in spheres of production. That agriculture is sometimes overlooked as a form of production, and that its nature seems both traditional and fragmented, makes it a challenging place to start this analysis. Nonetheless, food production and each of its constituent activities were scrutinized and redesigned by agriculturists and others in the post-1920 period. The upshot was creation of a more highly integrated agriculture, and a much tighter agricultural–industrial connection.

The Multipurpose Tractor   No single aspect of farming was more indicative of this shift than the tractor boom of the mid-1920s and after. Tractor demand had temporarily risen during World War I as farmers employed them in the Midwest and West to combat labor shortages. But post-Armistice sales dropped to prewar levels. The explanation was simple: tractor utility was physically circumscribed by intellectual constraints. It was in conception a single-purpose machine developed to capitalize on the uniqueness of Great Plains agriculture. Tractors could draw implements but not power them (tractors towed, for example, gasoline-engine-powered combines as well as plows), or could power mechanisms, such as threshers, while stationary. But tractors could not pull and power machinery simultaneously.

Only with a new understanding of the interrelatedness of farm endeavors did manufacturers seek to redesign tractors. Their development of "power takeoffs" in the 1920s transformed and popularized the tractor. Essentially splined drive-shafts extending from

tractor rears that could be mated to correspondingly shaped implement fittings to deliver power to these machines at any time, power
takeoffs created what advocates termed "general purpose tractors."
The possibilities of these vehicles seemed endless, and sales skyrocketed after 1923 when manufacturers perfected power-takoff
tractors suitable for row-crop cultivation. The industry quickly recognized the chaos generated by each manufacturer's policy of producing power takeoffs and compatible implements to its own
specifications, and agreed in 1927 to standardize industry-wide shaft
fittings and size, as well as the speed and direction of shaft rotation. These developments led an observer to conclude in 1932 that
the "idea of the integral design of the tractor and its implements,
the interdependence of the tractor and implements, has been fully"
realized, and as a consequence "the labor-saving possibilities of power
machinery may be more fully recognized."[8]

Adoption of general-purpose tractors (and later combines) with
power takeoffs was so swift that by about 1950 horses had virtually
disappeared from American farms. Farmers weighed the decision to
replace horses with machines in favor of tractors and combines,
but purchase of any particular piece of machinery rarely was clear-
cut. For these most significant investments, agriculturists needed
to balance a variety of factors, some of which were intangible. A
manufacturer's reputation, expertise, or price, which was appropriate for a different age, was an insufficient post-1920 determinant. Indeed, the idea of cost itself had become multifaceted. Before
choosing a machine, noted a commentator in 1931, farmers must
at least take into account "adaptability to local conditions; initial
cost; maintenance costs; power costs; design and construction;
suitability of construction materials; reliability and size of manufacturer; dealer's service record; ease of operation; ease of adjustment; ease of repair; appearance of machine; whether it needs
protection from the weather; ease of sharpening; availability of replacement parts."[9]

Inventing Crops to Fit the Machines   These general-purpose machines furnished motive power for many farm implements, but the
circumstances surrounding combination corn-pickers/huskers, which
were also products of the 1920s, reveal another aspect of agricultural change. These early devices sold poorly because they performed miserably. They shelled too thoroughly, husked inadequately,
had difficulty with stalks of different heights, and failed to strip
broken or blown-over stalks. Not until the late 1940s did these
machines gain popularity. Their acceptance then did not stem from

improved machine design, but from "design" of a machine-compatible corn, which was bred to meet the machine's needs as well as the limitations of the environment in which it was grown.

This multipurpose *hybrid* corn could be tailored to fit almost any environmental situation or set of physical parameters. It usually was bred for stronger stalks and roots, easy husking, disease resistance, yield, climatic adaptability, and uniform stands and ear size. Henry A. Wallace was a pioneer hybrid-corn industrialist, but credit for hybrid corn belongs to the Connecticut Agricultural Experiment Station's Donald F. Jones. Between 1918 and 1920, Jones both explained hybrid corn in a manner consistent with the post-1920 notion of systems and announced a mechanism to make it practical.

Americans had attempted to breed corn from the 1870s, but had focused on one variable, yield per acre. By the 1890s, corn breeders equated the corn's appearance with vigor and high yield, and chose seed kernels from ears deemed most nearly perfect. In 1896, Cyril Hopkins of the Illinois Agricultural Experiment Station refined this technique by devising the "ear-to-row" method of selecting potentially high-yield corn plants: planting only part of the seed from likely ears, and using separate rows for each ear's seed. Remaining seeds were stored for possible future use once each row had been harvested separately and the yield from different ears compared.

The ear-to-row method, like the others before it, rarely produced uniform high-yielding crops. Individual stalks and ears differed dramatically even for seeds planted from the same ear. George Shull, operating at the Carnegie Institution's facility at Cold Spring Harbor, New York, provided in 1908 an explanation for the diversity. He determined by inbreeding (self-fertilizing) corn plants that common corn varieties were in fact combinations of large numbers of strains; each kernel on an ear might be from a different strain. Only through "selfing" (self-fertilization) for several generations would "pure" corn breeds reveal themselves. And these breeds, most of which were very small, would breed true; their progeny would bear a remarkable resemblance to each other and their parents. Shull also several times crossbred two dissimilar purebred corn plants, sometimes with extraordinary results: favorable crosses would produce plants larger and more prolific than their parents, but these plants would not remain vigorous in succeeding generations.

Shull's experiments proved puzzling, but Jones deciphered them a decade later. To Jones, inbred corn's dwarfishness and lack of vigor stemmed from the transference of corn's hereditary factors in generally indivisible groups. That virtually precluded the possibil-

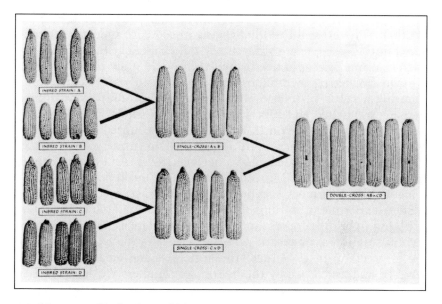

*Double-crossed hybrids could be created for almost any
environmental condition.*

ity of getting the most favorable characteristics in single plants;
traits were inextricably linked, and breeding for a desirable trait
most likely would also expose undesirable ones.

Yet Jones recognized inbreeding's utility. It revealed the root stock,
the good and bad traits, which were linked. He devised the double
cross, "which brings together the greatest number of different factors," as a method to rid plants of inferior characteristics because
"favorable growth characters tend to be expressed rather than unfavorable ones, whenever the two are paired"; inferior traits would
be suppressed by favorable ones. Jones's double cross used four
purebred lines and two generations. If these different lines were
each selected with a specific purpose in mind, high-yield hybrids
could be created that were adaptable to an extraordinary variety of
climatic conditions, were resistant to many common diseases and
pests, and produced similar stands and ears. [10]

Double crossing took time, skill, and capital because there was
no return while hybrids were being created. Moreover, hybrid seeds
were good for only one generation because hybridization's pluralistic vigor quickly disappeared. These factors dissuaded farmers from
producing their own hybrids. The first significant hybrid seed company, Wallace's Hi-Bred Corn Company, was created in the mid-

1920s to sell farmers hybrid corn seed; it was soon rivaled by Funk Brothers, Pfister, and DeKalb. The new seed-corn industry encountered initial resistance from farmers, who balked at the idea of paying premium prices for seed when they could grow their own. But shrewd marketing techniques, such as entering corn-yield tests or planting a patch of hybrid corn on farmers' land at company expense, established the superiority of hybrids and converted many farmers before World War II. Hybrid seeds accounted for more than 95 percent of American corn acreage by the mid-1950s. Average yields per acre had quadrupled since the 1920s. And farmers could harvest their crops with mechanical pickers/huskers.

Even before farmers accepted hybrid corn, seed companies, the USDA, experiment stations, and the Rockefeller Foundation were applying hybridization techniques to other crops. They developed hybrid sorghums, rices, and wheats, successes that led an enthusiast to proclaim in 1947 that "the principles and practices first discovered and developed by the hybrid-corn makers are destined . . . to banish hunger and want."[11]

The Genetic Manipulation of Animals   A similar reconceptualization of meat- and milk-producing animals, and how they could be adapted to enhance farm profitability, also occurred in the 1920s. As with corn, farmers had long artificially selected and bred animals. Until the 1920s, form and size (the yield per animal) predominated as farmers aspired to breed purebred animals conforming to idealized standards set up for these established breeds at farm shows. For example, Poland-China hogs proved a long-standing favorite, and farmers selected and bred those largest animals whose carcasses were essentially tubular in shape.

Farmers began to recognize in the 1920s that strict adherence to purebred criteria did not necessarily make economic sense. A number of tastes and needs characterized pluralistic America, and meeting each of these demands would increase profits. Farmers also realized that purebreds might not be the best-suited animals for particular environs. Some were more susceptible, for instance, to heat, drought, and pests. These considerations led farmers to embrace crossbreeding, and they looked to Europe for new purebreds. Danish hogs and British sheep, for example, were imported and crossbred with American varieties as farmers attempted to expand their markets and to develop "better" animals.

Agricultural scientists, led by Iowa State College's Jay Lush, provided a theoretical basis for the crusade and opened new possibilities. They redefined breeds as those animals most closely approximating statistical norms; breeds became aggregations of hereditary characteristics coalescing around specific points. This population-genetics approach produced new flexibility (animals could be bred around new, more suitable points), and scientists and farmers began to explore the creation of new "breeds." These efforts produced Saint Gertrudis, Beefmaster, Brangus, and McCan "purebred" cattle. A Swine Breeding Laboratory opened at Ames, Iowa, in 1936 to carry on this sort of work, while breeding research intensified at the USDA's Beltsville, Maryland, experiment station.

The introduction of artificial insemination techniques facilitated breeding, and became the technology that enabled these new breeds to become commercially viable. First employed systematically among New Jersey farmers, the technique spread quickly across America; where a bull normally could service thirty to fifty cows yearly, with artificial insemination he could service more than 2,000. From 7,539 cows artificially inseminated in 1939, the number grew to 1,184,000 in 1947, and to 7,500,000 in 1962.

New Agricultural Chemicals   After 1920, farmers engaged not only in genetic manipulation of animals and plants but also in environmental modification. Use of industry-produced chemicals and additives enabled agriculturists to create productivity-enhancing environments. A few agricultural chemicals had been employed decades earlier, but they were used only to combat extant pests, such as blights, not to modify environments. These after-the-fact chemicals gave way to environmental growth-promotants as farmers began to adjust environmental systems. Freeing environments of weeds (which competed with crops for space, sunlight, and nutrients, and which reduced crop quality, increased production costs, and decreased profits) was an early objective, but bore fruit only after World War II. Synthetic plant-growth hormones (auxins), which upset weeds' natural hormonal equilibrium, causing them to grow at uncontrolled rates and ironically to starve to death, constituted the most effective mid-century herbicides. Discovered in the late 1920s and developed as a by-product of chemical-warfare research, the first commercially available auxin, 2,4-D (2,4-dichlorophenoxy-acetic acid), went on the market in 1945. The chemical 2,4-D killed most broad-leafed plants, which included most weeds, and appeared harmless to plants useful to mankind, as well as to persons exposed to it through airplane spraying, dusting, or aerosols. As a scholar noted in 1948, "2,4-D seems to offer almost everything we ask of a weed killer."[12]

The insecticide DDT (dichloro-diphenyl-trichloro-ethane) offered

*Agricultural chemicals have become increasingly common
on farms. This mound of chemicals constitutes the average
amount used annually on this gentleman's 78-acre farm in
the early 1950s.*

more manipulative potential and seemed an even greater panacea.
Discovered in 1939 by a Swiss scientist, Paul Muller, it became
available in America three years later. DDT quickly gained fame
when the U.S. Army used it in occupied Italy to combat lice, and
then in the Pacific to combat malaria. After World War II, Americans employed it against more parasites than any previous insecticide. Like herbicides, DDT appeared to increase productivity without
harming desirable plants, animals, or humans. It spawned an entire
generation of related insecticides: chlorodane, toxaphene, gamma

BHC, dieldrin, and aldrin. These also were used massively and un-selectively.

Some of these insecticides were sprayed directly on animals to kill pests. That aspect of their use provided a complement for the new animal feeding techniques of the post-1920 period. To be sure, the discovery of vitamins in the century's first decades marked a major breakthrough, but it was consistent with a half-century-old quest to determine the nutritional elements necessary for normal growth. From the 1920s onward, however, feed companies, farmers, agricultural scientists, and pharmaceutical firms all sought means to manipulate animal growth through diet. They generally tried to promote accelerated growth and to free animals of their environmental limits. These investigators considered growth, health, and feed relative phenomena, their parameters adjustable to the purposes and situations at hand. This approach led to the introduction of sulfa drugs and then antibiotics as standard cattle-feed additives to prevent disease. The growth promotant, DES (diethylstilbestrol), first found use as a poultry-feed supplement, then as an implant, but left its mark as a cattle- and sheep-feed additive. Estimates suggested that it speeded growth some 30 percent. Rations were artificially constructed to utilize farm wastes. For example, corn cobs, molasses, and urea served as a cheap, nutritious cattle feed.

These feed initiatives, like the period's environmental and mechanical manipulations, depended on close links between agriculture and industry. Indeed, differences between these heretofore discrete forms of production became blurred as food producers engineered their environment and products. Demographic and economic changes accompanied industrial agriculture. Average farm size increased markedly, but the number of farmers declined precipitously. By 1960, only 3 percent of the nation's population engaged in agricultural pursuits. Each farmer could tend a larger area or more stock, but needed to do so to reduce unit production costs and thus justify the expense of the new manipulative methods.

Changes in Industrial Manufacturing after 1920

American manufacturing underwent conceptual changes comparable to those of agriculture as industrialists reconceptualized industry, industrial processes and organization, and marketing strategies. Rarely were these new developments trumpeted by individuals directly responsible. The self-aggrandizement of Singer, McCormick, or Ford seemed out of place in this new highly integrated industrial system. As important, many industrialists failed to grasp the radical nature of the changes they made. But industrial success after 1920 did not require manufacturers consciously to evaluate the differences in perception between themselves and their industrial forebears. It did necessitate, however, that they understand the milieu in which they operated and develop approaches in accordance with that understanding.

The Rise of General Motors   The dramatic rise of General Motors (GM) in the mid-1920s and the contrasting decline of the Ford Motor Company during the same period provides a compelling example of the new industrial system. Ford had conceived of his Model T as the car for the masses. Its single style enabled him to mass-produce cars, and to organize single special-purpose machines and machine tools to facilitate flow and generate economies of scale in his plants. Ford's rigidity stemmed from both his assumption that there existed a correct "American" approach to things (manifested by his Sociological Department, Americanization campaign, and other efforts) and his quest to perfect automobile manufacturing; he conceived of his corporation's role as the producer of "American" cars. But Ford's insistence on a single standardized automobile had become the butt of jokes in post-1920 America. Suggestions made the rounds that a person could purchase a Model T in any color so long as that color was black.

GM's Alfred P. Sloan, Jr., capitalized on the discontent embedded in such phrases to effect a revolution in automobile manufacture and sales in the 1920s and after. Arguing that a corporation's primary responsibility was to provide a high return to investors, Sloan rigorously maintained that corporations needed to determine and tap the market's demand rather than merely produce supplies, and that demand itself was plastic: it could be created, modified, or stimulated. What corporations required were concepts, strategies, and implementation. To Sloan, added volume alone spawned profits, and added volume came from carefully planned products superbly marketed. Product superiority and improved manufacturing techniques figured in corporate profits only incidentally, but they, too, could be stimulated. Corporate well-being was simply the consequence of effective management.

Although Sloan downplayed technological superiority as a crucial factor in corporate success, his GM stewardship was marked by a number of technical innovations: high-compression engines fueled by tetraethyl lead to prevent knock in the early 1920s, independent front-wheel suspensions to promote handling in the early 1930s, and automatic transmissions in the late 1930s. But Sloan's "better ideas" were tactical. He reorganized GM according to his precepts and introduced several programs consonant with his notion of demand's relative nature. He conceived of post-1920 automobile purchasers as primarily second-time buyers and contended that as a group their objectives differed from first-time owners. The latter desired basic transportation, while the former sought "comfort, convenience, power and style."[13]

Sloan helped these second-time buyers achieve their dreams. The General Motors Acceptance Corporation, established about 1920, provided loans at low interest so that persons could purchase higher-priced GM cars. Sloan also fostered creation of used car markets, which made it attractive for motorists to sell cars to purchase higher-ticket items. And Sloan was careful to see that GM offered a car "for every purse and purpose."[14] William Durant, GM's first chief, had had GM produce five different autos, but almost all targeted persons of upper-middle-class means. Sloan felt Durant's approach was unwarranted (it created competition within the company for the same market) and unfortunate (it conceded all other markets to other automobile manufacturers). He argued instead that each "General Motors car should be integral, that each car in the line should properly be conceived in its relationship to the line as a whole."[15] This continuous line of products would tap everyone, from the poorest American able to afford a car to the wealthiest, with no duplication. Each of the five vehicles would be sold in a different price range and highlight different features. This policy was in place by 1922. Chevrolet became GM's lowest-priced car and Cadillac its highest. The other three brands—Oakland, Oldsmobile, and Buick—aimed at various segments of the burgeoning middle class.

Sloan's pricing and product-line strategy was reflected in his reorganization of GM. He attempted to rid the company of unnecessary duplication of effort, but recognized that different markets required different approaches. Each of its five automotive divisions needed its own staff (executive, engineering, styling, production, marketing, sales, and the like), but overall GM must have "a system of co-ordination so that each part may strengthen and support each other part."[16] Sloan's reorganization, planned in 1920 and implemented in 1924, gave divisions most of the autonomy associated with independent companies, established interdivisional groups to harmonize activities and smooth out difficulties, and provided the corporation's chief executive with ultimate authority.

Nowhere was the dual "federal" nature of GM's organization structure more apparent than in the implementation of Sloan's greatest car-selling coup, the annual model change. The culmination of the ideas that resulted in continuous-line pricing, credit purchasing, and used-car markets, the annual-model-change concept emerged about 1925. Sloan decreed that there should be a single basic GM style each year, but left it to each division's stylists to modify that form to suit constituencies. For example, in 1927 Cadillac presented buyers with 500 color and upholstery options. Sophisticated planning went into predicting and stimulating consumer tastes; development of an annual model started two years prior to that model year. Divisions also kept close tabs on how each year's models were selling. Unlike Ford's static Model T, GM's annual models could not be marketed as new the following year. To prevent unprofitable model-year surpluses, GM's divisions required dealers to complete detailed sales questionnaires every ten days. Divisional inventories, purchases, and production rates were adjusted to approach each reporting period's estimation of demand. That flexibility permitted divisions to keep on hand smaller inventories of manufacturing materials. GM turned over inventories roughly twice yearly before adopting the annual model change; after 1925, the company turned them over an average of twelve times.

Annual models made change the normal state of affairs. This had profound manufacturing implications. Refitting facilities yearly with new special, single-purpose machines would be prohibitively expensive; wholesale adoption of Ford's mass-production techniques was incompatible with annual models. GM overcame that dilemma by introducing flexible mass production. It employed in its plants movable, general-purpose machines. Each had adjustable speeds and was powered by an individual electric motor. Workers resituated and set these machines at each model year's onset to form new lines facilitating product flow and mass production. GM also expedited model-year conversions by eschewing large central facilities for smaller, more specialized, and quickly modified plants.

GM adopted one final manufacturing feature. It strengthened fixtures to increase precision. More precise measurements enabled GM to extend interchangeability outside divisional boundaries. GM reconciled mass production with product variety by using some car parts in more than one division. Pontiac, established as a division in 1926, was perhaps the first beneficiary of this technique. Many of the parts used in high-volume Chevrolet also were used in the more expensive Pontiac. Higher-priced vehicles benefited from volume economies generated by lower-priced cars, and mass-production benefits spread throughout the entire GM product line.

GM's initiatives captured for the company the largest share of the American automobile market. Ford responded by abandoning his beloved Model T in 1927 and replacing it with the sportier but almost as static Model A. Ford's conversion to Model A production required him to shut down operations for six full months. In contrast, Chevrolet could gear up for a new model in a mere three weeks. But the Model A did not seize the post-1920 public imagination as had the Model T some two decades earlier. By 1933, Ford had stopped selling the Model A, adopted the annual model change, switched to flexible mass production, and embraced the concept of a product line aimed at a wide spectrum of buyers.

The Development of Plastics   Newer, less-well-developed industrial sectors also were changed, but often more subtly than the automobile industry. In the case of plastics, for instance, the crucial

Bakelite, the first useful non-cellulose-based plastic, was manufactured in ovens like this one from 1909. It and other special-purpose plastics gave way to general-purpose plastics, such as nylon, in the 1930s and after.

conceptual shift revolved around product definition. Pre-1920 plastic manufacturers presumed that plastics ought to be single-purpose substances that would replace a particular material in a certain set of related uses. This was true for celluloid and for Bakelite, the first useful noncellulose plastic. Compounded by Leo Baekeland, who had produced high-quality, easily handled photographic paper for Eastman, Bakelite was manufactured commercially from 1909. It was the first of a class of plastics produced by condensation of phenol with various aldehydes, and could be molded or used to coat or impregnate materials. Its ivorylike hardness made it suitable for billiard balls.

The announcement of Bakelite was greeted enthusiastically and added fuel to the search for special-purpose plastics. Researchers attempted to create a plastic for every specific function, or to find a precise function for every plastic substance created. Casein, furfural, acetate, and pentosan plastics emerged from those efforts. So, too, did rayon, which was the result of the textile industry's effort to produce an artificial silk. Composed of acified cellulose—generally wood pulp—treated with caustic soda and carbon disulfide, rayon was drawn and spun into threads, and then woven. Eight million pounds of rayon were produced in America during 1919, and the output increased more than sixteenfold a decade later. Chemical companies initially remained content to provide the textile industry the chemicals necessary for rayon, but after 1930 they aggressively manufactured the synthetic fiber themselves. In 1939, America produced more than 328 million pounds of rayon, almost all made by the chemical concerns.

Even before the chemical industry began large-scale rayon production, it had developed significant interest in and experience with plastics. Its studies of polymers and polymerization led to the development of polymerized fibers. Nylon, developed by the Du Pont Company, and the most important of these new plastics, was the first material marketed as a new, completely synthetic substance, designed not to replace one compound for one activity but many compounds for many activities; nylon was a general-purpose plastic. Its introduction followed from a carefully orchestrated Du Pont campaign. The company hired in 1928 Wallace Carothers, then instructor in organic chemistry at Harvard University, to head its organic-chemistry research group. Aware that certain polyesters and polyamides could be changed by cold stretching from random molecular arrangements into long, linear ones (like those found in natural silk and stretch-spun rayons), Carothers decided to study the synthesis of these high-molecular-weight polymers in hope of developing a multipurpose plastic. He and his postdoctoral associates isolated a fiber-forming polymer as early as 1929, but it readily melted. By 1934, Carother's group had produced nearly one hundred "superpolymers." The one labeled "66" seemed to hold the most commercial promise. Du Pont technicians investigated this polymer (hexamethylene diamine-adipic acid) for forty-four months and worked out means for its economical manufacture. It became known as Nylon when the company announced its creation in 1939. Nylon proved stronger than silk and much more adaptable. It soon replaced that natural fiber in women's stockings. As a consequence of the company's aggressive wide-ranging marketing, nylon had by 1941 replaced natural bristles in scores of household and industrial products: catgut in sports racquets, surgery, and musical instruments; gutta percha in wire insulation; some natural cements; steel machinery bearings; and leather. Umbrellas, shower curtains, and parachutes were among nylon's other early uses.

NEW MARKETING AND DELIVERY TECHNOLOGIES

Using Radio to Reach a National Market

Advertising played a prominent part in Du Pont's nylon marketing efforts, and the company took advantage of several intricate nationwide webs of radio stations to promote its product. This strategy did not, however, mark the chemical company as unique because radio had been a prime advertising medium for manufactured consumer products for more than a decade and a half.

Before the mid-1920s, entrepreneurs conducted large-scale advertising only through the mail or in local newspapers. Indeed, the earliest radio stations (which were owned by manufacturers) did not even accept advertising, but offered instead only public-service programming, which served radio manufacturers by establishing markets for radio receivers. Reduced prices for receivers had led to mass purchase of radios, which indirectly proved the medium's advertising potential. As early as 1922, WEAF in New York City (owned by Bell Telephone) allowed manufacturers to purchase air time to advertise their products; advertising quickly became the principal revenue source of radio stations. This thrust occurred despite opposition from then Commerce Secretary Hoover, who, as head of the federal department regulating radio, claimed that advertising undermined radio's public-service capabilities.

The new multipurpose radio received a boost in 1926 when Bell rented to stations long-distance wire connections. In that year, David Sarnoff, head of RCA, created the National Broadcasting Company (NBC) as an RCA subsidiary. A year later there were three major national networks: both the Blue and Red networks of NBC, and the rival Columbia Broadcasting System (CBS). Advertising could then be national as well as local, and more than $10 million was spent on it by 1928. No federal regulation existed of either the content or the frequency of ads, which increasingly accompanied programs ranging from entertainment to sports to news.

Congress made half-hearted efforts to oversee radio's development by enacting the 1927 Radio Act, which established the Federal Radio Commission, and the 1934 Communications Act, which replaced that agency with the Federal Communications Commission. Both bodies presumed public ownership of radio's frequencies, which were nevertheless allocated to private groups for short-term licenses renewable for proper operation and public service. In practice, few licenses were ever revoked, and most were therefore *de facto* long-term ones. In effect, regulation was primarily concerned with assigning frequencies, setting power-output limits, and controlling geographical access to radio facilities.

By 1941, on the eve of World War II, there were 660 American radio stations, 160 affiliated with NBC and 107 with CBS. Together these network stations comprised 86 percent of the licensed, and lucrative, nighttime broadcasting, and were invariably in key markets. In that year the federal government, under Supreme Court directive, forced RCA to dispose of one of its two networks, the Blue, which soon became the American Broadcasting Company (ABC) and a new competitor for advertising dollars. But networks and independent stations, while remaining relatively free of government regulation, were never without public scrutiny. Their advertising rates were directly proportional to the numbers of listeners they attracted, and those numbers were a function of their entertainment and public-service broadcasts. Networks and independent stations faced something of a dilemma; their profitability depended on determining before the fact what type of programming would draw the largest audience. In that sense at least, expected demand dictated actual supply. Mid-century radio programmers found it safer to copy a competitor's success (a practice that produced particular genres simultaneously on virtually all radio stations) than to initiate a new potentially risky form of programming. Indeed, new programming ideas generally were broadcast only when stations faced desperate dropoffs in advertising revenue.

The Railroads after 1920

Radio advertising created unprecedented nationwide awareness of manufactured consumer goods, which were often delivered to markets by railroads. In the 1920s railroads retained their traditional position as the nation's preeminent long-haul freight carriers, but they already faced increasing competition from newer modes of transportation and would ultimately be supplanted. The railroads did not begin to decline, however, as a result of refusing to adopt new technologies. In fact, railroads were quick to embrace them. Yet railroads were themselves at least in part responsible for their own demise. The technologies railroads incorporated were inconsistent with mid-twentieth-century desires, and so laid the seeds of their destruction; railroads made themselves anachronistic and eventually almost obsolete.

Railroads after about 1920 concentrated on becoming bigger, faster, and more efficient. Rails were repeatedly strengthened to absorb heavier loads, signals and traffic-control systems were made almost automatic, and air brakes were vastly improved to function under even the greatest burdens. Steam locomotives such as the Union Pacific 4000 class (1941) attained a weight of more than 600 tons and could travel at 70 mph. They began to be replaced in the 1940s by diesel locomotives, which were faster and lighter but more powerful and capable of transporting larger loads. Where many different companies produced steam locomotives, only a handful manufactured diesels, thereby allowing greater standardization of models and fewer total locomotives.

These technical factors enhanced railroad operation on high-volume, long-distance, point-to-point runs, but they severely hampered its utility for other forms of delivery. Mid-century railroads lacked the flexibility demanded by an America consciously attempting to decentralize manufacturing facilities and markets; they were in effect single-purpose carriers. Aside from the question of exceedingly high capital costs, huge locomotives maximized economies only when pulling a very large number of cars or very heavy cargo over considerable distance. That they were unable to offer a competitive price and to achieve a return justifying their great expense under other circumstances left a substantial void to be filled by other types of carriers.

New Delivery Technologies—Trucks and Buses

Trucks and buses capitalized on this opportunity, and also conveyed goods to areas inaccessible to railroads. Crude trucks and

Buses offered a flexibility unavailable from railroads.
Numerous small lines capitalized on this advantage by
merging to form the Greyhound system.

buses were built as early as 1900, but not until World War I was
their utility demonstrated both at home and abroad. Pneumatic tires,
developed by the Goodyear Tire and Rubber Company in 1916, prevented these heavy vehicles from destroying roadways, cargo, and
themselves through constant pounding. But pneumatic tires were
not the only technological innovation that helped trucks and buses
corner an increasingly impressive share of haulage. In 1927, the
Fageol brothers of Kent, Ohio, introduced the twin-coach bus; its
engines were placed underneath the passenger compartment to make
the vehicle's whole body suitable for heavier loads, as opposed to
earlier bus bodies that were simply placed on truck chassis. In the
1930s, Charles Kettering and others at GM improved existing diesel engines to make them more reliable and powerful for buses and
trucks alike. Larger trucks later separated their power unit from
the cargo units, which became semi-trailers that could be attached
or detached as needed.

Several railroad companies belatedly recognized trucking's potential, as well as their failure to form a flexible railroad network,
and ironically established their own trucking subsidiaries to combat independent trucking firms. Numerous small bus companies

sprinkled throughout the nation offered railroads a more direct challenge. They consolidated themselves into the Greyhound system, which from 1929 became America's largest bus line, delivering products and passengers nationwide. The foothold gained by these early trucking and busing haulers testified to the shortsightedness of railroads. These smaller, flexible carriers were able to capture a profitable share of the freight and passenger trade in an era of poor roadways and no interstate highway system (which was not inaugurated until 1956).

The Growth of Commercial Aviation

A more vigilant railroad industry probably could not have prevented development of yet another carrier, commercial airlines. It specialized in longer-distance haulage than many truck and bus concerns, and offered competition to railroads in that sector. Like railroads, the speedier airlines had huge capital costs.

Commercial aviation's roots go back to early air-mail service, which began in 1918 when the U.S. Post Office used army aircraft and young World War I veterans for the initial New York–Philadelphia–Washington route. Despite considerable loss of life, the Post Office had by 1924 established regular transcontinental mail flights. A year later, however, the federal government transferred mail service to commercial airlines, who were invariably eager for the contracts, given their generally precarious financial conditions.

Those air-mail routes eventually became parts of the first major airlines, including American, Delta, and United. These airline "families" emerged in the 1920s and 1930s. In that latter decade, flimsy cloth-covered biplanes gave way to all-metal craft, beginning with the Ford Trimotor. The Douglas Commercial, or DC-3, inaugurated in 1935, promptly became the principal civilian airliner (the American military also used it in World War II). Thousands were eventually built, and hundreds still operated as late as the 1970s. Its principal competitor, the Boeing 247, came from the company founded by Seattle lumberman Bill Boeing who, having secured an early air-mail contract for Chicago–San Francisco service, then began building planes to fulfill it.

Two events in the 1920s helped solidify the aviation industry's legitimacy. Charles Lindbergh's solo transatlantic flight in 1927 drew attention to the airplane's potential. The establishment of the Daniel Guggenheim Fund for the Promotion of Aeronautics in 1926 fostered aviation achievement. The fund enabled eight universities—Akron, Caltech, Georgia Tech, Michigan, MIT, New York

University, Stanford, and Washington—to form aeronautical engineering programs to train students and undertake research. Offshoots of Guggenheim-sponsored research included instrument-only flight, short-takeoff-and-landing aircraft, and weather reporting services.

In the next decade other technological advances followed. Planes powered by air-cooled, internal-combustion engines operated at speeds unheard of earlier. Consequent emphasis on increased altitude, size, and sturdiness engendered other mechanical modifications: stronger airframes; cockpit controls and instrument panels; electric, hydraulic, and de-icing systems; power plants; and pressurized cabins. The commercial implications were readily evident, and not just in the United States: in 1929, for example, 105 million passenger-miles were flown worldwide, while in 1939 airplanes logged more than 1,250 million passenger-miles. As air traffic increased, so did the need for support facilities; navigation and communications equipment; and airport controllers, groundcrews, and other technical personnel.

As competition for mail, freight, and passenger contracts intensified, and as charges of favoritism and even fraud surfaced, the federal government became more intimately involved. The 1930 Air Mail Act had tried but failed to establish large airline systems instead of small piecemeal carriers. Four years later, President Roosevelt ordered the army to resume air-mail service. By now ill-equipped for the task, its Air Service quickly lost pilots and planes, temporarily increasing postal rates. The 1934 Air Mail Act simultaneously restored air-mail service to private operators while separating commercial airlines from their manufacturers, thereby dissolving several aviation conglomerates. The 1938 Civil Aeronautics Act established a Civil Aeronautics Board (CAB), which in turn set up regular passenger and freight routes and continued subsidized mail service, thus virtually guaranteeing the industry profitable enterprises. Meanwhile the New Deal's Works Progress Administration provided funds for building and enlarging terminals and runways.

But commercial aviation's early history cannot simply be reduced to exclusively political or economic calculations. To be sure, extensive mail subsidies kept the industry aloft, and government regulation and assistance were crucial. But even these factors were insufficient to sustain commercial airlines in their competition with railroads. What fortified early commercial aviation and what entitled it to this unusual level of governmental support was its new modern character—in today's parlance, it was high tech. A "winged gospel" permeated America as the new technology's exciting, liberating potential captured public imagination. Commentators repeatedly maintained that airplanes would eliminate urban congestion, poverty, industrial competition, and even war. Aviation was to be a panacea; it would resolve almost all social questions. In that sense, aviation was comparable to the insistence by New Dealers and others that electricity and automobiles would pave the way for a new America "fit for life and the living."

MID-CENTURY HIGH TECH

The mid-century classification of airplanes, electricity, and automobiles as modern, high-technology initiatives was ironic. None was truly on the cutting edge: heavier-than-air powered flight had been accomplished as early as 1903, and automobiles and electricity were products of the previous century. Mid-century America did in fact have newer technologies, but they remained virtually unknown to the public. Only in the later twentieth century would these technologies gain substantial industrial or public constituencies. Yet these truly high-tech endeavors were as much a part of mid-century America as the TVA, for example. These new technologies were characteristically conceived from mid-century notions about the nature of systems and of parts within those systems. What was different about them, however, and what separated them from something like the TVA project was the locus of technological concern. Generally developed by physicists or electrical engineers laboring in universities or industry, these technologies concentrated not on system/part relationships in society, but on how they were manifested in the contemporary physics of indeterminancy, quantum mechanics, or electronics; the area of inquiry was often atomic or subatomic. Other technologists focused on the macroatomic level and strove to integrate several functions within single machines, taking care, of course, to establish compensatory mechanisms to balance interactions among elements. From these mid-century undertakings came photoduplication techniques, television, transistors, a new class of servomechanisms, and computers.

The Development of Computers

Events surrounding the early history of the computer demonstrate how mid-century organizational notions were reflected in that machine's conception and design. Computers were not merely calculating, measuring, or tabulating machines, which late–nineteenth- and early–twentieth-century Americans patented and built in great number. Rather, they were electronic, digital, integrated, multipurpose systems that harmoniously combined work done by several elements within a single device. The ENIAC (Electronic Numerical Integrator and Calculator), designed in the mid-1940s by J. Presper Eckert and John Mauchly at the University of Pennsylvania's Moore School of Electrical Engineering, was the first machine that met all the criteria and had general applicability. This machine incorporated ideas expressed in two other contemporary computing devices. In one instance, the Penn group apparently arrived at its determinations without consultation. In the other case, they borrowed heavily without acknowledgement or permission.

The pre-1920 calculator crusade ironically reached its zenith and limits in 1925, when an MIT group led by Vannevar Bush constructed a large-scale analogue calculator. It functioned like a slide rule, and was useful in solving differential equations with one variable. The Bush calculator was a mechanical, electric-powered device in which numbers were represented by directly measurable quantities computed by manually measuring the numbers of degrees specific gears rotated. A modified version of this machine found extensive use in World War II as the military employed it to compose artillery-firing tables.

Howard Aiken's Mark I was a different kind of machine. As early as 1937, Aiken expressly sought to join in a single machine the several tasks of storage, manipulation, and transfer of data. He conceived of an information-processing device—a true computing machine—not simply a calculator. Aiken's efforts were supported both by Harvard University, where he taught physics, and by International Business Machines (IBM). The Mark I (its official name was the IBM Automatic Sequence Controlled Calculator), which was completed in 1944 and integrated the three functions, was electromechanical. Its mechanical parts were controlled by inch-long electromagnetic relays, each of which opened or closed in 1/100 second. Some parts were adapted from existing IBM accounting equipment. The Mark I worked decimally, and without need of manual intervention. Programmers composed sequences of minute specific instructions, punched them on paper tapes, and fed them into the device. Aiken's machine was gigantic—more than fifty feet long, eight feet tall, three feet deep—and weighed more than five tons. It had 765,299 separate parts connected by 530 miles of wire. Operators were confronted by 1,444 dials.

At about the same time that Aiken conceived of the Mark I, a

*On the fiftieth anniversary of his invention, John V.
Atanasoff discusses the development of the computer with
a fellow academic.*

physicist at Iowa State College, John V. Atanasoff, and his graduate
student, Clifford Berry, began to build another, more revolutionary
machine. Atanasoff wanted to solve large, integrated systems of
linear algebraic equations (systems "frequent in the applied field of
statistics, physics and technology"), and felt that electronic machine power constituted "the only practical means" of solving them.
His prototype used vacuum tubes, not electromagnetic relays, and
was digital (binary), not decimal. It had a memory unit that was
composed of capacitors mounted on two rotating drums. Positive
charges at the condensers' ends corresponded to zero and negative
charges to one. Each drum could store thirty binary numbers, each
fifty characters long. Atanasoff added to this relatively small memory capacity by storing additional material in binary form on cards.
Machine-generated electric sparks marked the cards, which were
formatted similarly to the drums. His machine read the cards by
applying a voltage to them; the carbonized spots left by sparks would
be interpreted as ones, and the uncarbonized spots as zeros.

Atanasoff found that in about one case in 100,000 the sparker
failed to leave a carbonized spot. And since some of his equations
required millions of sparks, the occasional omission constituted a

significant problem. In late 1940, before he could remedy this difficulty, the federal government asked Atanasoff to undertake classified military research; ironically, he was given the task of analyzing analogue computing devices, particularly the Bush calculator. He later was called to Washington, D.C., for different, more extensive war-related research, and was forced to put aside his machine.

Atanasoff left Ames in 1942. About a year and a half earlier, he had the opportunity to meet Mauchly—then at Ursinus College—who was using analogue calculators to analyze weather statistics. Atanasoff discussed his machine with Mauchly, became a regular correspondent, and invited him to see it. Atanasoff also maintained that when perfected his device could outdo the Bush calculator. It took the Bush device months to determine a single artillery-firing table. Atanasoff estimated that his machine could solve the appropriate differential equation in two or three days.

Mauchly traveled to Ames in June 1941, and moved to the Moore School later that month. In September 1941, he wrote Atanasoff to ask if there was "any objection, from your point of view, to my building some sort of computer which incorporates some of the features of your machine." Mauchly also wanted to know whether, if he "got the Moore School interested," was "the way open for us to build an 'Atanasoff Calculator' (a la Bush Analyzer) here?" Atanasoff requested that Mauchly not develop Atanasoff's ideas until a patent application was filed. Atanasoff never filed that application. His war work took precedence,. In August 1942, Mauchly wrote his superiors at the Moore School, who were then also engaged in firing-table calculations, that "a great gain in speed of the calculation can be obtained if the devices which are used employ electronic means." He urged that they consider the merits of "the use of electronic circuits which are interconnected . . . for solution of difference equations."[17] He followed up this memo with a formal proposal, and permission was granted for an interdisciplinary research team, headed by Mauchly and Eckert, to construct such a device. The ENIAC was the result.

The ENIAC went into operation in 1946. It was as large as one box car and weighed thirty tons. Its 18,000 vacuum tubes worked 1,000 times as quickly as the Mark I's relays, though the Mark I was more reliable. The ENIAC was programmed by plugging cables from one of its parts to another, much like telephone switchboards. IBM punch cards carried data. A million cards were needed for its initial project, a simulation for the then untested hydrogen bomb.

The Pennsylvania group then joined with Princeton mathemati-

*The ENIAC required operators to trip thousands of
switches to program the machine for even rudimentary
calculations.*

cian John Von Neumann to build in 1950 the first stored-program
computer, the Electronic Discrete Variable Automatic Computer
(EDVAC). This government-financed computer had a memory much
larger than ENIAC and did not require manual programming. It
was also physically larger than ENIAC.

The Sperry-Rand Company's Universal Automatic Computer
(UNIVAC) was private enterprise's initial computer. Designed by
Eckert and Mauchly, the UNIVAC was about the same size as the
EDVAC and had many of the same features. It garnered priceless
publicity by predicting, an hour after the polls closed, Dwight Eisenhower's 1952 presidential election. IBM entered the computer
field with its Model 650 a few years later. It computed at about the
same speed as UNIVAC, but was smaller and cheaper and soon
outsold (or more precisely, outrented) its rivals. Yet total rentals
were modest; computers seemed useful only for large-scale scientific, military, and industrial projects.

The Development of Transistors

The transistors of the early 1950s seemed even less practical than contemporary computers. These tiny metallic devices were the product of a quest beginning in the 1930s to find adequate substitutes for amplifying-triode vacuum tubes (which were bulky, fragile, short-lived, and consumed large amounts of energy). Even as the triode search was commencing, however, researchers produced solid-state replacements for another type of vacuum tube, the rectifying diode. Indeed, World War II radar receivers employed small, very durable silicon–tungsten, copper–copper-oxide, and copper–selenium rectifiers. The components for these rectifiers required extensive purification, thus acquainting investigators with the properties of these substances in pristine and slightly altered states. Invention of these rectifiers—and ultimately solid-state amplifying triodes—depended on an understanding of the nature of subatomic systems that had been formulated in the previous decades and included solid-state and surface physics, Einstein's photoelectric effect, and quantum mechanics. Of particular importance was the notion that electrons occupied and were restricted to energy levels in solids called *bands*, that voids that could not be occupied by electrons stood between the bands, and that electrons could jump bands (without going through the voids) under certain conditions. This notion overthrew traditional conductor–insulator distinctions and established a continuum between them. From these assumptions came the concept of semiconductors, which made the idea of amplifying transistors appear logical and their development seem possible.

Bell Laboratories was among the earliest proponents of solid-state amplification. Long-distance telephony used huge numbers of vacuum-tube amplifying triodes, which were expensive to purchase and power, burnt out quickly, and disrupted service frequently. In the mid-1930s, Bell started to hire scientists and engineers explicitly to form a special interdisciplinary solid-state research team. As early as 1939, Bell researchers Walter Brattain and William Shockley installed a gridlike device within a semiconductor in an attempt to replicate a vacuum-tube triode, but their contraption failed to amplify current. The war interrupted the research, but Brattain and John Bardeen pursued it once they were freed of wartime responsibilities. They hoped to test Bardeen's theory that surface states immobilized current and made grid formation impossible. During their explorations of semiconductor surfaces they found that with two closely spaced electrode wires, a positive charge on one electrode would greatly enhance the semiconductor's capacity to carry current. Brattain demonstrated this amplifying effect on a germanium crystal to appropriate Bell staff on December 23, 1947.

Creation of the point-contact transistor resulted in Nobel Prizes for the three Bell employees. Shockley's subsequent sandwiching of minuscule impure semiconductors—the impurities gave the semiconductors positive or negative charges—yielded the junction transistor in 1951. It would prove more useful than the point-contact transistor. But neither seemed especially practical during the early 1950s. Both were limited in their applicability, were temperamental (easily damaged and not reliable), and were expensive to produce. Techniques to manufacture semiconductor material of sufficient purity required time and money; for the time being, vacuum tubes remained dominant.

TECHNOLOGY IN WORLD WAR II: SOLVING THE PROBLEM OF TOTALITARIANISM

World War II shattered the peace that had reigned since 1918 and placed dramatic new strains on the nation, but the country responded in a characteristically mid-century way. Indeed, Americans brought to the war those notions with which they had assaulted social questions and even scientific ones during the previous two decades. The war certainly did not disrupt America's optimistic assessment of technology's possibilities. This war was "total war"—"everybody's war"—and the country's technological capabilities promised to win battles at home and abroad. This integrated warfare extended beyond strictly military activities, for increased production and belt tightening within the United States, coupled with the introduction of new sophisticated weaponry, would enable America and its allies to defeat totalitarianism's forces. These heady sentiments about America's ability to engineer a social and technical, as well as military, victory translated into new institutions and policies. Development of new weapons figured prominently, and Roosevelt appointed a National Defense Research Committee in 1940 to prepare for war. Comprised of representatives from private foundations, universities, government, and industry, the committee established a contract system to encourage the nation's foremost scientists and engineers, regardless of geographic location, professional affiliation, or occupation, to undertake weapons research. Its functions were subsumed the next year by the Office

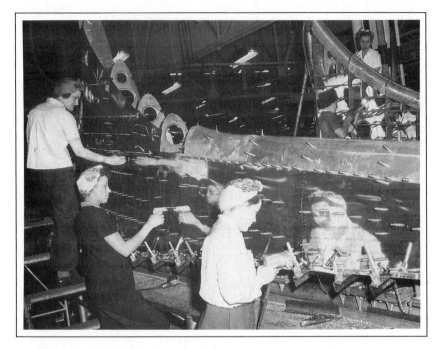

*Total war was everybody's war as these women assumed
factory jobs formerly reserved for men.*

of Scientific Research and Development, which was given the greater
charge of overseeing and coordinating all wartime research and
development.

Creation of the Atomic Bomb

The Office stimulated and funded production and development of
numerous new armaments, but most significantly it sponsored research on what would become the atomic bomb. It did not start
this research, however; several significant investigations predated
the Office's formation.

The earliest, most compelling work was done in Nazi Germany.
Two radiochemists, Otto Hahn and Fritz Strassman, announced in
1938 that they had successfully bombarded uranium with neutrons
to yield several lighter elements; the uranium's absorption of the
neutrons caused its nucleus to split, a process called fission, which
released enormous amounts of energy (200 million electron volts
versus the one or two per atom released by the most powerful con-

*America's production capabilities helped to combat the
totalitarian menace. This plant was converted to
manufacturing airplanes.*

ventional chemical reaction) with a corresponding decrease in matter. Their work reaffirmed contemporary atomic theory, which
hypothesized that matter and energy were continuous, interchangeable states ($E = mc^2$), that all elements were composed of
identical units (protons, neutrons, and electrons), and that elemen-
tal physical properties and characteristics were simply a consequence of the number of protons in the nucleus. These last two
postulates held out prospects for transmuting elements and for
creating new ones with particular properties—an important part of
bomb research—but it was the modest demonstration by Hahn and
Strassman of conversion of matter into energy that initially excited
scientists on both sides of the Atlantic. They quickly recognized
that fission reactions could release other neutrons, and contemplated possibilities of self-sustaining reactions—chain reactions—
whenever fissionable material achieved *critical mass* (the mass
necessary for a statistical guarantee that sufficient free neutrons
would strike other atoms). Neutrons could be slowed to yield con-

trollable chain reactions or left unimpeded to produce violent explosions. But only one uranium isotope (U-235) was uniquely suited to fission because whenever a neutron entered its nucleus, it released one or more neutrons.

These considerations, and Nazi Germany's documented efforts to develop nuclear power, led Albert Einstein, Leo Szilard, and other prominent scientists in 1939 to implore Roosevelt to establish a nuclear research program. Roosevelt responded cautiously. His National Defense Research Committee, and later the Office of Scientific Research and Development, authorized contracts for uranium-isotope-separation research, a crucial first step, because although U-235 is the only fissionable uranium isotope, it comprises only 1/140th of the world's uranium supply. Successful chain reactions require its separation from other uranium isotopes—especially the prevalent U-238—and its concentration. Gaseous diffusion was among the funded separation methods; investigators planned to diffuse the highly corrosive gas uranium hexafluoride through a permeable membrane and, because lighter molecules pass more quickly, separate U-235 from U-238. Small differences in molecular weight, coupled with difficulties in handling the gas, rendered this separation technique problematical.

Two years later, Office-supported University of California physicists used a particle accelerator to bombard the relatively plentiful U-238 and transmute it into a new man-made fissionable element, plutonium-239. This process for manufacturing nuclear material proved promising, a fact verified the following year when the Office notified Roosevelt of its confidence that a nuclear bomb could be developed in twenty-four months. Further support for this assessment occurred in December 1942. Enrico Fermi, an immigrant from fascist Italy backed by the Office, achieved the first successful controlled release of nuclear energy. Located under the University of Chicago's football stadium, Fermi's nuclear pile employed graphite latices and strips of cadmium and boron steel to slow and absorb neutrons.

Fermi's demonstration and the Office's strong recommendation helped convince Roosevelt of the bomb's immediacy. He authorized the Manhattan Project to keep bomb research secret, and charged it with building an atomic device within three years. General Leslie Groves headed the project, which in effect was a gigantic interdisciplinary research team composed of scientists from numerous disciplines and the Army Corps of Engineers. Project members secretly constructed plutonium-generating reactors at Hanford, Washington; a gaseous-diffusion facility at Oak Ridge,

Tennessee; and a physics research laboratory at Los Alamos, New Mexico. Under its director, Berkeley physicist J. Robert Oppenheimer, Los Alamos scientists and engineers designed and built bombs from the materials produced at Hanford and Oak Ridge.

The Development of Radar

The Manhattan Project effectively removed the Office from atomic research, and it then devoted all its efforts to conventional weaponry. The nexus between the Office and universities became especially pronounced as several interdisciplinary research centers were established on selected campuses to focus on particular types of weapons development. Investigators from across the nation, for example, were brought to the MIT Radiation Laboratory for radar research. Harvard spearheaded antiradar measures, and Johns Hopkins University's Applied Physics Laboratory became America's proximity fuse center. In most cases, the Office supported research that had been underway prior to the war. It also followed up preliminary work done by the army, the navy, or foreign powers.

Such was the case with radar (radio detection and ranging), which was as much a British as American invention and predated World War II. In 1935, Robert Watson-Watt, radio department superintendent of Britain's National Physical Laboratory, successfully transmitted radio waves at a moving aircraft and received their reflection or echo, which he timed to determine the plane's location. By 1939, round-the-clock radar stations twenty-five miles apart blanketed Britain's coast. They helped save the country during the Battle of Britain by enabling the Royal Air Force to meet incoming Nazi fighters; radar provided the number, range, direction, and altitude of enemy planes, day and night alike.

Prewar American developments were as extensive. The Naval Research Laboratory (NRL) produced the duplexer (a waveguide switch allowing single radar antennas to both transmit and receive) in 1936 and the XAF (a sea radar with a fifty-mile range), which it installed on two battleships in 1938. The Army Signal Corps contributed in 1937 the SCR-268, the first precision radar, which directed antiaircraft batteries. It also offered the SCR-270 model, and positioned it at Pearl Harbor, but the system was unfortunately deactivated as the Japanese attacked. The NRL also developed in 1941 the first airborne radar, the ASB, with homing, bombing, and search capabilities. Bendix, RCA, and Westinghouse shared the task of manufacturing more than twenty-six thousand ASB units during the war, which were installed in night fighters and torpedo bombers.

Modern radar began in 1940 when the British developed the high-power, multicavity magnetron, which was the first vacuum tube capable of producing enough power to make radar feasible at wavelengths of less then fifty centimeters (microwaves). It alone produced a strong pulse at superhigh microwave frequencies, which performed better than existing radar using long or medium (one- to two-meter) wavelengths. Indeed, all of the 150 distinct radar systems developed in America in World War II—ground, ship, and air varieties—operated on microwaves: from lightweight compact sets for fighter planes and PT boats to huge Early Warning Systems, the most powerful radar sets then built. Microwave radar required smaller antennas, distinguished more clearly two targets close together, and found especially low-flying aircraft more accurately.

Researchers at MIT's Radiation Laboratory improved the British magnetron by developing a more powerful version, the SCR-584, used for the first automatic tracking and gunfire-control radar. Other radar advances enabled pilots to drop payloads during poor visibility; to determine vessels' range and bearing (Plan Position Indicators, circular screens displaying vessels' locations); to obtain accurate fixes without emitting position-revealing signals (Loran—*long range aid to navigation*—developed at MIT and using pulsed shore-stations' waves to produce vast radiation grids); and to see territory ahead (the ASG, screens with radar maps).

Besides radar advances, radar countermeasures (techniques and devices to fool, jam, and confuse enemy radar) were actively developed and implemented. Frederick Terman, a Stanford electrical-engineering professor on loan to Harvard, headed the American effort. Mechanical reflectors, electronic jammers, and other means saved countless Allied soldiers and their equipment.

The Development of Sonar

The origins of sonar (*sound navigation and ranging*) were also found before World War II, but unlike radar, sonar was a predominantly American invention. American researchers had before 1920 transmitted sound underwater at higher-than-audible frequencies and obtained ranges of distant objects by timing the echoes. Devices that detected submarine propeller sounds or received echoes from hulls were developed in the following decades, but they worked well only under laboratory conditions. Wave motions, varying water temperatures, false echoes, or search ships' rolling and pitching disrupted them. Not until World War II were practical instruments developed. These included sound-range charts, underwater flares, underwater sound recorders, bathythermographs to measure water-temperature variations, improved depth charges and torpedoes, and sono-radio buoys to transmit underwater sounds to nearby ships and planes. These sonar-based devices cut down losses of ships and men, enabled the Allies to detect, track, and destroy German and Japanese submarines with unprecedented accuracy, and, given the difficulties of transoceanic warfare, proved an integral part of the Allied victory.

The Development of Proximity Fuses

Proximity fuses were an exception to the Office of Scientific Research and Development's practice of supporting preexisting weaponry. They were invented during World War II.

Ideally, fuses should detonate projectiles at points where maximum numbers of lethal fragments will pass through targets. Conventional fuses, however, especially timed ones, often exploded prematurely, belatedly, or inadequately. Proximity fuses were small, rugged radar sets placed in artillery shells, designed to detonate them at set distances from their targets. These proximity fuses made battlefield weaponry more effective. Industry, particularly Bell Laboratories, Crosley, Eastman Kodak, Raytheon, and Sylvania, worked with Johns Hopkins to surmount problems of firing shocks, temperature variations, and assembling complex small weapons with minimal errors and maximum safety. Acoustic and electrostatic proximity-fuses also were tried but proved impractical for quick military applications. Photoelectric and especially radio fuses worked much better, and the army and the navy adopted both. By the war's close, a quarter of America's electronics industry was devoted to producing them.

Proximity fuses were used in air-to-air, air-to-ground, ground-to-air, ground-to-ground, and ship-to-ship encounters, and against the Germans and Japanese alike. They were used to shoot down dive bombers, Japanese *kamikazes*, and German flying bombs, and were shot at enemy soldiers on attack or in foxholes with remarkable accuracy. The German V-1 flying bombs, or buzz-bombs, which traveled at 350 miles or more per hour, posed the greatest challenge; only the most determined Allied effort, using SCR-584 radar as well as advanced radio proximity-fuses, prevented a catastrophe.

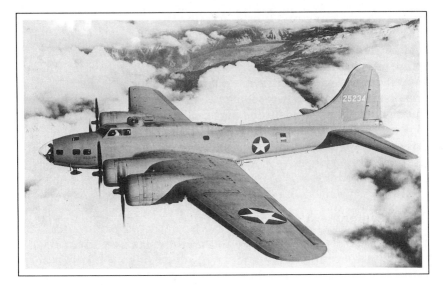

*The B-17 Flying Fortress practices for a bombing run as it
flies above the clouds.*

The Role of Aviation in World War II

As suggested by many of the uses of proximity fuses, airplanes played
a far larger role in World War II than in the previous conflagration,
and were integral parts of most military operations. They provided
cover for ground and naval assaults, protected land and sea convoys, diverted enemy fire, and transported troops, munitions, and
supplies. Air reconnaissance and bombing missions were essential
features of modern warfare. The B-17 "Flying Fortress," B-24 "Lib-
erator," and B-29 "Superfortress" bombers had been designed before America entered the war, but wartime production and
performance needs prompted repeated modifications, from leak-
proof fuel tanks to aluminum airframes (which increased speed,
capacity, range, reliability, and stability). In total, the nation produced more than 296,000 aircraft during the war.

Superior airpower and conventional weaponry as well as America's material resources helped decide the war, but the use of the
atomic bomb marked its formal conclusion. The first atomic weapon
was tested at Alamagordo, New Mexico, on July 16, 1945. Its core
of plutonium-239 was triggered by implosion; high explosives surrounding the subcritical volume of plutonium were detonated,
compressing the fissionable material into a smaller volume that

The atom bomb, "Little Boy," was detonated over
Hiroshima, Japan. It was 29 inches in diameter and 120
inches long. The force of its explosion was comparable to
20,000 tons of TNT.

exceeded the critical mass. Buoyed by the test's success, President Harry Truman issued Japan—the war with Germany had ended—an unconditional-surrender ultimatum that hinted at a new aerial weapon of unprecedented power. Japan's silence led to the dropping of "Little Boy" (an enriched U-235 bomb triggered by firing one subcritical mass into another) on Hiroshima on August 6, 1945. The blast killed 80,000 persons and destroyed or heavily damaged 96 percent of the city's buildings. Japan's continued silence produced the Nagasaki bombing (using "Fatman," which was similar in design to the Alamagordo bomb), which took 39,000 lives and leveled 40 percent of the city's structures. Each bomb contained more explosive power than 20,000 tons of TNT. Japan surrendered on August 15, eliminating an otherwise inevitable and costly Allied invasion.

AMERICAN USE OF TECHNOLOGY IN THE POSTWAR WORLD: SOLVING THE PROBLEM OF COMMUNISM

World War II reaffirmed America's technological strength, but the nation faced many challenges soon after its conclusion. Rapid demobilization strained the country's economy, while veterans' programs placed pressure on colleges and the housing industry. Communists would defeat China's nationalist forces and the Korean War would soon ignite. But America's technological superiority, national character, and continued vigilance appeared to offer protection from the "Red Menace." The Research and Development Board, formed in 1947, awarded contracts to universities for Defense Department research. Six years later Congress created two additional assistant defense secretaries—one for research and the other for development—to replace the Board and to extend the contract system. The National Science Foundation was established in 1950 to support basic research, some of which had military implications.

America's technological might also held out hope of nonmilitary solutions to the spread of communism; it would engage in a Cold War against communism. The nation's industrial prowess could inaugurate a worldwide era of plenty; it could stop and isolate communism by remaking the world in the United States' image. The Marshall Plan, resurrection and modification of Japan, foreign aid, and establishment of the United Nations were attempts by the United States to engineer a new world order. Americans conceptualized their world as an evolving whole (undeveloped, underdeveloped, developing, and developed countries constituted this evolutionary framework's vocabulary), and depended on the desire for material goods to carry that vision to fruition. Democracy would mean not simply the right to vote for the candidate of your choice, but also the right to material prosperity. More specifically, Truman attempted to halt the threat of communism abroad by "making the benefits of our scientific advances and industrial progress available for the improvement and growth of underdeveloped areas." Technical missionaries undertook irrigation and hydroelectric-power projects, programs to increase agricultural productivity, and disease-eradication campaigns in thirty-three nations during the early 1950s. America devoted its "full energy to help these countries grow and flourish in freedom";[18] technical development emerged as a vital component of foreign policy, much as it had formed a crucial aspect of New Deal domestic policy.

The nation, however, did not neglect military might. Turbojet engines and jet planes became air force staples; the F-80 fighter and the B-47 "Stratojet" and B-52 "Stratofortress" bombers were extensively used in the Korean War. Huge government defense contracts established Douglas, Boeing, and other airline manufacturers as integral parts of the military-industrial complex.

Atomic power grew in prominence after the Soviet Union exploded an atomic bomb in September 1949. Truman ordered a crash program to develop the hydrogen bomb, the dreaded next stage in nuclear weaponry. Several physicists active in the Manhattan Project, including Fermi and Oppenheimer, opposed Truman's initiative. The devastation of Hiroshima and Nagasaki concerned them, as did the vast destructive potential of thermonuclear weapons. Their advice was rejected, particularly once Edward Teller and others devised practical means to control fusion reactions. (Hydrogen bombs require nuclear reactions different from Fermi's fission reactions: not splitting heavy atomic nuclei, but rather fusing together the nuclei of light elements at extremely high temperatures to convert mass into energy.) The first American H-bomb was tested in 1954. Possessing the explosive power of fifteen million tons of TNT, it was more than 750 times as powerful as the A-bomb. A year later, the Soviet Union exploded a less powerful but similar device.

NOTES

[1] Quoted in Loren Baritz, *Servants of Power* (Middletown: Wesleyan University Press, 1960), p. 85.

[2] Baritz, *Servants,* p. 105.

[3] R. D. McKenzie, *The Metropolitan Community* (1933; New York: Russell and Russell, 1967), p. v.

[4] Thorstein Veblen, *The Engineers and the Price System* (New York: Viking, 1933), p. 135.

[5] Arthur Morgan quoted in Arthur M. Schlesinger, Jr., *The Coming of the New Deal* (Boston: Houghton Mifflin, 1959), p. 327.

[6] William J. Hale, "Farming Must Become A Chemical Industry," *Dearborn Independent,* October 2, 1926.

[7] Hale, *The Farm Chemurgic* (Boston: Stratford, 1934), p. ii.

[8] Archie A. Stone, *Farm Tractors* (New York: Wiley and Sons, 1932), p. 3.

[9] J. Brownlee Davidson, *Agricultural Machinery* (New York: Wiley and Sons, 1931), p. 372.

[10] D. F. Jones, "Selection in Self-Fertilized Lines as the Basis for Corn Improvement," *Journal of the American Society of Agronomy,* vol. 12 (1920), p. 81.

[11] A. Richard Crabb, *The Hybrid-Corn Makers* (New Brunswick: Rutgers University Press, 1947), p. 318.

[12] I. Bernard Cohen, *Science, Servant of Man* (Boston: Little, Brown, 1948), p. 100.

13 Remembered by Alfred P. Sloan, Jr., in his *My Years with General Motors*, edited by John McDonald and Catherine Stevens (Garden City: Doubleday, 1972), pp. 187–88.

14 Quoted in David A. Hounshell, *From the American System to Mass Production, 1800–1932* (Baltimore: Johns Hopkins University Press, 1984), p. 278.

15 Sloan, *My Years*, p. 74.

16 Sloan, *Adventures of a White-Collar Man* (New York: Doubleday, Doran, 1941), p. 135.

17 Atanasoff's and Mauchly's materials were reproduced in Arthur W. Burks and Alice R. Burks, "The Eniac: First General-Purpose Electronic Computer," *Annals of the History of Computing*, vol. 3 (1981), pp. 330–33 and *passim*.

18 Harry S Truman, "The Faith By Which We Live," *Vital Speeches of the Day*, vol. 15 (1949), pp. 227–28.

FOR FURTHER READING

Aitken, Hugh G. J. *The Continuous Wave* (Princeton: Princeton University Press, 1985). An excellent study of radio in twentieth-century America.

Akin, William E. *Technocracy and the American Dream* (Berkeley and Los Angeles: University of California Press, 1977). Examines the rise and fall of the Technocracy movement.

Arnold, Joseph L. *The New Deal in the Suburbs: A History of the Greenbelt Town Program, 1935–1954* (Columbus: Ohio State University Press, 1971). Chronicles the development and aftermath of the Greenbelt towns.

Baxter III, James Phinney. *Scientists Against Time* (Boston: Little, Brown, 1947). An examination of the role of scientists in World War II.

Bilstein, Roger E. *Flight in America, 1900–1983* (Baltimore: Johns Hopkins University Press, 1984).

Borth, Christy. *Pioneers of Plenty: The Story of Chemurgy*, rev. ed. (Indianapolis and New York: Bobbs-Merrill, 1942). A tribute to Hale and the chemurgic idea.

Boyce, Joseph C. *New Weapons for Air Warfare* (Boston: Little, Brown, 1947). A contemporary survey of the subject.

Boyer, Paul. *By the Bomb's Early Light* (New York: Pantheon, 1985). A social history of the early years of the atomic era.

Braun, Ernest and Stuart MacDonald. *Revolution in Miniature*, rev. ed. (Cambridge: Cambridge University Press, 1982). An excellent account of the development of transistors.

Burks, Alice R. and Arthur W. Burks. *The First Electronic Computer* (Ann Arbor: University of Michigan Press, 1988). The first book-length discussion of Atanasoff's contributions. First-rate on technical detail.

Conkin, Paul K. *Tomorrow a New World: The New Deal Community Program* (Ithaca: Cornell University Press, 1959). A discussion of social planning in the 1930s.

Corn, Joseph J. *The Winged Gospel* (New York: Oxford University Press, 1983). Examines the idea of flight in the American imagination.

Crump, Irving. *Our Army Engineers* (New York: Dodd, Mead, 1954). A tribute to the Army Corps of Engineers and their works.

Foster, Mark S. *From Streetcar to Superhighway* (Philadelphia: Temple University Press, 1981). Considers the role of the automobile in mid-twentieth-century America.

Friedel, Robert. *Pioneer Plastic* (Madison: University of Wisconsin Press, 1983).

Graham, Otis L., Jr. *Toward a Planned Society: From Roosevelt to Nixon* (New York: Oxford University Press, 1976). Highlights the idea of national planning.

Hallion, Richard P. *Legacy of Flight* (Seattle: University of Washington Press, 1977). Discusses the contributions of the Guggenheim Foundation.

Hawley, Ellis W., ed. *Herbert Hoover as Secretary of Commerce: Studies in New Era Thought and Practice* (Iowa City: University of Iowa Press, 1981). Investigates the role of the federal government in the 1920s.

Johnson, Charles W. and Charles O. Jackson. *City Behind a Fence: Oak Ridge, Tennessee, 1942–1946* (Knoxville: University of Tennessee Press, 1981). A portrait of the community during the uranium enrichment years.

Lewis, W. David and Wesley Phillips Newton. *Delta: The History of an Airline* (Athens: University of Georgia Press, 1979). An excellent history of the evolution of an airline.

Lowitt, Richard. *The New Deal in the West* (Bloomington: Indiana University Press, 1984). Surveys the various contributions of the New Deal agencies.

Lubove, Roy. *Community Planning in the 1920s* (Pittsburgh: University of Pittsburgh Press, 1963). An excellent examination of the "new" planning of the 1920s.

Lush, Jay L. *Animal Breeding Plans*, 3rd. ed. (Ames: Iowa State College Press, 1945). A classic text on practical population genetics.

McCraw, Thomas K., ed. *Regulation in Perspective: Historical Essays* (Cambridge: Harvard University Press, 1981). A valuable discussion of the question of regulation.

Morgan, Arthur E. *The Making of the TVA* (Buffalo: Prometheus Books, 1974). A retrospective assessment by TVA's leading proponent.

Schaffer, Daniel. *Garden Cities for America: The Radburn Experience* (Philadelphia: Temple University Press, 1982). A consideration of a garden city that was never completed.

Schlebecker, John T. *Whereby We Thrive: A History of American Farming, 1607–1972* (Ames: Iowa State University Press, 1975). A brief but rich survey of agriculture and agricultural technology.

Scott, Howard. *Science versus Chaos* (New York: Technocracy Inc., 1933). A contemporary document written by the founder of the Technocracy movement.

Simonds, William Adams. *Henry Ford and Greenfield Village* (New York: Stokes, 1938). A popular, uncritical study.

Soule, George. *A Planned Society* (New York: Macmillan, 1932). Extols the virtues of national planning.

Stewart, Irving. *Organizing Scientific Research for War: The Administrative History of the Office of Scientific Research and Development* (Boston: Little, Brown, 1948). A useful account of an important federal bureaucracy.

Williams, Robert C. *Fordson, Farmall and Poppin' Johnny* (Urbana: University of Illinois Press, 1987). A social history of tractors.

8

Technology as a Social Question: the 1950s to the Present

ASSAULTING THE ESTABLISHMENT AND ESTABLISHED NOTIONS

In 1956, William H. Whyte, Jr., published *The Organization Man*. Assistant managing editor of *Fortune*, Whyte criticized unthinking, uncritical worship of corporate organizations, and pilloried the gospel of scientism, the notion that expert determinations were always nonpartisan and not subject to error or debate. He railed against any organization's use of the concepts of belonging and togetherness to manipulate members and objectives, and called on his contemporaries to challenge the status quo. Whyte conceded that the organization was here to stay, and he did not advocate nonconformity. Rather, he urged men and women to examine the premises they took for granted, to give rein to their individual proclivities, and to redirect their various organizations. Individualism tempered by critical thought would awaken America from its blissful organization-inspired ignorance and produce a better tomorrow.

Whyte's book drew plaudits from many corners, became an

immediate best-seller, and remained popular for some two decades. Its cry for intraorganizational individualism struck a chord with its readers, as did its attacks on professional, technical expertise as the sole decision-making criterion. Americans have acted out their interpretations of Whyte's book during the thirty some years since its publication, although not always in the same ways, or even ways that Whyte might have appreciated. Tensions repeatedly erupted between the post-1920 organizational idea's implicit holism and the various guises of post-1950 individualism; the previous period's systematic notions appeared to hold true, but each contemporary event or occurrence became subjected to individualistic interpretation. And these interpretations were not necessarily of similar character. Freed from the restraint of a seemingly objective force—expertise—and replaced by the presumed validity of individual perspective, consensus became increasingly difficult.

Technology as the Driving Force Behind Modern Life

Technology, of course, did not escape this conceptual confrontation. To some, it was still a social benison, pristine in every way, while to others it had become a villain, a destroyer of communitarianism. Endowing technology with these superhuman properties absolved mankind of its sins, or posited blame on the handful who unleashed the monster. These views differed from earlier twentieth-century critiques, which generally had seized on a specific technological aspect, such as mass production, or sought to replace old technologies with new ones. But technology had rarely seemed a monolithic force. Most post-1950 Americans adopted neither position outright. They saw technology as a force, but not as monolithic, regarding it instead as a social question. Some technologies seemed harmful and others beneficial, but most appeared to combine aspects of the two extremes; technology was both a social solution and progenitor of social problems. Relationships between technology and society became more explicit in the post-1950s decades. What constituted the most crucial relationships in any particular case, however, was the individual's province to decide.

Technology's identification as a primary social determinant in an age of individual prerogatives made achieving majority opinions the most effective way to introduce major technological changes. Various civilian (as opposed to military or professional) oversight, consultation, or protest groups demanded fair hearings, as did industrial and governmental representatives. Antagonists regularly sought—and found—experts to sustain their differing positions. In-

This deluxe fallout shelter was equipped with five bunks, air mattresses, an air purifier, and a gasoline generator. It could sustain a family for nearly a week.

dividuals and groups disappointed by a majority conclusion sometimes sued. This process often made casualties of long-range plans. Majorities proved ad hoc and temporary, as individuals retained the right to modify their initial assessments. Public debate flourished, but the public interest was not necessarily served.

THE NUCLEAR AGE

No technology provided a more compelling example of the relationship between technology and society than nuclear weaponry. Although nuclear weapons were an essential facet of the nation's defense, disarmament protests began soon after the Soviet Union exploded its hydrogen bomb in 1955, and persisted into the early 1960s. A thriving bomb-shelter industry emerged, and movies, novels, and television offered visions of a world after nuclear war. Citizens' groups in America reacted passionately to the disclosure that atmospheric nuclear tests released large quantities of radioactive strontium 90, which found its way into cows' milk, and pressed the government to stop that practice. The Soviet Union and United States agreed in 1963 to end atmospheric tests, but continued underground testing as both nations sought technological breakthroughs and labored to increase nuclear "throw weight" and accuracy. One hundred–megaton bombs, each the equivalent of a hundred million tons of TNT, entered arsenals.

National Defense and Nuclear Debate

Public clamor intensified in the late 1960s and early 1970s as both nations built larger nuclear stockpiles and other countries developed nuclear weapons. Mutually Assured Destruction (MAD), Multiple Independently Targeted Reentry Vehicles (MIRVs), and "window of vulnerability" became catchwords of this nuclear age. America and the Soviet Union began Strategic Arms Limitation Talks (SALT), and reached two accords. SALT I (1972) established temporary offensive-weapon limits (subsequent treaties were to make final determinations) and prohibited sea- and space-based anti-ballistic missile development, testing, and deployment. SALT II (1979) was never ratified by the U.S. Senate, although both nations have roughly abided by its provisions. It controls the number and type of nuclear weapons and limits new system development and testing. Ultrasensitive electronic monitoring gear, as well as conventional intelligence gathering, are to guarantee both countries' compliance.

Americans in the mid-1970s carried the nuclear debate to a new level. Led by a disparate collection of technicians and scientists, opponents of nuclear weapons asked if the nation possessed the technical capability to undertake certain projects, and if those projects would fulfill their advocates' expectations. Disputes raged over the potential effectiveness of superhardened silos for intercontinental ballistic missiles (ICBMs), and the desirability of mounting ICBMs on mobile railroad cars. President Ronald Reagan's Strategic Defense Initiative (SDI), or "Star Wars," further focused debate in the 1980s. Many prominent scientists, engineers, and lay persons have condemned the plan to destroy enemy nuclear missiles in outer space—a nuclear shield—as unworkable, untestable, and outrageously expensive. SDI's defenders contend that the proposed system of hundreds of lasers, mirrors, and targeting instruments can be coordinated, that it will distinguish between weapons and decoys, and that it merits the expense.

Concerns also have arisen about nuclear war's aftermath. Some scientists maintain that a nuclear exchange would produce enough smoke and atmospheric debris to cause severe cold and darkness across the globe, creating a "nuclear winter" lasting several years. Others predict only a brief nuclear "autumn." Both rely on computer simulations to prove their points, but data fed into computers vary and, as a consequence, so, too, do conclusions. Similarly, some support the concept of limited nuclear war (limited in either the number of missiles launched or the geographical areas affected), while others worry about accidental nuclear war produced by the very systems designed to prevent such catastrophes. What have been portrayed as frightening fantasies in such movies as *Fail-Safe* and *War Games* seem to some painfully real. Still others dispute America's ability to respond to nuclear attacks and argue that a single warhead could disrupt the nation's power grid, wipe out computer memories, and create political, social, and economic chaos.

The Rise and Fall of Nuclear Power

The awesome force of nuclear energy has also affected peacetime nuclear-power policies. The 1946 Atomic Energy Act, America's first important legislation concerning nuclear technology, established the Atomic Energy Commission (AEC), which monopolized nuclear materials and thus stymied private initiatives even while taking formal control of atomic energy out of the military's hands. The ongoing Cold War made the AEC extraordinarily security-conscious and unwilling to risk civilian efforts. But newly elected President Dwight Eisenhower reversed this trend in 1953 by beginning an "Atoms for Peace" program that encouraged civilian construction of nuclear power plants. The first of these plants, in Shippingport, Pennsylvania, opened in 1957, followed by three others in 1959 and 1960: Rowe, Massachusetts (the Yankee plant); Morris, Illinois (the Dresden); and Indian Point, New York. These and several other early commercial reactors were primarily demonstration models intended to test different kinds of reactor designs. Nineteen types were seriously considered, and eleven finally constructed, before light-water varieties were deemed most practical.

Reactors built between the mid-1960s and 1979 (when central Pennsylvania's Three Mile Island plant [TMI] nearly suffered a "meltdown," or melting of the reactor core, with the consequent uncontrolled release of radioactive material into the atmosphere) were larger in size and capacity than the earlier models and, in theory, lower in cost. Rising coal and gasoline prices, and growing concerns over environmental pollution from coal-fired plants and over diminished fossil fuels, spurred these projects, as did federal financial assistance. For a time, half of all new power plants were nuclear, and what had been costly individual demonstration units became integrated into a seemingly viable industry. But their construction also engendered public concern.

Nearly sixty civilian nuclear reactors were operating when the Three Mile Island incident occurred. The inability or unwillingness of any technical expert or utility-company official on the scene to explain to the general public exactly what had happened, and what hazards might exist, proved more damaging than the incident itself. As two historians have observed, "what escaped at Three Mile Island was not only radiation, but, more importantly for the nuclear power industry, public confidence."[1]

To be sure, there had been plant shutdowns and other problems elsewhere, beginning with the 1952 Chalk River, Canada, accident (involving future president Jimmy Carter, who helped dismantle the reactor's core). And the TMI incident was considerably less se-

Lowering the core into the Shippingport, Pennsylvania, reactor inaugurated the nuclear power plant era.

This nuclear power plant at Three Mile Island,
Pennsylvania, was the site in 1979 of America's most
serious nuclear accident.

rious than the apparent meltdown in 1986 at Chernobyl in the Soviet Union. But the TMI accident helped sway American public
opinion away from support of nuclear power, which a majority now
views as both unsafe and too expensive. Especially after TMI,
lengthy, costly legal and environmental battles delayed and often
halted projects, including many already partially built, or even some
that had been completed after expenditures of billions of dollars
(though about 100 nuclear plants presently operate). The repeated
newspaper, magazine, radio, and television advertisements by the
nuclear industry (after TMI and then Chernobyl) stating that at least
American nuclear plants are inherently safe, if no longer so inexpensive to operate, have largely fallen on deaf ears. TMI itself is
still being decontaminated. Forecasts on the eve of TMI of a rise in
the percentage of American electricity produced from nuclear
power—from 10 percent in 1978, to 20 percent by 1985, to 50 percent by 2000—have proved rather optimistic.

In addition, concerns recently have arisen about earthquakes, airplane crashes, and sabotage destroying nuclear plants; about transportation and storage of hazardous nuclear-waste products; and about manufacture of reactors by unfriendly countries. Ironically, greater attention to energy conservation, combined with (temporarily) declining oil and natural-gas prices, have meant reduced need for the kind of endless electrical-power growth anticipated earlier and used to justify building most nuclear plants. Contributing as well to this change has been renewed reliance on alternative energy sources such as water, wind, sunlight, and geothermal sites. Indeed, electricity from newer coal plants has become cheaper than that from newer nuclear plants.

Yet it would be premature to write the nuclear-power industry's obituary. Need for ever more electrical power will rise again. Alternative energy sources are insufficient and fossil fuels remain finite. Moreover, new forms of nuclear power—through fast breeder-reactors and controlled thermonuclear fusion—might someday prove sufficiently safe and economic to be politically acceptable. The current principal variety of nuclear reactors—light-water fission—are inefficient in utilizing their uranium fuel, which in turn squanders the limited uranium resources in America and elsewhere.

Nuclear power controversies thus epitomize the declining contemporary faith in technical expertise, and the parallel concern over the potential implications of many technological pursuits. Few technologies, in fact, have been as widely associated as nuclear power with social and economic advance. Prophecies of not just virtually free electricity ("power too cheap to meter") but also the elimination of hunger, disease, normal aging, air pollution, slums, and even war were common after World War II. So, too, were forecasts of mobile "package" (nuclear) power plants, "miniature reactors" in every home and factory, "atomic engines" in every family car and truck and in every commercial ship and airliner, and "radioactive fertilizers" for every farm. Some envisioned small nuclear explosions to expedite various navigation and irrigation projects.

Symbolic, perhaps, of the controversies surrounding nuclear power is the ongoing dismantling and eventual burial of the Shippingport nuclear reactor, following its 1982 closing. Robots are dismantling the plant to reduce human exposure to radioactivity, and the radioactive sections then will be barged down the Ohio and Mississippi Rivers, through the Panama Canal, to a burial site in Hanford, Washington, where plutonium was first produced. The ground on which the reactor has stood eventually will be returned to its original contours and sodded—once it has been made free of man-made radioactivity. The cost of this process will far exceed the facility's original construction costs. Other later, larger reactors and several early nuclear-powered submarines will undergo similar procedures, with their radioactive sections likewise transported to Hanford or to other public or private burial sites. Simple demolition or abandonment of reactors or, with submarines, sea burial, would not contain the radioactivity and would thus pose environmental hazards. Indeed, many of the more than 135 reactors due to reach the end of their lives (generally, thirty years) by 2010 instead may be either entombed in concrete, or mothballed behind fences and guards for many years until their radioactivity diminishes, and only then be dismantled.

Not surprisingly, the federal government's 1986 announcement that America's first nuclear-waste dump would be built in the West, rather than the East or the Midwest, brought cheers from the latter two regions and sighs from the first—where nuclear power long has been most favored. Meanwhile, as experiments continue in New Mexico on potential burial techniques, the Hanford complex itself recently has been shut down because of various safety violations, thus disrupting plutonium production. Still, two University of Michigan nuclear engineers lament, "We suspect that if nuclear power is to be accepted and implemented, it will [nowadays] be only as a last resort, when the public realizes that all other options . . . have been exhausted or recognized as unsuitable or insufficient."[2]

SPACEFLIGHT

The history of spaceflight parallels in many ways that of the uses of nuclear energy. Both emerged from research begun during World War II, and both became the subject of superpower competition. Initially accorded overwhelming public support, both eventually lost much of their luster as various groups questioned their proponents' abilities, objectivity, and ambitions. They had become social questions.

The "Space Age" formally began on October 4, 1957, when the Soviet Union successfully launched the first artificial satellite, *Sputnik I*. This technological achievement by the supposedly backward Soviets shocked Americans, as did the launching a month later of *Sputnik II* (which carried a dog), and the failure another month later of an American satellite launch. Together they spurred an unprecedented American commitment to space exploration. Within a year President Eisenhower established the National Aeronautics and Space Administration (NASA). By January 31, 1958,

America's first successful satellite, *Explorer I*, had been launched, followed by *Vanguard I* on March 17, 1958. They in turn formally started the "space race" between America and the Soviet Union, which continues today.

The Origins of Spaceflight

American space efforts, however, predated these events. Clark University physicist Robert Goddard had achieved the first successful flight of a liquid-fuel rocket at Auburn, Massachusetts, on March 16, 1926. Goddard received little acclaim for his 184-foot flight, and died barely noticed in 1945. The German V-2 rockets (which were 5½-ton bomb-carrying missiles) acquainted the American military with liquid-fuel rockets during World War II, and both America and the Soviet Union utilized captured V-2 rockets in postwar years. In 1946 the army created the Atmospheric Research Panel to conduct V-2 research, but as the V-2 supply ran out, new rockets were developed: the Aerobee from Johns Hopkins University's Applied Physics Laboratory, and the Viking from the Naval Research Laboratory. These probed the upper atmosphere, or "near space," to altitudes of 150 miles, and made feasible earth satellites and space probes requiring greater rocket thrusts.

The 1958 legislation creating NASA declared that "it is the policy of the United States that activities in space should be devoted to peaceful purposes for the benefit of all mankind."[3] In 1967 sixty nations, America included, signed a pledge to use space only for peaceful endeavors. United Nations resolutions regarding space weapons, moon exploration, and astronaut welfare have helped create the space-law field. American–Soviet joint efforts have been modest and linked to overall relations and strategic needs; only the temporary spirit of détente, plus America's successful Apollo program, made possible in 1975 the primarily symbolic Apollo 18–Soyuz 19 flights, with their joint experiments and docking maneuvers.

The Space Race

More often the competitive spirit has prevailed. The space race was invigorated on April 12, 1961, when cosmonaut Yuri Gagarin in *Vostok I* became the first person to orbit the earth and return safely. On May 25, 1961, President John Kennedy stirred the nation with a pledge to land a man on the moon, before the Soviets, by 1970, a move that increased public interest and federal funding.

This control room at Cape Canaveral belies the individualism usually associated with spaceflight.

The seven original *Mercury* astronauts—described by some as the Cold War equivalent of medieval knights individually battling Godless communism—became national heroes as they ventured into space. Their most memorable flights were those of Alan Shepard (on May 5, 1961, aboard *Mercury 3*, the first American in space) and John Glenn (on February 20, 1962, aboard *Mercury 6*, the first American to orbit the earth). Following the Mercury program (1958–1963) came the Gemini program (1962–1966), which included the world's first successful orbital rendezvous and manned flights of record length. It was succeeded by the Apollo moon program (1961–1972), which competed with superior Soviet rocket thrusters by developing more powerful booster rockets and lunar landing vehicles. This program culminated with the *Apollo II* moon landing of July 16, 1969. As millions around the world watched, Neil Armstrong and Edwin Aldrin conducted experiments on its surface, collected rock and soil samples, and returned safely to earth.

Simultaneously, unmanned American expeditions surveyed ever more space mechanically, often ahead of Soviet efforts: *Mariner 2*

(1962) surveyed Venus; *Mariner 4* (1965), Mars; *Ranger 7* (1964), the moon; and *Surveyor I* (1966) landed softly on the moon. Later space probes such as *Pioneer 10* (1972) and *Voyager I* and *2* (1977) have transmitted data for several years about the planets farthest from earth and beyond.

Decline of Support for the Space Program

Public and congressional support for the space program declined sharply after the moon landing. It seemed the grand climax rather than a stage in ongoing "space-faring" activities. Indeed, public interest in this monumental event was short-lived; five successive moon landings attracted far smaller television audiences. In 1972 NASA began the space-shuttle program, not so much to match the Soviet Union, which wasn't truly competing here, as to provide practical means for further scientific exploration, future colonization, and commercial activities (especially carrying private communications satellites, including foreign ones, into space). NASA had long been involved in commercial ventures and in 1962 had launched Bell Telephone's *Telstar I* satellite. Constantly orbiting satellites are now routinely used for business communications and weather reports, as well as for military surveillance. The shuttle, however, promised to be even more practical than any of its predecessors: a reusable spacecraft—including its fuel tanks, orbiter, and, not least, booster rockets, which would fall into the ocean once the satellite was in space—that would save millions of dollars otherwise spent on new vehicles. The shuttle would make space transportation routine. It also would service the space stations that NASA deemed its next major project. Simultaneously, the shuttle's huge development costs, coupled with its appealing multiple uses, promised to keep NASA alive and well following the loss of purpose and support after the Apollo program ended.

Getting crucial military support required major design modifications, such as limiting efficiency for high-orbit launches, while obtaining President Nixon's approval necessitated severe funding cutbacks and in turn other design changes, such as reduced payload capacity. These alterations and budget cuts affected the shuttle's overall development and contributed to technical problems and cost overruns that plagued it from the start. Indeed, NASA cancelled other, less expensive projects of potentially more scientific value than the shuttle to compensate for those added expenditures. The shuttle's final cost of $11 billion doubled NASA's original estimate. Meanwhile, the European Space Agency, a cooperative orga-

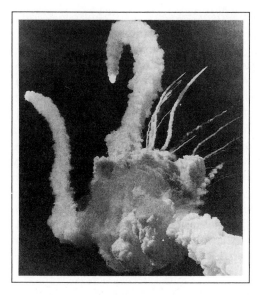

The Challenger *disaster reflected Americans' increasing disillusionment with the space program and, more generally, disappointment with modern technology.*

nization started in 1972, developed the largely French-built *Ariane* rocket booster at one-twelfth the shuttle's cost, using older and cheaper components to launch satellites at competitive international prices. Its first successful launch was in 1979.

The initial twenty-four shuttle missions, beginning with *Columbia* on April 12, 1981, were generally so successful that renewed public interest in the space program had faded once again by the time of the *Challenger* explosion on January 28, 1986. By then, it was widely assumed, shuttle missions were almost routine spaceflights (it was not accidental that America's first female astronaut, Sally Ride, went into space aboard a shuttle in 1983). That tragedy, however, not only claimed seven lives—among them that of the first "teacher in space," civilian Christa McAuliffe—but also postponed for several years further shuttle launches, as well as future NASA projects like space stations. NASA later decided to build a replacement shuttle despite widespread opposition.

The shuttle program, moreover, has not been free from political concerns, including (undocumented) speculation of pressure on NASA to launch *Challenger* despite cold weather to coincide with President Reagan's 1986 "State of the Union" address. Earlier speculation about the "teacher in space" search being a means of promoting both NASA and the president's image among skeptical professional educators appears well grounded.

The aftermath of the *Challenger* tragedy also publicized long-standing tensions between NASA and the military—which has its own ongoing space projects—regarding priorities, planning, oversight, and secrecy; competition between NASA and European cooperative ventures with European-made rockets; tensions within NASA over occupied versus unoccupied vehicles, and over large, expensive, showy projects favored by engineers versus smaller, cheaper, less visible ones favored by scientists; and tensions among NASA's top administrators and lower-ranking technicians over the safety of the shuttles themselves. The inability or refusal of key NASA officials to explain the situation to the public until forced to do so by government investigation, along with their extreme reluctance to take responsibility for their decisions, damaged public confidence in them. A 1976 assessment of America's space program still holds: "There is a base of public support . . . , but the case will have to be made for each new venture."[4]

TELEVISION AND THE FABRIC OF AMERICA

A similar questioning has characterized the nation's relationship in recent decades to television, a technology that has granted Americans unprecedented visual access to a wide range of events that they otherwise would have been unable to sample. This technology certainly was not greeted ambivalently by Americans in the late 1940s and early 1950s, for early television seemed a technical marvel. Americans rushed to purchase these huge vacuum-tube-based sets. Small screens, need for frequent repair, poor reception, pictures better described as green and gray rather than black and white, and long warm-up times did not detract from television's initial popularity. Families and others gathered together to view the new medium. Radio and movie audiences plummeted in size. During the next several decades, solid-state electronics, improved sound and picture quality, and color were added. Yet television's programming reputation failed to keep pace with the hardware. By 1961, many Americans agreed with FCC chairman Newton Minow's characterization of the medium as "a vast wasteland," a critique that persists to the present.

Criticism of Television Programming

Minow complained about the "procession of game shows, violence, audience participation shows, formula comedies about totally unbelievable families, blood and thunder, mayhem, sadism, violence,

An early color-television remote truck. Notice its tremendous size.

murder, western badmen, private eyes, gangsters, more violence and cartoons" broken only by "commercials—many screaming, cajoling and offending."[5] Later critics would add soap operas and "jiggle TV" (television programs featuring scantily clad men and women) to Minow's list. Minow and others maintained that television broadcasters had reneged on the obligation to elevate American society and had chosen instead to seek the lowest common denominator in a headlong rush for larger television audiences and greater advertising revenues. Others argued that contemporary social problems owed their genesis in part to television broadcasts; television's obsession with violence created a sick society. Children seemed particularly at risk. Senator Estes Kefauver's congressional committee on juvenile crime in the mid-1950s fingered television as a cause of juvenile delinquency, and Senator Thomas Dodd later chaired a Senate subcommittee in 1961 to investigate television violence. Dodd was especially concerned about syndicated shows, which would be repeated many times and corrupt the nation's youth. A special PTA committee targeted television violence. A subsequent committee of the same organization blamed television for the decline in children's reading skills.

Groups such as the Coalition for Better Television and Accuracy in Media broadened the critique. An organization of affiliated fundamentalist religious and conservative citizens groups, including the National Federation for Decency and the Moral Majority, the Coalition disliked television's pandering to sex and profanity, and organized boycotts of sponsors of offending programs. Accuracy in Media noted a pervasive liberal bias in television news and demanded equal time for conservative opinion. Representatives of various minorities and ethnic groups also objected to television's stereotypical or unflattering portrayal of them, claiming that the medium nurtured the seeds of intolerance.

Although most commentators see television as a purveyor of mass culture—much of it lowbrow or explicitly harmful—a handful reverse the indictment and attack television for fostering isolation and alienation. They worry that television viewing (which averages more than seven hours a day per household) decreases time formerly spent developing social skills and relationships, gaining experience, and maintaining familial attachments. The TV generation watches, it does not act, and it watches alone. The multi-television family has become standard. The result is passive, alienated individuals, a factor to those expressing such sentiments that seems of major importance to the republic's future. Not surprisingly, some parents have resorted to eliminating television entirely from their homes, while many more restrict and monitor their children's viewing time.

More recent television-related technologies seem unlikely to mollify television's critics. Communications satellites in stationary orbits paved the way for "superstations," cable television, and receiving dishes, which provide a smorgasbord of options able to placate even the most fragmented of societies. The videocassette recorder (VCR) of the 1970s has proven enormously popular and has been followed in the 1980s by the camcorder: a miniaturized lightweight camera for making home videotapes. Laser videodisc and compact-disc players, giant-screen TV, and video games have further personalized "home entertainment centers" as Americans further isolate themselves and rely more heavily on visual images.

THE ADVENT OF HIGH-TECH ELECTRONICS

The pervasive individuation of post-1950 American society, so forcefully reflected in the emergence of isolated television-viewing, also provided the context for the introduction of high-tech elec-

Three television sets enabled these men to watch three New Year's Day bowl games simultaneously. Keeping track of the audio must have been a problem.

tronics to the public. Pocket radios, cheap enough for almost everyone, were the first major transistor-based consumer product. Patrick Haggerty, president of Texas Instruments, deserves credit for understanding this potential market. He capitalized on the ascent of youth-oriented—individualistic—rock 'n' roll to establish a demand for TI's transistors, and persuaded a small company, Regency, to build transistor radios after skeptical major radio-makers refused Haggerty's initiative. They had good reason to do so. Early transistor circuits were time-consuming, difficult, and expensive to construct, and were easily broken. The task involved wiring together by hand thousands of components: transistors, diodes, resistors, and capacitors. Laborers, usually women, used tweezers to pick up miniature elements, connected them under magnifying glasses, and soldered them with toothpick-sized tools.

Development of the Microchip

Working independently in the late 1950s, Jack Kilby, a TI engineer, and Robert Noyce, a physicist at Fairchild Semiconductor, devised the "monolithic idea" that eventually became the silicon microchip. That these important inventors remain virtually anonymous in an age of celebrity is ironic. As they separately developed the idea, a complete circuit would consist of one part: a single block of semiconductor silicon containing all the components and connections of even the most complex circuit designs, with the wiring printed onto the chip during production; the individual part and the system became coterminus. Hence the "monolithic integrated circuit," which now is the core of not only computers and calculators but also cameras, clocks, pacemakers, toasters, typewriters, and countless other electronic devices. Where the earliest circuits contained a few transistors on a single microchip, the latest incorporate up to 500,000.

Noyce left Fairchild in 1968 to help found Intel. Beginning the next year, one of his engineers, Ted Hoff, developed an advanced form of integrated circuit, or "superchip," that brought together electronics and information processing. Using integrated circuits as memory devices, not just for logic operations, he had devised by 1971 a one-chip central processor unit now known as the microprocessor. Here the component *is* the computer, and the computer is a single component. Where custom-made chips, increasingly popular at the time, could not be used for any other product except the one for which they were designed, and posed financial and technological problems for their manufacturers, these general-purpose chips could be—and soon were—manufactured in huge quantities and then programmed for specific applications.

This "computer on a chip," as advertised by Intel, was rivaled also in 1971 by a genuine computer on a chip (which could process amounts of information large enough for many practical applications), developed by two engineers at Texas Instruments, Gary Boone and Michael Cochran. Soon other companies began producing microprocessors; this inevitably lowered prices and increased proficiency. In effect, microprocessors inserted intelligence and self-regulation into such now "smart" devices as elevators, car engines, gas pumps, traffic lights, typewriters, supermarket checkout stands, and clothes dryers. They did so through "transducers," devices linked to microprocessors that convert one form of energy into another. Future possible projects for transducers include implantable seeing-eye chips for blind persons (light-sensitive devices sending intelligible impulses to the brain), implantable electronic-ear chips for deaf persons (which might convert sound into intelligible impulses), and implantable speech-synthesizing chips for mute persons that might allow speech.

Pocket calculators represent the chip's triumph for many Americans. Individuation of American society was again the key, and Haggerty was again instrumental. Beginning in 1971, he directed Kilby to design a portable adding machine that, with the company's microchips, became the pocket calculator. It was the first major chip-based consumer product.

Chip manufacture in the mid-1980s took place in America and other industrialized countries, and was extensively automated. By contrast, assembling devices from these components generally occurred in Southeast Asia and required extensive labor, again frequently female. These workers used microscopes, which cause eyestrain, and were exposed repeatedly to caustic and toxic chemicals. Workers cut the silicon wafers into individual chips, discarded the nonworking chips, bonded the working chips, and soldered them to printed circuit boards according to templates. The finished circuit boards were then sealed in protective ceramic or plastic coatings.

Development of Computers—The Information Age

Adoption of this technology has assisted in making computers a fact of life for nearly all Americans. Eight million personal computers, five million programmable calculators, and a half-million large computers are currently in use, while millions more microprocessors are integrated into other devices. There is no need to know even simple mathematics or physics to operate these "user friendly" computers. The replacement of vacuum tubes with transistors and then microchip-based integrated electronic circuits was only one facet of the miniaturization and individualization of computers, of course. Magnetic storage tapes, and then hard and floppy disks and diskettes, replaced plugboards and paper tapes. High-speed line printers, plotters, and video terminals replaced electric typewriters and punch cards. Standardized program languages, such as IBM's Fortran (introduced in 1957) and the Defense Department's Cobol (developed in 1958), and canned programs and packages—often suited to particular tasks—have further increased access.

These developments have enabled Americans to enter, store, manipulate, and retrieve information in unprecedented amounts and at unprecedented speed. "Computer literacy" has become an integrated and (many would say) necessary element of a standard modern education. Parents, school systems, and businesses train their children, students, and employees. The YMCA and other organizations operate "computer camps." Nevertheless, these activities, and the use of computers in general, have their critics, who do not recognize the utopian "global village" and "electronic cottages" predicted, respectively, by Marshall McLuhan in the 1960s and by Alvin Toffler in the 1970s. Concern has been expressed about computer-generated unemployment and underemployment; psychological (alienation, boredom) and perhaps physical (excessive sedentary posture, eye strain, exposure to dangerous rays) maladies; invasions of privacy; transfer, scrambling or destruction of military, financial, legal, medical, and other confidential records by "hackers"; and above all, computer dependence or domination.

Lewis Mumford has argued that "no action, no conversation, and possibly in time no dream or thought would escape the wakeful and relentless eye of this deity: every manifestation of life [will] be processed into the computer and brought under its all-pervading system of control. This would mean," Mumford has concluded, "not just the invasion of privacy, but the total destruction of autonomy: indeed the dissolution of the human soul."[6] Few Americans, however, share Mumford's despair, and many treat their computers as nonthreatening friends, a trend understandably encouraged by computer manufacturers and distributors.

THE NEW AMERICAN MANUFACTURING

Computer manufacturers, as well as producers of other new devices, usually did not set up shop in America's traditional urban industrial heartland. Stretching from Boston and Baltimore to Chicago and St. Louis, this region of traditional concentration of American manufacturing firms and establishments had begun to decline following World War II. Not until the 1960s did the decline engender much comment. By then the industrialized Northeast and Midwest had lost thousands of companies and tens of thousands of jobs to other parts of the country, and especially abroad. The South and West had steadily attracted manufacturing companies through tax incentives, lower living costs, and opposition to organized labor. Western Europe, Latin America, and Asia offered strikingly lower operating costs. Social critics such as Daniel Bell detected the export of manufacturing jobs to foreign lands and decreed that America had entered the "post-industrial era" in which production and distribution of information would replace industrial goods. The na-

The development of geothermal energy resources provided an alternative to using fossil fuels for generating electricity in some areas of the country.

tion's new task was service. The preponderance of jobs rested in banking, government service, teaching, fast foods, and the like, not manufacturing. America was becoming "deindustrialized."

The inner-city riots of the middle and late 1960s, the Vietnam War, and the oil crisis of 1973 exacerbated the decline of the "frost-belt," as the urban industrial heartland was called. These three events challenged the long-standing assumption about the limitlessness of American material and human resources, which had seemed to translate into technologically generated plenty and might. The Great Depression, and the atomic bomb with its potential for vast devastation, had provided earlier inklings of the nation's possible limits, but the concern in the 1960s and after was very different. It produced strong new strains of selfishness in a society already infected with individuation; Americans took the perception of limits as a compelling justification to demand and actively to seek what they determined was their proper share of an apparently diminishing pie. President Nixon's New Federalism, which turned federal tax dollars and programs over to the states to use as they wished, was an explicit recognition of America's limitedness, as was California's Proposition 13, which drastically cut taxes and traditional services, and the growth of lawsuits between states over resource questions. But journalist Tom Wolfe's designation, "the me decade," more neatly captured the spirit of the 1970s and early 1980s. In this milieu, new manufacturing enterprises were publicly courted as states ardently competed for them. Those places that offered manufacturers the best terms (ease of transport, tax breaks, low overhead), and/or employees the greatest personal amenities (climatic, educational, cultural, recreational, monetary), were generally the victors.

The Growth of High-tech Industry

The "sunbelt"—states in the South and Southwest—was an early winner, welcoming hundreds of thousands of manufacturing jobs and millions of "frostbelt" refugees. But hard-hit states, such as Massachusetts and Michigan, soon counterattacked by moving aggressively toward what seemed a panacea. They sought to attract not traditional heavy-manufacturing facilities but "high-tech" industries, such as manufacturers of office and personal computers, fiber optics, lasers, video equipment, robotics, machine-vision systems, and genetic-engineering firms. This form of technology-based production presumes a superiority to prior forms, if largely through its cleanliness, efficiency, and small size (as compared with conventional factories and power plants), as well as through its "paperless" communications systems. And it fits neatly within the notion of a postindustrial society, given high-tech industries' foundation on information collection, analysis, and distribution.

High-tech industries have brought new jobs to frostbelt economies, and unemployment rates have consequently declined. Massachusetts was perhaps the foremost example of this turnabout in economic fortunes, and its city of Lowell exemplified that trend: from one of the nation's highest urban unemployment rates, dating back to the decline of the textile mills that made the community famous, to one of the nation's lowest, thanks to Wang Laboratories and other high-tech local industries. Other states did not enjoy the educational and cultural climate that, in part, had made Massachusetts so attractive; nor did other states enjoy its diversity of high-tech industries—as exemplified by those along Route 128, a freeway running through suburbs ten miles from Boston. Nor is it likely that high-tech industries will prove a permanent solution to long-term economic problems. The history of Silicon Valley, America's earliest locus of high-tech industries, presents a very different picture.

An Example of High-tech Industry—The Silicon Valley

Silicon Valley is the popular name for the center of the nation's semiconductor industry. It is located in Santa Clara County, south of San Francisco, and until 1940 was a quiet, productive farming area.

The Stanford Connection   Silicon Valley owed its spectacular rise to many factors, but foremost among them were the efforts and vision of one man: Frederick Terman, a Stanford University electrical-engineering professor and administrator. Terman nurtured Stanford's electrical-engineering program, the heart of semiconductor research, into one of America's foremost. No less important, he encouraged his many graduating students to establish infant electronics companies near the Palo Alto campus; set up for them and others the innovative Stanford Research Institute in 1946, and the nation's first industrial park, Stanford Industrial Park, in 1956, both adjoining the campus; and obtained numerous governmental research contracts for electronic components during World War II and after. In effect, Terman pioneered new university–business–government relationships that have since become models for other educational institutions and corporations.

Terman's achievements were made possible by, among other things, the federal government's ever-increasing need for new weapons systems, especially ballistic missiles; by Stanford's proximity to various military and naval installations; by its proximity to San Francisco, the West's financial center, and in turn to venture capital; by its abundance of good land and talented students for university–business expansion; and by the area's pleasant climate and cultural attractions. Local industries readily moved from agriculture to defense. Soon new ones arrived and prospered, beginning with Hewlitt-Packard, started in 1938 by two Stanford graduates who manufactured custom-made audio oscillators in their Palo Alto garage shop. In 1948 the two Varian brothers, pioneers of the Klystron tube for radar and microwave communications in Stanford's own laboratories a few years earlier, established Varian Associates, which later became Stanford Industrial Park's first tenant. In 1955 William Shockley left Bell Laboratories to set up Shockley Transistor Company in Palo Alto. That began the dramatic growth

Merging the campus ideal and the industrial, Stanford's
Industrial Park in California became the model for later
high-tech efforts.

of the semiconductor industry and, equally significant, the spinning off to nearby locales of new firms from older ones—as was
the case with the eight Shockley employees who, in 1957, left to
establish Fairchild Semiconductor (among them future silicon-mi-
crochip co-inventor Robert Noyce).

The Rise of Silicon Valley   By 1970 Silicon Valley had become the
world's densest concentration of high-tech enterprises, while Santa
Clara had become one of America's wealthiest counties. Large corporations such as General Electric and Kaiser had set up branch
manufacturing plants in the area, and they and others, such as IBM
and ITT, also had established research centers at Stanford's Industrial Park and elsewhere. The Park leased space only to high-tech
firms beneficial financially and intellectually to the university, and
provided a model of both landscaping and architecture for later industrial parks: low-lying, clean, attractive campus-like settings
antithetical to the huge ugly factories and sooty smokestacks of
traditional big industries.

As the complexity of semiconductor research and development grew, and as the need arose for custom-made integrated circuits for major defense contractors, Silicon Valley's critical role as military and aerospace subcontractor expanded. So, too, did the number and size of companies producing materials and equipment for manufacturing semiconductors. Not surprisingly, no other part of the country obtained so many defense contracts.

The Silicon Valley's Decline In the 1980s, however, Silicon Valley was confronted with saturated markets and fierce competition both domestically and abroad. Many of its companies were unable to sell their product to sufficient consumers to stay profitable or autonomous, and a number merged or folded. Symbolic of these upheavals were the separate departures from Apple Computer (the legendary "overnight success" in personal computers) of its two founders, Steve Jobs and Steve Wozniak, as they were replaced by professional managers from nontechnical backgrounds.

Simultaneously, new large high-tech concentrations sprang up throughout the country, and in Western Europe and Japan. Competition further undermined Silicon Valley's economic stability. High tech's vaunted ability to set up shop almost anywhere had, ironically, siphoned off jobs from this former glamour spot.

Robotics and Labor Relations

Silicon Valley's glamour, and that of other high-tech spots, does not always result in glamourous products. For example, high-tech produced robots supply labor for some dangerous manufacturing operations but not always with salutary results. The National Institute for Occupational Safety and Health reported sadly that on July 21, 1984, a 34-year-old machine operator "went into cardiorespiratory arrest and died after being pinned between the back of an industrial robot and a steel safety pole."[7] This robot, like other industrial varieties, resembled an insect more than a human, with its blunt beetle jaws and long needle nose. Driven by computer-operated motors, it could be programmed for many tasks. Where robots had been used primarily for loading, unloading, and manufacturing large, heavy objects, they more recently have been used in fine assembly work. These newer precision robots have become popular, with thousands being produced and sold annually. There now exists a Robot Industries Association, based in Dearborn, Michigan (site of both Henry Ford's home and the world's biggest auto plant, the Rouge facility). Manufacturers have widely praised robots as additions to or replacements for (part of) their human work forces. Unlike humans, robots do not get sick, take vacations, join unions, or criticize management.

Robots rarely pose a physical menace to workers, but often threaten their jobs. Commentators predict "workerless factories" operated by centralized computer systems and incorporating computer-aided design (CAD), computer-aided manufacturing (CAM), and complete automation processes. Many displaced workers would be compelled to seek lower-paying service-sector jobs. Critics are even less optimistic about employees who remain once robots are installed. They contend that robots are designed not merely to replace some workers but also to control others by dictating the pace of operations, and cite General Motors' new heavily computerized and automated Saturn division as an apt example. Promoted as an experiment in reinventing the manufacturing process, Saturn will avowedly attempt to reduce union control over work rules, and will link wages directly to productivity. To these observers, robotics will permit GM to integrate manufacturing in one location and control the work force; human workers, ironically, will be treated much like robots.

Some evidence exists that such dire projections may miss the mark. Other automotive manufacturers and some industries relying on robots have compensated workers for loss of income and skills by establishing "quality of worklife" programs. Workers share decision-making power with management on not only work processes and materials but also pricing, profits, and plant location. Moreover, as one high-tech proponent observes, "robots can't run factories," and their presence doesn't "eliminate the need for human skill and judgment."[8] Nevertheless, the extent of the shared decision making remains to be seen.

Transportation for the New Manufacturing

No matter by what means products are manufactured, and no matter what the product, they still require transport to market. Manufacturers in the post-1950 period increasingly scorned railroads, and that carrier's percentage of the nation's freight haulage dipped well below 30 percent by 1980. The much more flexible trucks and buses continually seized a greater share of transportation, and passage of the Federal Aid Highway Act in 1956 was instrumental. That act specified that funds collected from taxes on gas and tire sales, plus truck use, would go into a new Highway Trust Fund, which would pay for the largest building program in American history: 41,000 miles of interstate highways connecting 90 percent of America's cities with populations more than 50,000, including 5,000 miles in urban areas, with 90 percent of the cost to be paid by the federal government, and the states paying the rest. By 1973, 82 percent of the interstate-highway system had been completed (although the project's actual costs have far exceeded original estimates) and 16 percent was under construction. As of 1986, only 1,000 miles remained to be built. Temporary reduction of the official speed limit on interstate highways from 70 to 55 mph (an energy-conservation measure in wake of the oil embargo) and the rapid rise of petroleum fuel costs did not shake manufacturers' faith in trucks and buses or lead to a rebirth of rail transport. Nor did the federal government's creation in 1975 of Conrail (Consolidated Rail Corporation) out of the Penn Central Railroad and other failing northeastern freight railroads, and billions in federal loans, stem railroad decline. Those railroads that have remained profitable, such as the Norfolk and Western and Union Pacific, have done so through diversification; freight haulage constitutes only a small and not generally very successful part of their business interests.

Air transport also contributed to the railroad's demise, becoming the foremost transporter of mail and passengers over long distances even before Boeing introduced the first successful commercial jet, the 707, in 1958. These jet engines added to air's advantages by proving easier to maintain than conventional internal-combustion engines, and by permitting greater passenger and cargo capacity as well as speed. The deregulation of airlines in the late 1970s further enhanced air transport's long-distance haulage prospects. With marketplace control over fares and routes, stiffer competition and lower fares distinguished popular markets, while airlines cut expenses by reducing or abandoning service in unpopular ones.

HIGH-TECH AGRICULTURE

"The economic and social superiority of modern farming as a way of life is a fiction; as a technology it is destructive; as a Green Revolution hopeless,"[9] wrote one critic of American agriculture, which has been a source of persistent debate for nearly thirty years even as agricultural yields skyrocketed. Most attacks have focused on the prevalence of larger agricultural units (Harvard's John H. Davis coined the term "agribusiness" in about 1955) and the research system and methods that seem to support them. Critics contend that the United States Department of Agriculture (USDA) and land-grant universities have produced an agriculture that is needlessly capital intensive and dependent on hazardous chemicals, expensive machinery, and costly hybrid seeds. Small family farms have been unable to compete, and their land has been gobbled up by agribusiness concerns. Consumer groups often adopt a different perspective. They maintain that American agriculture is hopelessly antiquated; its bias towards smaller, inefficient production units drives food prices up and keeps them high.

Some Results of Pesticides, Mechanization, and Hybrid Crops

Rachel Carson was among the first to bring public attention to bear on indiscriminate agricultural pesticide and herbicide use. A fish and wildlife biologist, Carson's classic *Silent Spring* (1962) made connections between pesticide overuse and misuse and the (1) killing of natural predators normally controlling pests, (2) development of pesticide resistance in pests, (3) nonpests' transformation into pests, (4) death of many traditional wildlife species, unable to reproduce in sufficiently healthy numbers, and (5) the sudden rise of cancer, leukemia, hypertension, and cirrhosis-of-the-liver rates in persons living in areas of heavy pesticide use or eating foods contaminated by them. Although many (especially the pesticide industry) initially disputed Carson's findings, subsequent studies suggested that at least in the case of DDT, she failed to go far enough. Fat-soluble DDT and many of its descendants have an affinity for living tissue and cannot be readily broken down. They are stored in body fat, the effects of which may not be realized for years or generations. Some chemical solutions, such as the herbicide 2,4,5-T, apparently cause birth defects in animals and humans. A biochemical dream became a nightmare.

Carson's initiative, and protests, lawsuits, and legislation by others, have resulted in a virtual DDT ban. Other, presumably safer, pesticides and herbicides continue in use. Integrated pest management (considering the environment as a whole), sometimes without chemicals, has made inroads, as have more traditional solutions, such as the introduction of pest predators, lures, and drainage.

Critics of agricultural mechanization have had less success. A USDA pronouncement in 1968 that "machines are not made to harvest crops; in reality, crops must be designed to be harvested by machines" has become a standard dictum.[10] The rubbery-tasting hard tomato is a case in point. Two University of California, Davis, staff members, Jack Hanna and Coby Lorenzen, marketed in 1961 a mechanical tomato harvester and a strain of tomato plant that would bear its fruit within a limited period, ripen simultaneously, hold on the vine for thirty days without deteriorating, survive mechanical harvesting, and withstand shipment. But such a tomato was necessarily hard and tasteless. By 1967, nevertheless, the hard tomato and mechanical harvester combination was used on 90 percent of California's tomato acreage, easily a majority of the nation's crop. An estimated 32,000 migrant tomato-pickers lost their jobs. Although California tomato acreage actually increased, three-quarters of the state's tomato farmers left the business, selling their land to larger growers. Formerly significant tomato production in the states of New Jersey and Ohio was virtually eliminated. Similar harvester–produce combinations are now employed for lettuce, grapes, and other crops, resulting in similar "technological unemployment."

Use of hybrid seed has drawn less attention. Most debate has centered on agricultural practices, often accompanying hybrid seed use: heavy fertilizer, pesticide and herbicide application, irrigation systems, aquifer and ground-water pollution, and the like. Few have questioned that hybridization itself might invite catastrophe. Those who have predicted such an event cite hybrids' greater genetic uniformity compared to the native strains they have replaced. This similarity makes them susceptible to a single plague. A new disease strain or insect pest could wipe out a year's harvest; the 1970 corn-leaf blight killed a fifth of the South's corn crop. Other commentators worry about loss of distinctly different germ plasms as natural varieties are broken into their constituent pure breeds. They fear that seemingly nonproductive purebreds will be destroyed, but that those pure strains might have proved essential in the future; such plants could have a crucial resistance to a new disease or pest. The pioneer hybrid-seed manufacturer, Henry A. Wallace, was among the earliest to voice this concern. Partly because of Wallace's efforts, and because it makes economic sense, most industrial hybridizers have established germ-plasm banks in which they keep seeds of nonproductive and otherwise unused purebreds, purebreds that eventually may become important and profitable. Also, recent scholarship has demonstrated that many of the Living History Farms that dot America and grow representative crops from the nation's past were conceived of as germ-plasm banks by industrial hybridizers, and have been financially underwritten by these plant manufacturing interests. Recent critics have not been persuaded by these ad hoc practices and demand the creation of federally controlled germ-plasm repositories.

THE GROWTH OF BIOTECHNOLOGY

Hybridization, pesticides, and herbicides were pre-1950 technologies that became social questions in the 1950s and after. Biotechnology was born after mid-century and was immediately a social question. It consists of two basic processes: genetic engineering and fermentation. In the early 1950s, scientists uncovered the genetic "code" by which deoxyribonucleic acid (DNA) controls cell reproduction in living organisms. Through replication of DNA molecules, hereditary information passes from generation to generation. By the early 1970s, researchers could alter that code by "recombination": cutting a piece of DNA and splicing it with another piece. This gives the microorganism specific qualities not found in nature, and is therefore called genetic engineering. The recombinant DNA (rDNA) molecules then can be inserted into a host microorganism—whether a bacterium, fungus, protozoon, plant, or mammal—which can reproduce itself. This is fermentation.

To be sure, manipulation of heredity is hardly recent. But investigators no longer have to wait long periods for species to propagate. They instead can modify plant and animal genetic makeup to their own timetable, and can create new individual organisms. Nor need they depend on chance errors in reproduction for new species variation to emerge. Unlike with conventional breeding, genes now can be moved from one species to another or individual units can be cloned—made genetically identical.

Biotechnology is at least as revolutionary a phenomenon as computers, microchips, and microprocessors. As one scholar observes, "Biotechnology is the projection onto an industrial scale of a new view of nature [itself] as programmed matter"—or a change in "what the very word *life* means."[11] In recent years, hundreds of domestic and foreign companies have used fermentation to produce food proteins, insulin, interferon, and hemoglobin. Despite serious setbacks—such as certain organisms successfully fermented experimentally losing their desired characteristics in large-scale fermentation—there is enormous potential for genetic engineering not only for agricultural and pharmaceutical companies but also for chemical and energy firms: fragrances, paints, plastics, resins, and solvents, for example, all could be produced by genetic engineering. In addition, scientific-instrument concerns have begun producing sophisticated computerized machines that make artificial strands of DNA, and others that decode genetic instructions.

As with other forms of high tech, the participants in these various enterprises range from huge multinational corporations to tiny local firms. Like those other high-tech industries, too, key participants include academic scientists and engineers, particularly biologists and biochemists. That many academics have become financial as well as research partners in biotechnology firms raises questions of conflicts of interest and mixed loyalties. Similarly, MIT's acceptance in the early 1980s of $125 million from businessman Edwin Whitehead to establish an institute for life-sciences research raised questions of institutional conflicts of interest: Who would control this semi-autonomous MIT affiliate, from its academic appointments and evaluations to its patents and profits? Perhaps to ward off public interference, recombinant DNA researchers themselves issued guidelines in 1974 for their own work, and proscribed risky research areas. Under the auspices of the National Academy of Sciences, they sought voluntary compliance, but limited their focus to the activities of the scientific community. In the 1970s, the concern was whether experimental transfer of DNA molecules from one organism to another posed hazards through accidental release into the environment. By the 1980s this had been judged a minor concern, but was replaced by worries by some about the *deliberate* (if, so far, limited) release of laboratory-produced genetic materials into the environment by commercial genetic-engineering firms. Here the guidelines that were established in 1976 by the National Institutes of Health (NIH) were peripheral (they dealt only with accidental release), given new ethical concerns about genetic engineering and biological warfare and new public consciousness of biotechnology's scope.

Yet the changing industrial and political climates in the 1980s favoring deregulation, particularly for enterprises that might revitalize American industry that is facing stiff foreign competition, limited further action by federal and state governments. And the NIH guidelines did not apply to industry anyway, only scientists. A few municipalities with local biotechnology research facilities, such as Cambridge, Massachusetts, did issue their own updated guidelines. And a few lower-court cases, and Environmental Protection Agency and Department of Agriculture rulings, temporarily halted agricultural field-testing. By and large, though, public debate and action were modest. In 1986, however, President Reagan signed a comprehensive package of rules, guidelines, and definitions that will establish America's regulatory policies regarding biotechnology for a decade or more. In general, these will determine how five federal agencies will assess benefits and risks for biotechnology experiments and products.

Whatever the risks involved, the social consequences of utilizing biotechnological knowledge are surely far-reaching. In medicine and public health, for example, biotechnical advances either can already, or in the future, may provide screening for genetic diseases of (prospective) parents or offspring; provide new vaccines or new cells to fight diseases; and provide new microbes to digest sewage, detoxify wastes, and turn the rest into food substances. And in agriculture, for instance, biotechnology may shift the principal ability to produce food from farmers to scientists, and in turn to agribusiness, through the genetic manipulation of seeds, creation of artificial seeds, engineering of livestock and plant embryos, and the use of chemicals multiplying crop yield on demand, while avoiding traditional and, as currently seen, controversial pesticides, herbicides, and insecticides. That dormant utopian visions of prolonged healthy lives and endless plenty should spring up again amid such prospects is hardly shocking. Simultaneously, however, persistent dystopian nightmares of mankind playing God, as contemporary descendants of the obsessed and so ill-fated Victor Frankenstein of Mary Shelley's 1818 novel, also have been revived.

THE ENVIRONMENT AND ENGINEERS

Attempts to regulate biotechnology, pesticides, and herbicides were manifestations of a more generalized concern about the quality of American life, which led to creation of numerous national and state regulatory agencies in the 1960s and 1970s. Each was designed to tackle a facet of this larger, more nebulous problem. Unlike earlier regulatory mechanisms, their mandates were not restricted to specific industries. More than twenty such instrumentalities were formed by Congress in the 1970s alone. The Environmental Protection Agency (1970) and the Occupational Safety and Health Administration (1973) were among the most powerful and active.

Establishment of these agencies indicated that technological progress no longer was automatically equated with social progress; concerns for the environment, peace, and other quality-of-life issues seemed to hold at least as much appeal as improvements in hardware. This technology–quality-of-life union, now always subjected to individual scrutiny and interpretation, was a repudiation of technical expertise as the sole evaluation criterion, and a rejection of engineers and scientists as the most appropriate decision makers. Reports of unsafe automobiles, chemicals, nuclear power plants, and the like reinforced the suspicions of a wary public, and further tarnished the image of American engineers. Technological issues seemed far too important to be left to engineers and scientists, who in the post-1950 era appeared increasingly aloof from ethical concerns, and were often portrayed as similar to those technocrats who so disillusioned Thorstein Veblen back in 1933.

The Response of Engineers to Criticism

Engineers and scientists have responded to these thrusts in very different ways. Some have stressed that technical experts do indeed worry about ethics and the quality of life, and have formed such organizations as the Union of Concerned Scientists and Physicians for Social Responsibility to act on them. Others have chosen to meet the challenge head on. They question not just the competence of the general public to understand what they deem complex technological issues like nuclear power but the very right to decide those issues through local referendums and legislative processes. Thus the two University of Michigan nuclear engineers, asked whether "democracy is a viable form of government in a technological era," imply that it is not.[12]

Criticism of the Army Corps of Engineers   The critique of expertise has stung engineers at every level. The Army Corps of Engineers, the nation's largest engineering organization with more than forty thousand employees (most of them civilian), has been hit especially hard. Best known for its longtime endeavor to control the Mississippi River, the Corps had developed more than twenty-five thousand miles of inland waterways and spent more than $8 billion constructing dams, reservoirs, levees, dikes, floodwalls, and river-channel diversions and enlargements. Beginning in the late 1960s, however, public support for the Corps weakened and the kinds of projects that had been widely hailed as technological achievements in earlier decades now became sources of contention. Resistance by the Corps to criticism, especially from "nonexpert" outsiders, hurt not merely its image but also its operations.

Among the most serious criticisms came those from TVA's first chairman, Arthur Morgan, himself an internationally acclaimed civil engineer who had personally supervised numerous large-scale projects akin to those of the Corps. In a 1971 book entitled *Dams and Other Disasters,* Morgan documented long-term Corps opposition not only to environmentalists, Indians, and other citizens who got in the way of Corps enterprises but also to basic civil-engineering practices that likely would have lessened floods and costs alike: reservoirs, jetties at river outlets, cutoffs across river bends, and other alternatives to dams and levees. Moreover, the Corps, Morgan showed, repeatedly claimed that it had supported technical advances that it had actually long rejected. The West Point emphasis on tradition, on avoiding originality, and on applying short-term military solutions to long-term, nonmilitary engineering problems, Morgan concluded, must finally cease.

The initial major confrontation between the Corps and environmentalists took place in the late 1960s and early 1970s over a proposed flood-control project along Illinois' Sangamon River. Opponents ultimately managed to stop the project by providing convincing technical evidence of the project's unsuitability, not on traditional aesthetic and conservation grounds but precisely on economic and health grounds. Later controversial Corps projects were also sometimes halted, or at least modified, through similar arguments that challenged the Corps' very expertise. In general, moreover, only those controversial projects begun before the late 1960s, or those enjoying strong political support, have escaped modification or termination.

Indeed, the Corps, like other government agencies, has been belatedly involved in water resource development and environmental studies, especially since passage of the 1965 Water Resource Planning Act, the 1968 Wild and Scenic Rivers Act, the 1969 National Environmental Policy Act (requiring environmental impact statements for every proposed Corps project), and the 1972 Federal Water Pollution Control Act. The last law gave the Environmental Protection Agency regulatory powers regarding water quality that deprived the Corps of some of its traditional authority, while simultaneously requiring the Corps to inspect and upgrade its dams in the name of both health and safety. Moreover, the Corps is now actively concerned with quality-of-life issues and the social, cultural, economic, and demographic effects of its proposed projects, not just the environmental effects. If, on the one hand, the Corps has finally institutionalized comprehensive planning, it has on the other hand rejected its hitherto automatic commitment to undertaking huge, expensive public-works projects as the only water-resource-management measure. Thus the Corps, for example, currently provides leisure facilities for more Americans than any recreational or sports agency. (The Corps also remains involved with natural and man-made disaster relief efforts.)

Criticism of Urban Planners   Engineers and planners modifying the urban environment also suffered from this type of criticism. Until the late 1960s, a "bulldozer mentality," epitomized by Robert Moses, who spent more than $25 billion refurbishing New York

*In many cities, late–nineteenth-century tenements are
demolished and the poor moved elsewhere to make way for
young urban professionals.*

City, characterized urban-renewal efforts. City areas were razed to
create room for new, more modern housing or business facilities,
and freeways, bridges, and tunnels were constructed to make
downtown areas more accessible to suburban residents. Beginning
in the 1960s, those living in or near razed neighborhoods—these
were generally poor people—publicly objected to destruction of their
neighborhoods and formed associations to combat it. Each organization assumed the mandate of controlling the development or
change within that neighborhood; it was the neighborhood's right
to decide what constituted a quality life within that neighborhood,
not experts serving suburban or corporate interests, a stance that

frequently pitted individual associations against city government and its engineers (and even against other associations). These associations got a boost in the late 1960s, and particularly following the 1973 oil embargo, when former suburbanites and young suburban-bred professionals reversed a century-old trend and settled in increasing numbers inside city limits. These new well-to-do urbanites cherished the architectural diversity and potential convenience of inner cities and packed economic clout sufficient to make city engineers and elected officials listen. The result was "gentrified" neighborhoods in which shells of old buildings were restored and preserved, while the insides were refitted for new uses. However, this process was sometimes accompanied by concerted drives to force long-standing "undesirable" residents out of their neighborhoods.

Infatuated with urban life, these "yuppies" led a resurgence of inner cities, and demanded urban mass transit to suit their needs. San Francisco, Washington, and Atlanta bowed to the pressure and opened new rail/subway systems in 1972, 1976, and 1979 respectively. Several western cities are currently reviving or building new modern versions of the electrified street-railways abandoned decades ago for automobiles. And Los Angeles is even building a subway. These initiatives require tearing up some parts of the urban environment and almost always run afoul of the interests of some neighborhood associations—especially those that have not undergone gentrification—which protest and often sue.

THE TECHNOLOGY OF POLICY

As evinced by the difficulties inherent in trying to placate each urban neighborhood association affected by an urban-renewal project, attempts to analyze, plan, or predict in an era of individual prerogative may seem hopeless, but several methods and establishments have sprung up to undertake these tasks. These organizations generally incorporate representatives from several different fields of knowledge, and set out to develop rationales for certain actions. They explicitly acknowledge relationships between technology and society and hope their determinations will shape opinion and serve as the basis for public policy. Rarely are they as successful in this last endeavor as they would like. The absence of the aura of expertise in post-1950 America undercuts their efforts. These consortiums also frequently disagree; their perspectives range over the entire ideological spectrum. Individuals tend to ignore those studies and suggestions with which they do not agree, confident that another group will provide "documentation" that supports their point of view. Indeed, it is the sense that the impact of technology on society is such a profound social question that underlies these efforts; it generates these sorts of activities in spite of their highly dubious character.

Systems Analysis

Systems analysis originated as a British military technique in World War II to incorporate radar and other unconventional equipment into the air defenses, and is the grandparent of the modern technology of policy. It remains an influential and popular technique for "solving"—and, sometimes, anticipating and even preventing—problems of modern technological societies like the United States. Far from being limited to strictly technical or material problems, systems analysis is frequently applied broadly to social problems as well. Health, welfare, education, transportation, and crime are among the areas that have applied systems analysis. Thanks to computers and additional components of the contemporary information revolution, systems analysis now can be applied to ever more problems in government, industry, universities, and elsewhere. Systems analysis is akin to and sometimes overlaps with cost/benefit analysis, planning-program-budget analysis, risk analysis, and technology assessment and forecasting.

However labeled, systems analysis presumes that all problems are fundamentally technical in nature and thus quantifiable. Moreover, even if problems are defined in nontechnical ways, this theory states that they can be reduced to terms subject to systems analysis. Problems that are supposedly not reducible—philosophical or moral problems—are, by definition, not genuine problems; they must be either reformulated or abandoned. Systems analysis also presumes group efforts, usually teams of experts from various disciplines (not just engineering) who work together to devise the best solutions, or range of alternative solutions, to the problems at hand. Akin to the parts of well-run machines, the members of the group represent all the "parameters" to be considered: engineering, economics, physics, chemistry, management, and so forth. As social critic Theodore Roszak observes, "the good systems team does not include poets, painters, holy men, or social revolutionaries, who, presumably have nothing to contribute to 'real life' solutions."[13] Like the Technocrats of an earlier era, systems analysts see the world as chaotic and in need of the kind of order only they can provide: efficient, honest, nonideological, and apolitical. Like the

Technocrats, too, they readily employ jargon, as demonstrated by the words of Simon Ramo, a founder of the giant defense-related corporation TRW: "As the end result, the approach seeks to work out a detailed description of a specified combination of men and machines—with such concomitant assignment of function, designated use of matériel, and patterns of information flow that the whole system represents a compatible, optimum, interconnected ensemble for achieving the performance desired."[14]

In recent years, systems analysis has been criticized as not only elitist, self-justifying, self-perpetuating, and narrowly conceived but also inefficient, inadequate, and unscientific. Providing a range of alternative solutions to problems is one thing, but providing *the* solution is quite another matter. Nor does systems analysis have a particularly stellar record in the field in which it initially was embraced: the military. There, a seemingly endless progression of poorly made weapons and vehicles, and enormous cost overruns, have many Americans equating military expenditures with waste and greed.

Technology Assessment

Technology assessment (TA), another form of the technology of policy, is the systematic study of the possible impact on a society of a particular form of technology. It is especially concerned with effects that are unintended, indirect, or delayed. The term was coined in the late 1960s by Connecticut congressman Emilio Daddario. His efforts led to passage of the Technology Assessment Act of 1972, and in turn to the creation that same year of the United States Office of Technology Assessment (OTA). Its establishment reflected, on the one hand, America's fading faith in technological progress and, on the other hand, its desperate desire to anticipate and so perhaps to prevent or solve problems stemming from new technologies.

Closely related to TA and the OTA is the environmental impact statement (EIS), the mandatory use of which, since 1969 (when the National Environmental Policy Act became law), has been widespread and significant. Though in some cases such statements may limit or even stop technological developments like dams and power plants, in others they may actually speed the development's completion. While the EIS is usually restricted to a specific geographical site, TA often treats a technology that could be situated in many sites. And while EIS is usually limited to "go/no-go" decisions, TA often considers several alternative strategies.

Understandably, TA is ordinarily concerned with present and future phenomena alone. Assessments have covered such technological issues as auto safety, consumer product safety, occupational health and safety, and nuclear power safety. Some practitioners dismiss past technological developments as irrelevant to the consideration of current ones. Others, however, have begun including historical perspectives in their public-policy considerations. That in turn has led to "retrospective technology assessment" (RTA), a term coined in 1974 by Joseph Coates, then of the National Science Foundation.

Some RTA projects draw analogies between past predictions of the impact of technological developments on society and current ones; the railroad, for example, has been studied in conjunction with the space program. Others have examined the same general phenomena—pollution, for instance, or telephones—in both earlier and more recent times in order to utilize these past expectations and policies to illuminate present and future ones, and so to minimize "future shock."

Direct connections among the past, present, and the future are never easy to draw, except superficially, and historians who engage in RTA face dilemmas about not oversimplifying their findings while still making them meaningful. Paradoxically, growing knowledge of technology's rich and varied history makes it ever harder for historians to generalize too boldly about what really happened, much less what might yet happen, in any given situation. Paradoxically too, TA and RTA require assumptions (about events or processes being repetitive or cyclic) that historians normally reject, yet which are necessary for any practical applications. Finally, TA and RTA have been criticized by conservative social critics as hopelessly naive and have been likewise attacked by radical social critics as facades shielding modern technology from fundamental, systematic scrutiny.

Futurecasting

Forecasting of future trends often employs systems analysis, but uses other methodologies, too. By definition it looks only to the future, where technology assessment can look to the past and present as well. Moreover, forecasting need not be anywhere as rigorous as that other enterprise; nor need it be as restrained, insofar as forecasting is not as frequently tied to governments and thus has no immediate agency waiting to implement it. Although prophecies are probably as old as civilization, "futurecasting" is now a virtual industry in itself. Its mission is not merely to anticipate the future but indeed to shape it. Thanks to computers and other components of the information revolution, forecasting, like systems analysis, is ever more popular, even in the face of evident weaknesses.

Forecasts of whatever variety reflect at least some dissatisfaction with the present; they are not found in contented, stagnant societies where the citizens hope that the future will resemble the present. Forecasts are by definition intended to instill either hope or concern for the future, and so to spur improvements or prevent problems, or both. Given the increased possibilities of realizing those predictions, the traditional gap between prophecy and fulfillment has narrowed and sometimes disappeared.

Forecasters routinely employ mathematical and social models of growing complexity, and construct elaborate possible scenarios. Many utilize massive data banks as their starting point, then extrapolate specific developments and general trends. "Social indicators" often are used to predict broader changes, as is the "Delphi method" for reaching consensus among various experts; the latter entails mailing anonymous written questionnaires, rather than arranging face-to-face contact, to achieve more objective conclusions free from personality conflicts and group pressures.

The drawbacks of such approaches are similar to those of systems analysis: value-laden assumptions and methods; conservative presumptions of continued social stability (lest the forecasts be wholly inaccurate); inability to accommodate radical and unexpected changes; self-fulfilling prophecies; and aversion to nonquantifiable issues and dilemmas. No less important, forecasting is frequently and simply wrong: some of the technological wonders predicted come about sooner than, and in forms other than, those predicted; unanticipated technological advances prove significant; and, not least, the impact of technology on society—and vice versa—is different from that of the forecast. Examples are endless, but the computer, to take a notable one, was not predicted by any of the fifty most prominent American forecasters writing between 1890 and 1940, while the immediate successors of those forecasters expected that a mere handful of computers and operators would suffice for the foreseeable future.

Weather forecasting, an enterprise presumably sheltered from most of these problems, is nevertheless a telling example of forecasting's limitations. The National Weather Service, part of the Commerce Department, employs not only advanced "supercomputers" and mathematical models but also sophisticated radar, instrument-filled balloons, buoys, and, not least, satellites (the first of which was launched in 1960). Data are also collected from hundreds of offices in cities and towns across America, and from thousands of ships and planes throughout the world. By now, predictions with an approximately 80-percent accuracy rate are possible for from four to seven days, where a few years ago a mere two-day forecast could be that reliable. According to one student of these efforts, "Meteorologists are convinced that if only they could design the perfect mathematical model—one that represents the atmosphere in all its complexity—they could forecast the weather with real precision."[15] Yet the complexities of weather conditions today—some of them ironically affected by technology itself—are such that that expectation is painfully naive.

Similar assumptions govern economic forecasting, the record of which for accuracy pales beside even that of weather predictions. Indeed, relatively short-term economic forecasts are *intended* to be revised as the dates to which they apply near. Yet economic forecasters ply their trade with unrestrained determination—not, however, as economist John Kenneth Galbraith has wryly put it, because they know, but because they're asked. And their forecasts are indeed influential and affect policies in the public and private sectors alike. Forecasters using various models base their predictions primarily on the economy's past behavior, then integrate, with the help of computers, those analyses with visions of future actions by the government and by industry. The limitation of such models is their inability to measure the subsequent changes that do occur because of limitations imposed on their vision by what those groups themselves expect will happen. Paradoxically, then, economic forecasts look too far backward while not looking sufficiently ahead: hence, their constant need to be revised.

Lack of Consensus Among Forecasters The president of the World Future Society, a leading professional prognosticator, lamented George Orwell's limited success in predicting the shape of the world in 1984 from the perspective of 1948, when he completed his influential novel *1984.* That Orwell failed to anticipate numerous developments is indeed true, and he does not receive good marks on the scorecard popular with such full-time futurists. But the larger—and elementary—truth missed by this "expert" is that *1984* was written precisely to prevent its grim story from coming true, and that Orwell had in mind his own day—in England, in the Soviet Union, and in the Socialist movement—as much as 1984. Like all serious utopians or dystopians, Orwell was criticizing his own world more than literally describing ours. Such a misreading of *1984* is surprising, and reveals that many contemporary futurologists are blissfully ignorant of the dependence on the present.

Perhaps most illuminating, however, of the problems confronting modern forecasting is the lack of consensus among alleged experts regarding America's, and the world's, very future. The boundless optimism of a Buckminster Fuller and the stark pessimism of a Jacques Ellul *(The Technological Society)* are repeated not only in other speculative writings but also in the more technical treatises of, say, the upbeat Herman Kahn *(The Next 200 Years* and *The Coming Boom)* and Julian Simon *(The Ultimate Resource)*, and the downbeat Club of Rome *(The Limits to Growth)* and *The Global 2000 Report* (to President Jimmy Carter). Using presumably the same data, these forecasters disagree completely about such crucial areas as future population growth, food supply, energy sources, and environmental conditions. Indeed, the Club of Rome reversed its pessimistic findings just four years later, and with barely an apology or explanation. The facts, then, fail to speak for themselves, and are themselves in dispute. No Delphi methods, no matter how prolonged, would create any consensus here. What Kahn and Simon together (in *The Resourceful Earth*) say about *The Global 2000 Report* indicates this gap: "The original *Global 2000* is totally wrong in its specific assertions and its general conclusion. It is replete with major factual errors, not just minor blemishes . . . . Many of its arguments are illogical or misleading . . . . Our statements about the future . . . are intended as unconditional predictions . . . . We feel no need to qualify these predictions upon the continuation of current policies, as *Global 2000* claimed to do . . . ."[16]

Such disputes have spurred less technocratic visions, which may be termed "alternative futures." "Appropriate" technology is part of this trend. In general, these visionaries, by contrast to the professional prognosticators like Kahn and the World Future Society, are less elitist and dogmatic about their views and more interested in others' views, and are as concerned with cultural and psychological conditions as with economic and social ones. They adhere implicitly to the conservationist Sierra Club's motto: "Not blind opposition to progress, but opposition to blind progress."

Growth of the Ideology of Appropriate Technology

The growing acknowledgement, beginning in the mid-1960s, of the various limits to—and costs of—unrestrained technological growth

have led to the concept of "appropriate" or "intermediate" technology. Its leading advocate, E. F. Schumacher, a prominent British
economist, was not against technology; he was no contemporary
Luddite, or machine-breaker. Rather, he and his associates
throughout the world (not only in the industrialized West) have
sought smaller, more personal, and decentralized forms of technology "appropriate" to different societies. Less affluent and less industrialized societies do not necessarily require, and may not benefit
from, the same kind of complex and large-scale tools, machines,
business enterprises, and management practices that flourish in the
United States. For example, inexpensive and easily constructed
windmills, water tanks, water pumps, latrines, spinning wheels,
solar cookers, and small industries have been successfully introduced in Africa and Asia and have simultaneously raised living
standards and preserved existing ways of life. Their comparative
cheapness and simplicity are, in engineering terms, their elegance
and strength. These devices and processes often have been accompanied by the renewed use of indigenous raw materials, of energy-
generating waste products, and of small animal-powered or hand-
operated machinery. Such labor-intensive schemes employ local
workers and lessen the exodus from the countryside to over-
crowded cities.

Unlike earlier academic proponents of "modernization," who be-

*Harnessing the wind to
provide electric power is
one means of reducing
America's dependence on
fossil fuels. This wind
turbine generates electricity
sufficient for five hundred
homes. Each blade is one-
hundred-feet long.*

lieved that vastly different foreign cultures could readily absorb Western technology—and Western institutions and values—eventually to become model industrial democracies, Schumacher and others recognize the intrinsic diversity of nations, cultures, and geographic regions. Smaller-scale, more decentralized technology may appear more "appropriate" to societies less industrialized than the United States, but even here advocates question the traditional assumption that bigger is automatically better, and more efficient. The widespread preference in recent years for smaller, safer, and more fuel-efficient automobiles, and for solar, wind, water, and even coal energy sources as opposed to oil and natural gas are examples of appropriate technology at work—if often for economic more than social or cultural reasons. So, too, are the increasing demands for improvements in mass-production processes, particularly group-task alternatives to the monotonous and unfulfilling single-task assembly-line process—and here *not* necessarily for economic reasons, given the high wages and generous benefits of automobile workers, for instance.

Searching for the Good Life   The recent popularity of organic farming and of "natural" foods are further examples of appropriate technology at work in America—and here for various reasons. Spurred by disenchantment with contemporary urban industrial life, thousands of young Americans, most of them born and raised in cities and suburbs, became farmers in the 1960s and 1970s. In search of a happier, more satisfying daily existence, they eagerly tried their hand at what two of their heroes, Helen and Scott Nearing, called (in a 1959 book) *Living the Good Life.* Indeed, the Nearings' formerly obscure work now became a best-selling guide for aspiring agrarians, as did their complementary *Maple Sugar Book* (1950). Few, however, were as willing as the Nearings to dispense with the perceived benefits of modern technology, from machine tools and kitchen appliances to stereos and televisions. And fewer had the simple, even ascetic, tastes and firm discipline of the Nearings, who moved from rural Vermont to coastal Maine in 1952 when the former locale lost its rusticity. Many of these converted farmers became disillusioned and returned to more familiar surroundings and occupations.

A great many Americans have become concerned with the harmful effects on their health of chemical and synthetic fertilizers, pesticides, and food additives. They have sought to replace chemical methods of food growth, processing, and preparation with "natural," biological ones. In addition to eating low-cholesterol, high-fiber foods (sometimes avoiding meats altogether) and exercising regularly, they avoid processed foods like refined sugar and flour. Few produce their own food, and those who do generally restrict themselves to backyard gardens. Yet many have become avid readers, if not practitioners, of the organic-gardening techniques especially espoused by Rodale Press, the profits from which have enabled it to establish a farming research center near Emmaus, Pennsylvania, and to manufacture equipment for organic gardeners elsewhere. Some "natural" food enthusiasts also have become severe critics of giant agribusiness corporations. These alternative consumers also recycle newspapers and other household items and often purchase wood-burning heaters. Their practices are usually not, as with the Nearings, alternatives to affluence, but rather its by-products. Growing up in a highly technological society, they, like other practitioners of appropriate technology, wish not to abandon technology but to reform it.

Not all these alternative developments will persist. Their popularity may be temporary, as with solar energy, or may fluctuate, as with compact cars. Moreover, what is appropriate to less-industrialized countries may be utterly inappropriate to "postindustrial" ones like the United States. As a sympathetic critic of appropriate technology observes, "small is not always beautiful, local is not always better, and labor-intensiveness is not always desirable."[17] Indeed, defining appropriateness is perhaps a futile exercise, insofar as the term is both vague and relative.

AMERICANS AND THEIR TECHNOLOGIES

Alternative technology advocates, like most other post-1950 Americans, consider technology as a cause of and a solution to social problems. That approach reduces society and culture to objects upon which technology acts. It removes a vital human dimension from technology even as it considers technology's presumed human implications. And as long as this discourse persists, Americans will lambast or glorify (or lambast *and* glorify) technology. But for all the fearsome scenarios that this sort of analysis conjures up, consideration of technology as a forceful, perhaps the most forceful, causative agent of social and cultural change or constancy is ultimately quite comforting; it permits blame for the present to be deposited on technology, the few who "created" it, or "the organization" that sustains it. Each of these explanations insulates those considering the issue from responsibility for the present state of affairs.

This other-directed technological determinism provides solace for the present but raises the question of the ability of Americans to control future technological developments. Within this context, the idea of individual prerogative as the sole means of assessing technologies becomes a double-edged sword. Although it makes achievement of a consensus on any particular technological issue virtually impossible, individual prerogative does offer an ironic satisfaction. It encourages individuals to participate publicly in deliberations about technologies, which (as with the Hawthorne experiments) provides participants with a sense of belonging, of being part of a group. And it is this sense of belonging, not the assessment of technologies, that is crucial. The establishment of networks of individuals—being part of the group—becomes the end result of these deliberations, not the means to adjudicate technologies. No matter what the outcome of any assessment, no matter what the ultimate decision, those who had voiced their opinion on any side of the issue can revel in knowing that theirs was a job well done; they had proven themselves responsible and eased their consciences by contributing their input. In effect, concern about specific technologies functions as a way of providing individuals a group identity in an individuated society, and enables them to renounce responsibility for future technological developments even as they make a show of exercising that "responsibility."

This paradox stems directly from post-1950 American notions. So, too, does the assumption that technology is both a social solution and cause of social problems, as well as its corollary, that technology is the progenitor of society and culture. An examination of the history of technology in America, however, reveals the flaws in that assumption and corollary. Technologies and their applications in America's past have been the consequence of cultural notions and, within that context, social desires, needs, and ambitions. Rather than a progenitor of society and culture, technology has stood as a manifestation or reflection of cultural and social perceptions; it is a human product. And as these notions and desires have differed from one period to another in the nation's past, technologies also have changed as Americans have understood and employed them in a manner consonant with their contemporary ideas. This has led to an abandonment of some technologies as no longer viable, a reinterpretation (often subtle) of other existing technologies, and creation of new technologies. Indeed, technology has been an integral facet of American experience, but as a mechanism through which cultural notions and social desires were expressed or achieved. It remains so.

NOTES

1 Phillip L. Cantelon and Robert C. Williams, *Crisis Contained: The Department of Energy at Three Mile Island* (Carbondale: Southern Illinois University Press, 1982), p. xi.

2 James J. Duderstadt and Chihiro Kikuchi, *Nuclear Power: Technology on Trial* (Ann Arbor: University of Michigan Press, 1979), pp. 196–97.

3 *United States Statutes* 72 (1958): 426.

4 Jon D. Miller, "The Impact of Two Decades of Space Exploration on the Development of American Attitudes toward Science and Technology," in Joel A. Tarr, ed., *Retrospective Technology Assessment-1976* (San Francisco: San Francisco Press, 1977), p. 288.

5 Quoted in Erik Barnouw, *Tube of Plenty: The Evolution of American Television* (New York: Oxford University Press, 1975), p. 300.

6 Lewis Mumford, "The All-Seeing Eye," in John G. Burke and Marshall C. Eakin, eds., *Technology and Change* (San Francisco: Boyd and Fraser, 1979), p. 66.

7 Quoted in "Robots at Work," *Boston Phoenix*, July 2, 1985, p. 1.

8 Larry Hirschhorn, *Beyond Mechanization: Work and Technology in a Postindustrial Age* (Cambridge: Massachusetts Institute of Technology Press, 1984), p. 1.

9 Quoted in Jim Hightower, *Hard Tomatoes, Hard Times*, rev. ed. (Cambridge: Schenkman, 1978), p. 330.

10 Raymon E. Webb and W. M. Bruce, "Redesigning the Tomato for Mechanized Production," *USDA Yearbook* (1968), p. 104.

11 Edward Yoxen, *The Gene Business* (New York: Harper and Row, 1984), p. 14.

12 Max Gates, "U Experts Try to Clear the Air on N-Power," *Ann Arbor News*, November 25, 1979, p. C–7.

13 Theodore Roszak, "The Citadel of Expertise," in Burke and Eakin, eds., *Technology and Change*, p. 85.

14 Simon Ramo, "The Systems Approach," in Burke and Eakin, eds., *Technology and Change*, pp. 77–78.

15 Gordon Williams, "The Weather Watchers," *Atlantic*, vol. 257 (March 1986), p. 72.

16 Quoted in Albert H. Teich, ed., *Technology and the Future*, 4th ed. (New York: St. Martin, 1986), p. 193.

17 Witold Rybczynski, *Paper Heroes* (Garden City: Doubleday Anchor Books, 1980), pp. 165–66.

FOR FURTHER READING

Almond, Gabriel A., Marvin Chodorow, and Roy Harvey Pearce, eds. *Progress and Its Discontents* (Berkeley and Los Angeles: University of California Press, 1982). Essays by leading scholars on the ideology of progress today.

Bell, Daniel. *The Coming of Post-Industrial Society* (New York: Basic Books, 1973). A sociological account of the changes created by deindustrialization.

Bilstein, Roger E. *Flight in America, 1900–1983* (Baltimore: Johns Hopkins University Press, 1984).

Boguslaw, Robert. *The New Utopians* (Englewood Cliffs, NJ: Prentice-Hall, 1965). Early critique of systems analysis.

Boroush, Mark A., Kan Chen, and Alexander N. Christakis, eds. *Technology Assessment* (New York: Elsevier North Holland, 1980). A comprehensive introduction to the topic.

Braun, Ernest and Stuart MacDonald. *Revolution in Miniature*, rev. ed. (New York: Cambridge University Press, 1982).

Burns, Alan. *The Microchip* (New York: Wiley, 1981). Thorough examination of the social consequences of microchips.

Corn, Joseph J., ed. *Imagining Tomorrow* (Cambridge: Massachusetts Institute of Technology Press, 1986). Readable essays on the fate of various technological advances.

Deken, Joseph. *The Electronic Cottage* (New York: Morrow, 1981). Popular introduction to personal computers.

Doyle, Jack. *Altered Harvest: Agriculture, Genetics, and the Fate of the World's Food Supply* (New York: Viking, 1985). A critique of contemporary agricultural practice and biotechnology.

Dunlap, Thomas R. *DDT: Scientists, Citizens, and Public Policy* (Princeton: Princeton University Press, 1981). Explains the crusade against DDT in terms of values and public policy, not scientific expertise.

Flink, James J. *The Car Culture* (Cambridge: Massachusetts Institute of Technology Press, 1975). A provocative critique of the car as a negative force in American culture.

Fuller, R. Buckminster. *Utopia or Oblivion* (New York: Bantam Books, 1969). Popular essays by the leading contemporary technological utopian.

Graham, Frank, Jr. *Since "Silent Spring"* (Boston: Houghton Mifflin, 1970). Considers the rise of the environmental movement in the 1960s.

Hall, Peter and Ann Markusen, eds. *Silicon Landscapes* (Boston: Allen and Unwin, 1985). Essays on the economic and social aspects of high tech.

Henderson, Hazel. *Creating Alternative Futures* (New York: Berkley Windhover Books, 1978). Essays by a leading advocate of humanistic, nontechnocratic futurism.

Hoos, Ida R. *Systems Analysis in Public Policy*, rev. ed. (Berkeley and Los Angeles: University of California Press, 1983). Powerful critique of the overuse and misuse of systems analysis.

Howard, Robert. *Brave New Workplace* (New York: Viking, 1985). Popular indictment of the contemporary workplace.

Kuhns, William. *The Post-Industrial Prophets* (New York: Weybright and Talley, 1971). Easily readable study of contemporary technological visionaries and critiques.

Lappe, Mark. *Broken Code: The Exploitation of DNA* (San Francisco: Sierra Club Books, 1984). An examination of the possible effects of biotechnology.

Lewis, W. David and Wesley Phillips Newton. *Delta: The History of an Airline* (Athens: University of Georgia Press, 1979). Pioneering history of a major American airline.

McDougall, Walter A. *The Heavens and the Earth: A Political History of the Space Age* (New York: Basic Books, 1985). Award-winning history of the politics behind space programs worldwide.

McLuhan, Marshall and Quentin Fiore, *War and Peace in the Global Village* (New York: Bantam Books, 1968). Excellent example of McLuhan's visionary writing.

Mahon, Thomas. *Charged Bodies: People, Power, and Paradox in Silicon Valley* (New York: New American Library, 1985). Short popular biographies of Silicon Valley's leading figures.

Malone, Michael S. *The Big Score: The Billion-Dollar Story of Silicon Valley* (Garden City: Doubleday, 1985). Readable history of Silicon Valley.

Mazlish, Bruce, ed. *The Railroad and the Space Program: An Exploration in Historical Analogy* (Cambridge: Massachusetts Institute of Technology Press, 1965). A useful example of retrospective technology assessment.

O'Neill, Gerard K. *The High Frontier: Human Colonies in Space* (Garden City: Doubleday Anchor Books, 1982). An optimistic glimpse of future spacefaring.

Reid, T. R. *The Chip* (New York: Simon and Schuster, 1984). Readable tale of the simultaneous invention of the silicon microchip.

Roland, Alex, ed. *A Spacefaring People* (Washington, D.C.: NASA, 1985). First-rate essays on the history of space.

Rose, Mark H. *Interstate: Express Highway Politics, 1941–1956* (Lawrence: Regents Press of Kansas, 1979). Discusses the origins of the interstate highway system.

Roszak, Theodore. *The Cult of Information* (New York: Pantheon, 1986). Powerful indictment of the computer age.

Schumacher, E. F. *Small is Beautiful* (New York: Harper and Row, 1973). A classic by a leading figure in the appropriate-technology movement.

Snow, C. P. *The Two Cultures: And a Second Look* (New York: Cambridge University Press, 1964). Classic account of the gap between scientists and humanists.

Stares, Paul B. *The Militarization of Space* (Ithaca: Cornell University Press, 1985). A study of the use of outer space for various weapons systems.

Toffler, Alvin. *The Third Wave* (New York: Morrow, 1980). Very popular vision of high technology.

Turkle, Sherry. *The Second Self: Computers and the Human Spirit* (New York: Simon and Schuster, 1984). Important contemporary study of the social and psychological consequences of computers.

Wiener, Norbert. *The Human Use of Human Beings: Cybernetics and Society* (Boston: Houghton Mifflin, 1950). Pioneering examination of the computer's social consequences.

Winner, Langdon. *Autonomous Technology: Technics-out-of-Control as a Theme in Political Thought* (Cambridge: Massachusetts Institute of Technology Press, 1977). Significant critique of contemporary technology.

ILLUSTRATION CREDITS

11 National Museum of American History, Smithsonian. 18 Museum of American Textile History. 21 M. Diderot's *Encyclopédie*. 23 Saugus Iron Works, Saugus, Massachusetts. 26 M. Diderot's *Encyclopédie*. 28 Courtesy of the New York Historical Society, New York City. 36 The Historical Society of Pennsylvania. 39 Courtesy of the Chicago Historical Society. 44 Smithsonian Institution. 61 Courtesy of American Society of Mechanical Engineers (both). 63 Courtesy of the New-York Historical Society. 64 The Bettmann Archive. 66 The Bettmann Archive. 71 (top) American Antiquarian Society; (bottom) The Edward W. C. Arnold Collection, lent by the Metropolitan Museum of Art; photo courtesy of the Museum of the City of New York. 76 Library of Congress. 80 The Bettmann Archive. 97 Library of Congress. 99 The Bettmann Archive. 110 Smithsonian Institution. 112 Culver Pictures. 115 Smithsonian Institution. 118 State Historical Society of Wisconsin. 119 HBJ Collection. 120 Smithsonian Institution. 123 Department of Manuscripts and University Archives, Cornell University, Ithaca, New York (both). 125 The Bettmann Archive. 138 California Museum of Photography, University of California, Riverside. 147 Culver Pictures. 149 Culver Pictures. 151 The Bettmann Archive. 153 Library of Congress. 154 The Bettmann Archive. 157 San Diego Gas & Electric Company. 160 The Bettmann Archive. 162 Courtesy of AT&T. 184 Courtesy of the Chicago Historical Society. 187 Standard Oil Company (New Jersey). 191 Reprinted with permission, Caterpillar Inc. 193 Reprinted with permission, Caterpillar Inc. 200 Reprinted with permission, Caterpillar Inc. 203 Culver Pictures. 206 Culver Pictures. 208 Culver Pictures. 213 Department of Special Collections, Stanford University Libraries (both). 215 The Bettmann Archive. 222 GE Hall of History. 223 California Museum of Photography, University of California, Riverside. 224 Culver Pictures. 228 National Archives, Signal Corps. 234 From the Collections of Henry Ford Museum & Greenfield Village. 237 From the Collections of Henry Ford Museum & Greenfield Village. 239 International Museum of Photography at George Eastman House (detail). 242 California Museum of Photography, University of California, Riverside. 243 California Museum of Photography, University of California, Riverside. 246 California Museum of Photography, University of California, Riverside. 251 The Boeing Company Archives. 265 Culver Pictures. 271 From the Collections of Henry Ford Museum & Greenfield Village. 279 Reproduced through the courtesy of the Botanical Museum of Harvard University. 282 The Bettmann Archive. 287 The Bettmann Archive. 292 Culver Pictures. 297 Courtesy of the author. 299 The Bettmann Archive. 302 The Boeing Company Archives. 303 The Boeing Company Archives. 308 The Bettmann Archive. 309 UPI/Bettmann Newsphotos. 317 UPI/Bettmann Newsphotos. 320 Charles A. Ruch, Historian, Westinghouse Historical Collection. 321 UPI/Bettmann Newsphotos. 325 NASA. 327 AP/Wide World Photos. 329 UPI/Bettmann Newsphotos. 331 UPI/Bettmann Newsphotos. 335 Southern California Edison photo. 338 Stanford News Service. 349 UPI/Bettmann Newsphotos. 357 UPI/Bettmann Newsphotos.

INDEX

A.G. Spalding and Brothers, 205
Adams Act, 192
Adams, Herbert Baxter, 194
Adams, John, 85
Adams, John Quincy, 56
Adding machines, 227
Addressographs, 227
Adz, 11, 25
Aerodome, 248
Agricultural colleges, 192–94
Agricultural societies, 82, 116, 191
Agriculture, 187–88; changes after
    1920, 276–83; chemurgy
    movement, 272–75; colonial,
    13–15, 29; cultivators, 121–22;
    drills, 121; genetic manipulation of
    animals, 280–81; harrows, 15, 121;
    high-tech, 341–43; hybridization
    of plants, 278–80, 343; machine-
    based, 116; McCormick reaper,
    117–19; need for roads, 193–94;
    pesticides, 281–83, 342; railroads
    and, 188–89; systematizing the
    farm, 189–94; uniform, 115–23. *See
    also* Deere, John; Plows; Tractor
Aiken, Howard, 296
Aircraft, 293–95, 310; B-17, 308; B-24,
    308; B-29, 308; B-47, 310; B-52, 310;
    Boeing 247, 293; Boeing 707, 241;
    Douglas Commercial, 293; F-80,
    310; Ford Trimotor, 293. *See also*
    Aviation
Airmail, 251, 294; beginnings, 293
Air Mail Act (1930), 294; of 1934, 294
Alamogordo, New Mexico, 308
Aldrin, Edwin, 325
Almy, William, 74
Alteneck, Hefner, 143
Amateur Athletic Union, 202
Ambrotype, 93
American Automobile Association,
    210
American Broadcasting Company
    (ABC), 290
American Cyanamid, 177
American Electrical Company, 144
American Federation of Labor, 240–41
American Fusee Company, 136

American Home Economics
    Association, 221
American Institute of Chemical
    Engineers (AICE), 172–73;
    organized, 165
American Institute of Electrical
    Engineers, 229; organized, 165
American Institute of Mining
    Engineers, 165
American Kitchen Magazine, 221
American Machinist, 225
American Motor League, 210
American Revolution, 1, 34
American Road Makers, 208
American Society of Civil Engineers
    (ASCE), 165, 166–69
American Society of Mechanical
    Engineers (ASME), 171, 235–36
American System, 72, 126
The American Woman's Home, 220
Ampere, Andre M., 98
Apollo 18-Soyuz 19, 324
Apollo moon program, 325
Apple Computer, 339
Archer, Frederick Scott, 93
Arian, 327
Arkwright, Richard, 46, 47
Armstrong, Neil, 325
Army Corps of Engineers, 53, 55–56;
    criticism of, 347–48; Manhattan
    Project, 304; public works projects
    in 1930s, 265–66; railways, 67;
    Sangamon River project, 348
Articles of Confederation, 34, 37, 38,
    41, 43, 45
Atanasoff, John V., 297–98
Atomic bomb, 304–8; H-bomb, 311,
    317; Fatman, 309; Little Boy, 309
Atomic Energy Act (1946), 319
Atomic Energy Commission (AEC),
    319
AT&T, 177
Automatic coupler, 188
Automatic telegraph, 137
Automobile, 268; automotive
    institutions, 210; Buick, 211, 285;
    Cadillac, 285; Chevrolet, 285, 286,
    287; General Motors Acceptance

Automobile (*continued*)
  Corporation, 285; General Motors
  Company, 211–12, 284–87; Model
  A, 287; Model T, 232, 286, 287;
  Oakland, 285; Oldsmobile, 285;
  operator licensing, 211; Pontiac,
  286; systematic production,
  232–33; tapping of the middle
  class, 209–10; United States
  Motor Company, 212
Automobile Club of America, 210
Aviation, 251; advent of powered
  flight, 247–49; commercial, 293–
  95. *See also* Airmail; World War I:
  air technology; World War II:
  aviation

Baekeland, Leo, 288
Bakelite, 288
Baltimore and Ohio Railroad, 67, 99,
  100
Bardeen, John, 300
Barnum, P.T., 146
Baruch, Bernard, 245
Baseball, 203, 204
Beecher, Catherine, 220–21
Beecher, Lyman, 85
Beer, 9; colonial manufacture of,
  15–16
Bell, Alexander Graham, 162, 163,
  164; development of telephone,
  158–60; Great Centennial
  Exhibition, 137. *See also* Bell
  Telephone
Bell, Daniel, 334
Bell Laboratories, 300, 307
Bell Telephone: long-distance service,
  162–63; telephone and radio,
  164–65; underground cables, 161
Bendix, 305
Benz, Karl, 209
Bergman and Company, 148
Berry, Clifford, 297
Bessemer, Henry, 105–6, 198
Bicycling, 193, 207–8; bicycle
  manufacture, 230–31
Bigelow, Jacob, 83–84
Biograph projector, 214
Biotechnology, 344–46
Blacksmithing, 12, 14

Blake, Francis, 159
Blast furnace, 24
Boeing, Bill, 293
Bond, William C., 126
Bonneville Dam, 266
Boone, Gary, 332
Borden, Gail, 126
Boulton and Watt, 60
Boy Scouts, 194
Brady, Mathew, 129
Brattain, Walter, 300–301
Bricks, 29
Brickworkers, 29
Bridges, 2, 59, 65
Briscoe, Benjamin, 211, 212
British Marconi, 164
Brown and Sharpe Universal Milling
  and Grinding Machine, 137
Brown, Smith, 74
Brush, Charles, 148, 158; arc lighting,
  143–44
Bureau of Reclamation, 265
Burnham, Daniel, 186; Chicago Plan,
  186
Burr, Aaron, 37
Buses, 291–93
Bush Calculator, 296, 298
Bushnell, David, 60
Bush, Vannevar, 296

Calhoun, John C., 56
Calotype, 92, 93
Camcorder, 330
Camera, 91–92, 195
Campfire Girls, 194
Canalboats, 62
Canals, 2; building of, 62–66; Erie
  Canal, 63–65; expansion of
  railroads, 91; Gallatin's report on
  roads and canals, 55–56; George
  Washington and the Potomac River
  Company, 39–40; locks, 65–66
Carding engines, 71, 73
Carding machines, 46, 74
Carey, Matthew, 85
Carlyle, Thomas, 84–85
Carnegie Foundation for the
  Advancement of Teaching, 170–71
Carothers, Wallace, 288–89
Carpenters, 17, 27, 73

Carson, Rachel, 342
Carter, Jimmy, 320; *The Global 2000*, 356
Caulkers, 27
Cement. *See* Concrete
Central Pacific Railroad, 101, 102
Chalk River, Canada, 320
Challenger, 327–28
Channing, William Ellery, 80
Charles River Bridge decision (1837), 94
Chemistry, 272–74
Chemurgy, 272
Chernobyl, USSR, 321
Chesapeake and Ohio Canal, 67
Chicago Railway Exposition, 152
Chicago World's Fair, 273–74
Civil Aeronautics Act (1938), 294
Civil Aeronautics Board (CAB), 294
Civil War, 91, 129; railroads after the, 101–2
Clay, Henry, 73
Clermont, 59
Clipper ships, 124–25
Club of Rome. *See The Limits to Growth*
Coal industry, 103–5, 198
Coalition for Better Television and Accuracy in Media, 330
Coates, Joseph, 353
Cobol, 333
Cochran, Michael, 332
Colliers, 23, 24, 104
Colonial architecture, 11
Colonial government, 6, 7–8; and interstate commerce, 37–38
Colt, Samuel, 114
Columbia Broadcasting System (CBS), 290
Columbian Exposition, 156, 186
Columbia space shuttle, 327
The Coming Boom, 356
Communications Act (1934), 290
Compact disk player, 330
Compton, Karl T., 274
Computer-aided design (CAD), 340
Computer-aided manufacturing (CAM), 340
Computers: Cobol, 333; development of, 295–99, 333–34; EDVAC, 299; ENIAC, 296, 298; FORTRAN, 333; IBM Model 650, 299; manufacturing, 334; Mark I, 296; systems analysis, 351; UNIVAC, 299. *See also* Computer-aided design; Computer-aided manufacturing
Concrete, 182–83
Conestoga wagon, 39
Conrail, 341
Constitutional Convention, 41
Continental Congress, 34
Cooke, Morris, 267
Cooke, William F., 98
Cooper, James Fenimore, 80
Cooper, Peter, 68
Coppersmiths, 29, 30
Cordage, 26
Corliss Engine, 137, 140, 199
Corliss, George, 137
Corporations: early–nineteenth-century, 54–55; mid-nineteenth-century, 94–95
Cort, Henry, 103
Coughlin, Father, 263
The Council of National Defense (CND) 244, 245
Coxe, Tench, 46–47
Crockett, Davy, 76
Crompton, Samuel, 47
Crosley, 307
Crystal Palace, 126, 136
Cultivators. *See* Agriculture: cultivators
Curtiss, D.S., 189

Daddario, Emilio, 352
Daguerre, L.J.M., 91–92
Daguerreotype, 91, 92–93
Daimler, Gottlieb, 209
Dams and Other Disasters, 347
Daniel Guggenheim Fund for the Promotion of Aeronautics, 293
Daniell, J. Frederic, 98
Davis, John H., 341
Davy, Humphry, 98, 143
Day and Newell safe, 126
DDT, 281–82, 342
Dearborn, Michigan, 270
Declaration of Independence, 37, 79, 135

Deere, John, 120–21
De Forest, Lee, 163–64
DeKalb, hybridization of corn, 280
Dick, David, 126
Dickson, William K.L., 214
Diode, 163
DNA, 344, 345
Dodd, Thomas, 329
Draper, William, 92
Drawknives, 12
Dreadnaught, 246
Drills. *See* Agriculture: drills
Du Pont, 177, 288, 289
Durant, William Crapo, 211–12
Dynamos, 142–44, 228

Eagle plows. *See* Plows, Eagle
Eastman, George, 195, 215, 216
Eastman Kodak, 177, 307
Eaton, Amos, 82, 83
Eckert, J. Presper, 296, 298, 299
Edison Company for Isolated Lighting,
    148
Edison Electric Illuminating
    Company, 148; Light Company, 148;
    Tube Company, 148
Edison Lamp Works, 148
Edison Machine Works, 148
Edison, Thomas, 141, 149, 150,
    163, 176; AC power, 155;
    cinematography, 214; electric
    traction, 152; Great Centennial
    Exhibition, 137; incandescent
    lighting, 145–48; kinetoscope, 214;
    Naval Consulting Board, 244; Pearl
    Street Station, 147; phonograph,
    145; stock ticker, 145; telephone,
    145, 159
EDVAC, 299
Einstein, Albert, 304
Eisenhower, Dwight, 299, 319, 323
Electricity, 133; AC current, 149–50,
    155–58; DC current, 142, 148, 150;
    electric power and the factory,
    229–30; electrification of America,
    141–58; General Electric, 155–56;
    generators, 142; Pearl Street Station,
    147; rural electrification, 266–68;
    traction, 151–55; TVA, 266–68. *See*

also Dynamos; Electric traction;
    Lighting; Telegraphy
Electric traction, 151–55, 181
Electromagnets, 142
Electronics: high-tech, 330–34;
    integrated circuits, 332; microchip,
    332–33; Texas Instruments, 331;
    transistors, 300–301. *See also*
    Computers
Ellicott, Thomas: *The Young Mill-
    Wright and Miller's Guide*, 52–53
Ellul, Jacques, 356
Emerson, Ralph Waldo, 89, 220
Emigrant Friends' Society, 69
Emmaus, Pennsylvania, 359
Engineering Magazine, 225
Engineering societies. *See individual
    society title*
Engineer's Council for Professional
    Development, 173
ENIAC, 298, 299
Environmental Impact Statement
    (EIS), 352
Environmental Protection Agency,
    346
Equalizer lever, 68
Erie Canal, 63–65, 96
Erie Railroad, 100
European Space Agency, 326–27
Evans, Oliver, 79, 81; steam engines,
    59, 62; *The Young Mill-Wright and
    Miller's Guide*, 52–53
Everett, Edward, 76
Ewbank, Thomas, 126
Explorer I, 324

Factories, nineteenth-century, 106–8
Fageol Brothers, 292
Fairbain, William, 128
Fairchild Semiconductor, 338
Faraday, Michael, 128, 142
Farmer, Moses G., 142, 146
Farmer's Advance, 18
Farm Securities Administration, 272
Fatman, 309
Federal Aid Highway Act (1956), 340
Federal Communications
    Commission, 290
Federal Power Commission, 267

Federal Radio Commission, 290
Fermi, Enrico, 311; nuclear pile, 304
Fessenden, Reginald, 163, 164
Field, Cyrus W., 127–28, 152
Field, S.D., 152
Fillmore, Millard, 126
Film, 91, 92, 93, 195–96; motion
    picture, 214, 215, 216
Fireproof tiles, 182
Firestone Tire and Rubber, 269
Fission, 302–4
Fitch, John: steamboat, 40, 41, 59
Fitts Road Steam Engine, 137
Five Dollar Day, 236
Flanders, Walter, 212, 232
Flaxbrake, 17
Fleming, John Ambrose, 163
Floor anchor, 27
Football, 203–5
Ford English School, 236, 238
Ford, Henry, 210, 284, 287; chemurgy,
    274–75; Eagle boats, 247; General
    Motors bid for takeover, 212;
    industrial sociology, 236–38;
    moving assembly line, 233;
    systematic techniques, 232–33. See
    also Ford Motor Company; Ford
    Sociological Department; Ford
    Village Industries
Ford Motor Company, 274, 284
Ford Sociological Department,
    236–38, 271
Ford Village Industries, 270–72
Forest Products Laboratory, 201
Forest Service, 265
Fortran, 333
Franklin, Benjamin, 40; colonial
    printing, 31–32
Franklin Institute, 99, 174
French and Indian War, 33
Fritsche, Carl B., 274
Froe, 12
Fuller, Buckminster, 356
Fullers, 19
Fulton, Robert, 62, 79; North River,
    59; submarine, 60
Funk Brothers, hydridization of corn,
    280
Fusion, 322
Futtocks, 27
Future casting, 354–56

Gagarin, Yuri, 324
Galbraith, John Kenneth, 355
Gale, Leonard D., 98, 99
Gallatin, Albert: 1808 report on roads
    and canals, 55
Gamble, James, 113
Gambrel, 113
Garden cities, 268
Gaulard-Gibbs, 150
Gaulard, Lucien, 149
Gauss, Carl, 98
General Electric, 164–65, 177; AC
    power, 156; formation of, 155
General Film Company, 215
General Motors Acceptance
    Corporation, 285
General Motors Company, 177, 212,
    292; decline in urban mass transit,
    269; rise of, 284–87; Saturn
    division, 340; technical
    innovations, 284–85. See also
    General Motors Acceptance
    Corporation
General Survey Act of 1824, 56
Generators. See Dynamos
Geological Survey, 265
Gibbs, John, 149
Gilbreath, Frank, 235
Gilman, Daniel C., 174
Girl Scouts, 194
Glenn, John, 325
The Global 2000 Report, 256
Goddard, Robert, 324
Gompers, Samuel, 241
Good Housekeeping, 221
Good Roads Movement, 93, 188
Goodyear, Charles: The Great
    Exhibition, 126
Goodyear Tire and Rubber Company,
    137, 177, 292. See also Goodyear,
    Charles
Gramme, Z.T., 143
The Grange, 187
Gray, Elisha, 159
The Great Centennial Exhibition,
    136–40, 207; Centennial
    Corporation, 135; Centennial
    Medal, 137
The Great Exhibition, 126
Greenbelt, Maryland, 268
Greendale, Wisconsin, 268

Greenhills, Ohio, 268
Greyhound Bus Lines, 293
Grove, William, 98
Groves, General Leslie, 304
Guild system, 78
Gunpowder, 35
Gutta percha, 127, 128

Haggerty, Patrick, 331, 333
Hahn, Otto, 302–3
Hale, William J.: chemurgy, 272, 273–74, 275
Hall, E.J., 249
Hamilton, Alexander, 43–46
Hanford, Washington, 304
Hanna, Jack, 342
Hargreave, James, 46
Harmonic telegraph, 158–59
Harrison, John, 68
Harrows. See Agriculture, harrows
Hatch Act, 192
Hawthorne Works. See Western Electric: Hawthorne Works
Henry, Joseph, 99, 126, 142
Henry, Patrick, 41
Hertz, Heinrich, 163
Hewlitt-Packard, 337
Hi-Bred Corn Company, 279–80
Highland Park, 232, 236
High-tech industries, 336; Silicon Valley, 337–39
Highway Trust Fund, 340
Hoff, Ted, 352
Holbrook, Josiah, 83
Home Furnishing Goods Exhibition, 223
Home Furnishing Review, 223
Home Insurance Building, 185
Hoover, Herbert: President, 261–62, 266, 267; Secretary of Commerce, 289
Hopkins, Cyril, 278
Howe, Elias, 109
Howe, Frederick C., 186
Hunt, Walter, 109
Hussey, Obed, 117

IBM, 296, 299
IBM Automatic Sequence Controlled Calculator. See Mark I

Illinois Agricultural Society, 122
Illinois Canal, 65
Illinois Central Railroad, 102
I.M. Singer and Company, 109–11
I.M. Singer and Company Gazette, 109
Indentured servant, 22
Industrial Workers of the World (IWW), 236
Institute of Radio Engineers, 165
Intel, 332
Intercollegiate Athletic Association of the United States, 205
Intercollegiate Foot Ball Association, 204
Interstate Commerce Commission, 188
Iowa State Agricultural College, 221
Iron: colonial, 22–25, 28; increase in production, 141; mining, 197; nineteenth-century, 103–5; railroads and, 102–3; smelting, 198–99; systematizing production, 197–99; use of coal, 104. See also Steel: Bessemer process
Ironworkers, 25

Jackson, Andrew, 56, 76, 79
Jacobi, M.H., 98
Jeffersonian Embargo, 72
Jefferson, Thomas, 40, 43, 45, 55
Jenney, William Le Baron, 185
Jersey Homesteads, 272
Jervis, John B., 68
Jobs, Steve, 339
Joiners, 27
Joint Committee of Engineering Education of the National Engineering Societies, 170, 171
Joint-stock companies, 5
Jones, Donald F., 278–79
Journal of Home Economics, 221
Journeymen, 13

Kahn, Herman: The Coming Boom, 356; The Next 200 Years, 356
Kefauver, Estes, 329
Kelly, William, 105–6
Kennedy, John F., 324
Ketchum, William F., 117

Kettering, Charles, 292
Kilby, Jack, 332, 333
Kinetoscope, 214
King, James T., 223
Kitty Hawk, North Carolina, 247, 248
Klystron tube, 337
Knox, Henry, 43

Land Ordinance (1785), 38
Langley, Samuel Pierpont, 248
Laser videodisc, 330
Latrobe, Benjamin, 79, 81
League of American Wheelmen, 208, 210; formed, 207
Leather, colonial manufacture of, 19–20
L'Enfant, Pierre, 48
Lighting, 200, 226; arc, 137, 143–45; gas, 143, 146; incandescent, 145–50; Pearl Street Station, 147–48. *See also* Brush, Charles; Edison, Thomas; Electricity
Lilienthal, Otto, 248
The Limits to Growth, 356
Lindbergh, Charles, 293
Little, Arthur D., 172–73, 261
Little Boy, 309
Liverpool and Manchester Road, 66, 68
Living the Good Life, 358
Living History Farms, 343
Livingston, Robert, 60
Locomotives, 152, 291; Rainhill Trials, 68; steam, 68–69
Lodge, Oliver, 163
Loew, 216
Logging industry, 199–210
Long, Huey, 263
Looms: colonial, 17–18; power, 71, 74–75
LORAN, 306
Lorenzen, Coby, 342
Los Alamos, New Mexico, 305
Louisiana Purchase, 60
Lowell, Francis Cabot, 74–75
Lowell mills. *See* Textiles: Lowell, Massachusetts
Lowell Offering, 76
Luddites, 84
Lush, Jay, 280
Lusitania, 244

McAdam, John, 58
McAuliffe, Christa, 327
McCormick, Cyrus, 94, 120, 121; Great Centennial Exhibition, 126; reaper, 117–19
MacDonald, Thomas, 194
Machine shops, 70, 77; metal working, 71; textile, 73
McLuhan, Marshall, 334
Madison, James, 41, 42, 45, 56
Manhattan Project, 304
Mann, Charles R.: report on American engineering education, 171–73
Manny, John H., 117
The Maple Sugar Book, 358
Marconi, Guglielmo, 163
Mariner 2, 325, 326
Mariner 4, 326
Mark I, 296, 298
Marshall, John, 40
The Marshall Plan, 310
Marshall, Thomas, 47
Marston, Anson, 193
Maryland Farmer, 189
Mason, 73
Massachusetts Assembly, 32
Massachusetts General Court, 33
Master-Apprentice system, 12, 69
Mather, Increase, 15
Mattock, 14
Mauch Chunk Road, 66
Mauchly, John, 296, 298
Maxim, Hiram, 146, 248; machine gun, 245
Maxwell-Briscoe, 212
Mayo, Elton, 258, 259
Mechanical Engineering Teachers Association, 170
Mechanical News, 255
Mechanics institutes, 53, 81–82
Menlo Park, New Jersey, 147
Mercantilism, 1–2, 28, 38, 182; beyond mercantilism, 52–53; colonial large-scale enterprise, 21–22; colonial social structure, 5–7; United States Constitution, 1
Mesabi Range. *See* Iron: mining
Mexican War, 91, 124
Microchip, 332
Middlesex Canal, 75
Midvale Steel, 137

Mill River Dam collapse, 168
Millikan, Robert, 274
Mills, 4, 48, 62, 70, 75; bark, 20;
  colonial, 9–11; flour, 29; fulling, 4,
  16, 19; grist, 10–11, 19, 62;
  Hollander, 31; horizontal, 10;
  logging, 199–201; Norse, 10; paper,
  9; querns, 10; rolling, 25; sawmill,
  1, 19, 25, 27, 62; spinning, 75;
  splitting, 25; textile, 73–74;
  watermill, 53; weaving, 75; wind,
  9, 29
Millwrights, 12, 70, 107–8
Mimeographs, 227
Mining, 201–2; copper, 196–97. See
  also Iron: mining
Minow, Newton, 328–29
Model A. See Automobile: Model A
Model T. See Automobile: Model T
Monroe, James, 56
Moral Majority, 330
Morgan, Arthur E.: Dams and Other
  Disasters, 347–48; TVA, 266
Morse Code, 163
Morse, Samuel F.B.: Great Centennial
  Exhibition, 135; photography,
  91–92; telegraphy, 98–99;
  transoceanic telegraph, 128. See
  also Morse Code
Moses, Robert, 348
Motion Picture Distribution and Sales
  Company, 216
Motion pictures, 212–16;
  kinetoscope, 214; zoopraxiscope,
  214
Motion Pictures Patent Company
  (MPPC), 215–16
Muller, Paul, 282
Mumford, Lewis, 334
Münsterberg, Hugo, 238
Mushet, Robert, 105–6
Muybridge, Eadweard, 212–14

NASA, 324; created, 323; space
  shuttle program, 326–28. See also
  Spaceflight
National Bank, 45
National Broadcasting Company
  (NBC), 290
National Bureau of Standards, 273

National City Lines, 269
National Conference on City
  Planning, 186
National Defense Research
  Committee, 301, 304
National Electric Light Association,
  229
National Environmental Policy Act
  (1969), 348; becomes law, 352
National Farm Chemurgic Council,
  275; created, 274
National Federation of Decency, 330
The National Institute for
  Occupational Safety and Health,
  339
National Institutes of Health, 345
The National Manufactory, 47
National Research Council (NRC),
  257; created, 244
The National Road, 55, 56
The National Science Foundation, 310
National Weather Service, 354–55;
  public works projects, 265
The Naval Consulting Board (NCB),
  244
Naval Research Laboratory, 305
Nearing, Helen and Scott, 359; Living
  the Good Life, 358; Maple Sugar
  Book, 358
Neat's Foot Oil, 113
Neuman, John von, 299
Newcomen engine, 40, 41
New Deal: Army Corps of Engineers,
  265–66; chemurgy, 275; critics of,
  263–64; garden cities, 268;
  Roosevelt and, 262–70; rural
  electrification, 266–68; TVA,
  266–68; urban mass transit, 269;
  WPA, 294
New York Assembly, 33
New York Central Railroad, 100
The Next 200 Years, 356
1984, 355–56
Nixon, Richard, 326, 335
Norris, Tennessee, 266
Norris, William, 69
Northwest Ordinance (1787), 38
Noyce, Robert, 332
Nuclear power: Atomic Energy
  Commission, 319; critical mass,
  303; decline of, 322–23; Fermi's

nuclear pile, 304; Hanford, Washington, 304; Los Alamos, New Mexico, 305; Nazi Germany's effort, 304; Oak Ridge, Tennessee, 304–5; successful bombarding of uranium, 302; Three Mile Island, Pennsylvania, 319–21. See also Atomic bomb; Fission; Weapons: nuclear weaponry
Nuclear weapons. See Atomic bomb; Weapons: nuclear weaponry
Nylon, 288

Oak Ridge, Tennessee, 304
Occupational Safety and Health Administration, 346
Oersted, Hans Christian, 98, 142
Office of Scientific Research and Development, 301–2, 305; proximity fuses, 307; uranium-isotope-separation research, 304
Ohio Canals, 65
Ohm's Law, 99, 145
Olds, Ransom, 209–10
Omnibus, 151
Oppenheimer, J. Robert: atomic bomb, 305; opposed to hydrogen bomb, 311
The Organization Man, 315–16
Orwell, George: 1984, 355–56
Otis elevators, 137
Otis, James, 32
Otto, Nikolaus, 208–9
Owen, Robert, 75

Pacific Guano Company, 136
Panhard and Levassor, 209
Panic of 1837, 56
Papermaking, colonial, 31–32
Paramount Pictures, 216
Parliament, British: taxes on American colonies, 32–33
Pearl Street Station, 147–48
Pennsylvania Assembly, 42
Pennsylvania Railroad, 100
Pennsylvania Society for the Encouragement of Manufacturing and the Useful Arts, 46
Pewters, 29, 30
Pfister, hybridization of corn, 280
Phonoautograph, 158

Phonograph, 145, 214
Photography, 91–93, 195. See also Motion pictures
Phrenologists, 93
Physicians for Social Responsibility, 347
Physiognamist, 93
Pig's Foot Oil. See Neat's Foot Oil
Pilgrims, 26
Pinchot, Gifford, 201
Pioneer 10, 326
Pittsburgh Survey, 186
Plastics, 287–89
Plows: colonial, 14–15; Eagle, 120; Prairie Breaker, 120–21; shovel, 14, 121, 122; steam, 122–23. See also Deere, John
Pneumatic Steel Company, 106
Pneumatic tires, 292
Pope, A.A., 193, 207, 231
Populist party, 187
Pork industry, 111–14
Portland cement. See Concrete
Post, George B., 185
Potomac Company, 40
Pratt and Whitney, 137
Prince Albert, 125–26
Printing, colonial, 31–32
Printing telegraph. See Stock ticker
Procter and Gamble, 113
Procter, William, 113
Produce Exchange Building, 185
Promontory Point, Utah, 101
Proposition 13, 336
Providence Tool Company, 114
Public Utility Holding Company Act, 267
Pullman sleeper car, 137
Puritans, 15, 26
Putting-out-system, 38, 77

Quern, 10
Quincy Railway, 66

Radar, development of, 305–6
Radio: ABC, 290; beginnings of, 163–64; Bell Telephone and, 163–64; British Marconi, 164; CBS, 290; Federal Communications Commission, 290; Federal Radio

Radio (*continued*)
  Commission, 290; General Electric,
    164–65; marketing with, 289–90;
    NBC, 290; pocket size, 331; RCA,
    164–65, 290; WEAF, 289
Radio Act (1927), 290
Railroads, 2, 111, 141; after 1920, 291;
    agriculture and, 188–89; ascendancy
    of, 96–106; automatic coupler, 188;
    building of, 67–68; Conrail, 341;
    cow-catchers, 68; early–nineteenth-
    century, 66–69; equalizer lever, 68;
    government investment in, 101–2;
    iron industry and, 102–5; logging
    industry and, 199; mining and, 196;
    Rainhill Trials, 68; Railroad
    Valuation Act (1913), 188; steel
    industry and, 105–6; swivel truck,
    68. *See also* Telegraphy: railroads
    and; Locomotives
Railroad Valuation Act, 188
Rainhill Trials (1829), 68
Ramo, Simon, 352
Randolph, Edmund, 43
Ranger 7, 326
Ransome, Ernest L., 183
Raytheon, 307
RCA, 164–65, 290, 305
Reagan, Ronald, 318, 327, 345
Reaper. *See* McCormick, Cyrus:
    reaper
Rensselaer School, 82–83
The Research and Development
    Board, 310
The Resourceful Earth, 356
Retrospective technology assessment
    (RTA), 353
Reverberatory furnace, 103
Ride, Sally, 327
Rittenhouse, David, 43
Robotics, 339–40
Robotics Industries Association, 339
Rockefeller Foundation, 280
Rocket, 68
Roebling, 137
Roosevelt, Franklin D., 266, 267, 272;
    air mail service, 294; endorsement
    of chemurgy, 273; Manhattan
    Project, 304; National Defense
    Research Committee, 301. *See also*
    New Deal: Roosevelt and

Roosevelt, Nicholas, 60
Roosevelt, Theodore, 206, 248
Roszak, Theodore, 351
Royal Society of Arts, 125, 142
Rumsey, James, 60, 62; death of, 59;
    paddlewheel-powered pole boat,
    40, 41
Runkle, John D., 170–71
Rural Electrification Administration,
    267
Russell Sage Foundation, 186

St. Louis's Manual Training School,
    241
SALT I (1972), 318
SALT II (1979), 318
Salt manufacture, colonial, 9
Sanitary Home, 221
Sarnoff, David, 290
Satellites. *See* Spaceflight: satellites
Sawyers, 27
Schneider, Herman, 171
Scholfield, Arthur, 74
Scholfield, John, 74
Schumacher, E.F., 357–58
Scientific American, 225
Scott, Howard, 263
Scott, Walter Dill, 239, 259
Seward's Folly, 124
Sewing machine, 109–11
Shelley, Mary, 346
Shepard, Alan, 325
Shipbuilding, 21–22, 29; colonial,
    25–28. *See also* Shipping;
    Shipsmiths; Shipwrights
Shipping, mid-nineteenth-century,
    124–25
Shippingport, Pennsylvania, 319,
    322
Shipsmiths, 25, 26, 27
Shipwrights, 27
Shockley Transistor Company,
    337–38
Shockley, William, 300, 337
Shreve, Henry M., 62
Shull, George, 278
Silent Spring, 342
Silversmiths, 29–30
Simon, Julian: *The Ultimate
    Resource*, 356

Sinclair, Upton, 263
Singer, Issac M., 108–111
Singer Sewing Machine Company,
    136
Skyscraper, 183, 185
Slater, Samuel, 47, 74
Sloan, Alfred P., Jr., 284–86
Smith-Hughes Act (1917), 243
Smith-Lever Act, 192
Society for the Advancement of
    Management, 236
Society for Establishing Useful
    Manufactures, 47
Society of Industrial Engineers, 236
Society for the Promotion of
    Engineering Education, 170
Society for the Promotion of the
    Science of Management, 236
Sonar, 306–7
Spaceflight: Apollo moon program,
    325; *Challenger* disaster, 327;
    Columbia, 327; decline of support
    for, 326–28; European Space
    Agency, 326; *Explorer I*, 324;
    Gemini program, 325; Mercury
    program, 325; origins of, 324;
    satellites, 323–24, 325–26; shuttle,
    326–28; space race, 324–26;
    Sputnik I, 323. *See also* NASA
Space shuttle, 326; *Challenger*,
    327–28; *Columbia*, 327. *See also*
    Spaceflight
Spalding, Albert G., 205–7
Speaking telephone. *See* Telephone
Speedwell Ironworks, 99
Spinning jacks, 74
Spinning jenny, 46, 71, 74
Spinning wheel, 16–17
Sports, 202–6
Sprague, Frank J., 153–54, 155
Sprengel, Herman, 145
Sprengel pump, 146
Springfield Armory, 114
Sputnik I, 323
Sputnik II, 323
Stamp Act, 32
Standard Oil of California, 269
Stanford Industrial Park, 337, 338
Stanford Research Institute, 337
Stanley, William, 149, 150
Steamboat, 40–41, 59–62

Steam engine, 40, 152, 157; Corliss
    engine, 137
Steam locomotive. *See* Locomotives
Steam plow, 122–23
Steam ships, 124–25
Steam shovel, 197
Steam thresher, 122
Steam traction, 152, 190
Steam turbogenerators, 157
Steel industry, 141; Bessemer process,
    105; railroads and, 105–6;
    skyscrapers, 185; systematizing,
    197–99. *See also* Iron
Steinhill, Karl A., 98
Stephenson, George, 66; *Rocket*, 68
Stephenson, Robert, 128
Stevens, John, 59
Stock ticker, 145
Stockton and Darlington Railway,
    66, 68
Stone, General Roy, 193
Stowe, Harriet Beecher, 220–21
Strassman, Fritz, 302–3
Strategic Arms Limitation Talks. *See*
    SALT I; SALT II
Strategic Defense Initiative (SDI),
    318
Streetcars, 151–52
Submarines, 60, 246–47
Subways, 269
Sugar Act, 32
Surveyor 1, 326
Swan, Joseph, 145
Sweeney, Orland R., 272–73, 274
Swine Breeding Laboratory, 281
Sylvania, 307
Systems analysis, 351–52
Szilard, Leo, 304

Talbot, William Henry Fox, 92
Talking telegraph recorder. *See*
    Phonograph
Tanner, 20
Taylor, Frederick W., 235, 267
Taylor, William R., 189
Technical education, 82–84
Technocracy movement, 263–64, 352
The Technological Society, 356
Technology assessment, 352;
    retrospective, 353

Technology Assessment Act (1972), 352
Telegraphic printer, 214
Telegraphy, 141, 180; competition from telephone, 158–59; railroads and, 97, 98–101; transoceanic, 127–28; *See also* Morse, Samuel F.B.; Western Union
Telephone, 133, 141, 145, 180; development of, 158–63; Great Centennial Exhibition, 137; radio and, 163–65; Western Union's interest in the, 160. *See also* Bell, Alexander Graham; Bell Telephone
Television, 328–30
Teller, Edward, 311
Telstar I, 326
Tennessee Valley Authority (TVA), 266–68, 275
Terman, Frederick, 306, 337
Tesla, Nikola, 155–56
Texas Instruments, 331
Textiles, 47, 94; cloth, 38; colonial manufacture of, 16–19; cotton, 16, 73–74; linen, 17, 18, 38; Lowell, Massachusetts, 75–77; textile mills, 73; use of machinery in manufacture of, 71–75; wool, 9, 16,17, 38, 73, 74. *See also* Looms
Thomson, Elihu, 144, 146
Thomson-Houston Electric Company, 144, 145, 155; DC lighting, 150; electric traction, 154
Thomson, William (Lord Kelvin), 128
Three Mile Island, Pennsylvania, 319, 320–21
Threshing, 15
Throstles, 71, 73
Thurston, Robert, 171, 176
Tintype, 93
Toffler, Alvin, 334
Tomato harvester, 343
Townshend Acts, 32
Tractor, 190–91; multipurpose, 276–77
Trade associations, 78
Transistor, 300–301
A Treatise on Domestic Economy, 220
Treenails, 25
Triangular trade, 26

Triode, 164
Trip hammer, 19, 25
Trucks, 291–93
Truman, Harry, 310; atomic bomb, 309; hydrogen bomb, 311
Trumbull, John, 133
Turner, Frederick Jackson, 194
Turnpikes, 2, 96; building of, 57–59; building bridges for, 59; highways in twentieth century, 340–41; National Road, 56
Twentieth Century Fox, 216
Typewriter, 137, 227

The Ultimate Resource, 356
Union of Concerned Scientists and Physicians for Social Responsibility, 347
Union Pacific, 101, 102
Unions, 240
United States Bureau of Public Roads, 194
United States Constitution, 1, 49; a commercial document, 41–42
United States Department of Agriculture (USDA), 192, 194, 341; chemurgy, 273; hybridization, 280–81
United States Electric Lighting Company, 144
United States Golf Association, 202
United States Lawn Tennis Association, 202
United States Military Academy, created by Congress, 53
United States Mint, created by Congress, 43
United States Motor Company, 212
United States Office of Technology Assessment, 352
Univac, 299
Universal Film Manufacturing Company, 216
Urban planning, 348–50

Vacuum cleaner, 225
Vail, Alfred, 99
Van Buren, Martin, 99
Van Rensselaer, Stephen, 82

Vanguard I, 324
Varian Associates, 337
Varley, C.F., 128
VCR, 330
Veblen, Thorstein, 263, 347
Victoria (queen of England), 125
Vincent, J.G., 249
Virginia Assembly, 9, 33
Virginia Company of London, 22
Volta, Alexander, 98
Vostok I, 324
Voyager I, 326
Voyager 2, 326
Vulcanized rubber, 126

Wabash Canal, 65
Walker, Timothy, 84–85
Wallace, Henry A., 278, 343; backing Orland Sweeney, 273; Hi-Bred Company, 279
War of 1812, 71, 74
War Industries Board, 245
Warner Brothers, 216
Washing machine, 223–24
Washington, George, 31, 39, 60; Potomac Company, 40, 41
Water frames, 71
Water Resource Planning Act (1965), 348
Waterwheels: horizontal, 10; overshot, 10; undershot, 10. *See also* Mills
Watson, Thomas A., 158, 159
Watson-Watt, Robert, 305
Weapons: canons, 36; flintlock, 34; Gatling gun, 137; gunsmithing, 34–35; ICBM, 318; interchangeable parts, 114–15, 231; MAD, 318; manufacture of, 34–36, 127; Maxim's machine gun, 245; MIRV, 318; Model 1841 percussion rifle, 114; Model 1842 percussion rifle, 114; nuclear weaponry, 317–19; Pennsylvania long rifle, 34–35; proximity fuses, 307; Robbins and Lawrence rifle, 126; SDI, 318; Sharps rifle, 137; tank, 246; torpedo, 246. *See also* Atomic bomb; Aviation

Webster, Daniel, 85
West, Benjamin, 60, 98
Western Electric, 159, 161, 259; Hawthorne Works, 257, 258, 260; Light Company, 144
Western Union: formed, 101; transcontinental telegraph, 128; and telephones, 159, 160, 161
Westinghouse: airborne radar, 305; radio, 164–65
Westinghouse air brake, 149
Westinghouse, George, 154, 156; AC lighting, 148–50; AC powered street railway, 155; Great Centennial Exhibition, 137; Niagara Falls project, 157; railroad mechanisms, 149
Westinghouse Lighting Company, 148
Weston, Edmund, 144, 146
Wheatstone, Charles, 98, 128
Wheeler and Wilson Manufacturing Company, 109
Wheelwright, 25
White City, Columbian Exposition, 186
Whitehead, Edwin, 345
Whitesmiths, 29
Whitney, Eli, 79, 81
Whitney, Willis, 177
Whyte, William H., Jr.: *The Organization Man*, 315–16
Wildlife Service, 265
Wild and Scenic Rivers Act (1968), 348
Willcox and Gibbs, 109
William, Charles, 158
William Sellers and Company, 137
Wilson, Allen B., 109
Wilson, Woodrow, 244
Winthrop, Governor John, 15
Winthrop, John, Jr., 22
Wisconsin Agricultural Convention, 189
Wolcott, Alexander, 91, 92
Wolfe, Tom, 336
Woodward, C.M., 241
Works Progress Administration, 294
World Future Society, 356
World War I, 243–47; air technology, 247; communications in, 164; land

World War I (*continued*)
  technology, 245–46; munitions
  production, 249–50; sea technology,
  246–47
World War II: aviation, 308–9;
  postwar world, 309–11; technology
  during, 301–9. *See also* Atomic
  bomb; Nuclear power; Radar;
  Sonar; Weapons
Wozniak, Steve, 339

Wright Brothers, 247, 261; first
  successful flight, 247

Yale locks, 137
YMCA, 203
*The Young Mill-Wright and Miller's
  Guide*, 52

Zoopraxiscope, 213, 214